Alexander Supalov
Inside the Message Passing Interface

Alexander Supalov

Inside the Message Passing Interface

—

Creating Fast Communication Libraries

DE
—
G

PRESS

ISBN 978-1-5015-1554-5
e-ISBN (PDF) 978-1-5015-0687-1
e-ISBN (EPUB) 978-1-5015-0678-9

Library of Congress Control Number: 2018952175

Bibliographic information published by the Deutsche Nationalbibliothek
The Deutsche Nationalbibliothek lists this publication in the Deutsche Nationalbibliografie;
detailed bibliographic data are available on the Internet at http://dnb.dnb.de.

Published by Walter de Gruyter Inc., Boston/Berlin
Printing and binding: CPI books GmbH, Leck
Typesetting: MacPS, LLC, Carmel

www.degruyter.com

To my beloved wife and wonderful children

About De|G PRESS

Five Stars as a Rule

De|G PRESS, the startup born out of one of the world's most venerable publishers, De Gruyter, promises to bring you an unbiased, valuable, and meticulously edited work on important topics in the fields of business, information technology, computing, engineering, and mathematics. By selecting the finest authors to present, without bias, information necessary for their chosen topic *for professionals*, in the depth you would hope for, we wish to satisfy your needs and earn our five-star ranking.

In keeping with these principles, the books you read from De|G PRESS will be practical, efficient and, if we have done our job right, yield many returns on their price.

We invite businesses to order our books in bulk in print or electronic form as a best solution to meeting the learning needs of your organization, or parts of your organization, in a most cost-effective manner.

There is no better way to learn about a subject in depth than from a book that is efficient, clear, well organized, and information rich. A great book can provide life-changing knowledge. We hope that with De|G PRESS books you will find that to be the case.

DOI 10.1515/9781501506871-203

Acknowledgments

I would like to thank Stuart Douglas and Jeffrey Pepper of De Gruyter for approaching me with an offer that prompted the writing of this book. My thanks also go to Jeffrey Pepper and Jaya Dalal for their efforts on the manuscript, to Mary Sudul for diligent copyediting, and to MacPS for their production work.

Jeff Squyres and Kenneth Raffenetti helped me with certain aspects of Open MPI and MPICH, respectively, saving quite a bit of time that would have otherwise gone into issue investigation.

Many anonymous reviewers provided valuable feedback on my conference submissions, helping me fine tune the text of the book. I extend my gratitude to them, too.

DOI 10.1515/9781501506871-204

Contents

Preface

Implementation of the *Message Passing Interface* (MPI) is deemed one of the more exalted of systems programmer's occupations: somewhat below the presumed complexity of an operating system (OS) with all its drivers, akin to a compiler and its runtime libraries, and more demanding than application programming. This is partly explained by the nature of the MPI itself: as just another software library in form, it sits between the MPI application (and thus the application level) and the hardware (and thus at and *below* the operating system level), striving to achieve the best possible performance. Of course, the important services of the operating system are called upon when necessary, however, now and then an MPI implementor must go down to the wire, bypassing the operating system either directly or with the help of other specific systems libraries that in turn talk directly to the respective hardware components.

This book seeks to demystify this endeavor, often perceived as open to just a few exceptional daredevils who dwell in ivory towers, spending uncounted taxpayer money running expensive custom built parallel instal-lations for purposes far too remote for anyone in the street. Well, going backward through this statement, this is all wrong. First of all, High Performance Computing (HPC), of which the MPI is an integral part, has come to affect everyone, anytime, and everywhere. Looking up tomorrow's weather forecast, driving a car, boarding a plane, searching the Web, or buying goods online—you rely on HPC and likely the MPI more often than you may think. Second, parallelism has reached the computer you use for your everyday business and leisure activi-ties. The laptop of today surpasses supercomputers of the 1970s, and kids' toys carry around more computing power nowadays than the whole Apollo 11 mission that landed men on the Moon back in the 1960s! And third, any skill, however exalted, can be learned given the will and capacity to learn.

Communication permeates the modern world. This is why learning how to create fast, efficient, compact, and scalable communication libraries is important now and is going to be even more important in the future. Even though MPI per se may be unique in its laser focus on cutting edge technical HPC, the methods and ap-proaches used during MPI design, implementation, and testing are going to be pertinent in all other areas of communication software. This book will provide you with a better understanding of the MPI internals and their if's and why's. It will also let you extract more performance out of your HPC applications. Beyond that, it will show you how to approach comparable tasks in any programming language on any device, whether that is a supercomputer running Linux, a laptop running Windows, or your smart-phone running Android or iOS.

DOI 10.1515/9781501506871-206

Introduction

This is a hands-on guide to writing a Message Passing Interface (Message Passing Interface Forum, 2015) implementation from scratch. This book also discusses the major MPI implementations, best optimization techniques, application relevant usage hints, and a historical retrospective of the MPI world—all based on the experience of a quarter century spent inside the MPI. It offers:

- Intimate knowledge of the MPI internals presented in a hands-on fashion
- Optimizations to achieve the maximum performance possible in the hardware
- Deep understanding of the MPI advantages and drawbacks

Who Should Read This Book

This book is intended for MPI implementors (existing and hopeful), creators of alternative communication interfaces looking beyond what's on their plate, advanced MPI users seeking deeper understanding of the MPI internals, and programmers willing to learn more about software design, development, and optimization.

What You Need to Know

This book is neither an introduction to the HPC in general or the MPI in particular, nor a description of a particular MPI implementation, nor a vendor-specific MPI usage manual. There are a number of good information sources covering the aforementioned areas that you can read in order to get up to speed if necessary.

In addition to this, you should have a rather good grasp of systems programming, communication and networking technologies, and the MPI standard itself. Knowledge of the C programming language as well as of the Linux operating system is essential. Knowledge of the Bourne shell and Fortran programming languages may occasionally be helpful, too, as well as practical experience of creating sizeable software products and using the associated code management techniques and the program building tool chain.

Notation and Conventions

I tried to keep this book as informal as possible. All you need to know is that source code will be highlighted using this font, and that code excerpts will look as follows:

```
This is a source code snippet
```

DOI 10.1515/9781501506871-207

If a bit of the source code needs to be skipped to smoothen the narration, you will see an ellipsis (...) in the listing. You will find the missing bits either later on in this book or in the source code repository provided at the book's web page https://www.degruyter.com/view/product/488201 under Supplementary Information. In order to facilitate your search, every listing will carry an exact file name that should be consulted for details.

Commands will show up as Linux shell command strings:

```
$ which mpiexec
```

Here, the initial dollar sign ($) is not to be entered because this is a shell command prompt. If a command line is so long as to exceed the line width of this book, a sequence of backslash (\, to be entered by you) and a greater-than sign (>) at the beginning of the next line (output by the shell instead of the dollar sign) will be used to keep the formatting correct, fully in line with the interactive shell conventions. Program output will be shown verbatim using the same font, however, long lines may be wrapped around without any special mark-up, just as if they were displayed in the terminal window.

Program invocation syntax may possess optional arguments that will conventionally be marked by square brackets [ ] as follows:

```
$ mpiexec -n number_of_processes program [arguments]
```

Note that variable parts of the run string will be shown in fixed width *italics*. If there are variants to choose from, they will be surrounded by curly braces {} and separated by the vertical bar (|), for example:

```
$ mpiexec -{n|np} number_of_processes program [arguments]
```

Of course, there is more than one way to cut the cake:

```
$ mpiexec -n[p] number_of_processes program [arguments]
```

References to the system calls and functions will be given in the usual man page format, like memcpy(3), where the number denotes the part of the Linux manual. Our own functions will be referred to without any number in the parentheses, like loh_wtime().

We will rely on MPI standard terminology, using the spelling agreed on by the MPI Forum. So, you will meet *nonblocking communication*, *elementary datatypes*, an *MPI implementor*, and a few other oddities. By the way, you can see here that important terms will be shown in *italics* when they are introduced.

Finally, we will be careful about the units of measurement used for memory size. When dealing with the main memory and caches, as well as networking, we will use binary size units, e.g., kilobyte (KiB) as 1024 bytes, megabyte (MiB) as 1024 times 1024 bytes, and so on. When dealing with the disks and other external media, we will use decimal units, e.g., kilobyte (KB) as 1000 bytes, megabyte (MB) as 1000 times 1000 bytes, and so on. *Message latency* will be measured in seconds or fractions thereof, e.g., in nanoseconds (ns) equal to 10^{-9} second, microseconds (μs) equal to 10^{-6} second, and milliseconds (ms) equal to 10^{-3} second. *Message bandwidth* will be measured by the number of bytes transferred per second.

How to Read This Book

You should read along and do all the exercises (at least mentally) in order to profit most from this book. It will help enormously to have a good Linux workstation handy, and even better to have access to a reasonably sized Linux computing cluster with more than an Ethernet network to it. Still better will be to know how to use them all, because in exchange for giving expert advice, this book expects from you a certain level of professionalism. Ask around: people are always there to help you out with the local customs and eternal wisdom.

The following awaits you inside:
- *Chapter 1: Overview*—Revisit the basics: parallel hardware, MPI and related standards, as well as sound software development principles.
- *Chapter 2: Shared memory*—Create a simple MPI subset capable of basic blocking synchronous point-to-point communication over shared memory.
- *Chapter 3: Sockets*—Create an MPI subset capable of blocking and nonblocking point-to-point communication over Ethernet and other IP capable networks.
- *Chapter 4: Extensions*—Learn how to add essential MPI features including communicator management, datatypes, collective operations, error handling, and language bindings.
- *Chapter 5: Optimization*—Optimize the extended MPI subset internally using multiple fabrics, advanced implementation techniques, and available special hardware.
- *Chapter 6: Tuning*—Further improve performance by changing relevant MPI settings depending on the target platform and/or target application.
- *Chapter 7: And the rest*—Learn about the most important remaining MPI features and their impact upon the MPI internals.
- *Chapter 8: A look ahead*—Understand MPI limitations to better assess its past and future.

In addition to this, there are *References*, *Index*, and *Appendix A: Solutions* with easy answers to all those nagging questions.

NB: All opinions expressed in this book are mine unless backed by an explicit authoritative reference—your mileage may vary. Good luck!

Chapter 1
Overview

The implementation process of a Message Passing Interface library is guided by laser focused dedication to performance. This dedication permeates the MPI standard, it drives the advices to implementors contained therein, it forces an implementor to think thrice about every bit and byte, every conditional operation, and extra argument added to the source code, because all seemingly desirable extras result in additional nanoseconds in the critical code path. Yes, this is not a typo. We are going to deal in nanoseconds in this book. A thousand of them will make a whole microsecond, and this will soon seem to you like walking instead of flying, because characteristic latencies associated with the shared memory and modern networks lie in the area of a couple hundred nanoseconds per message transferred. Just for comparison, if you blink, it takes your eyes between 300 to 400 milliseconds, or 300 thousand to 400 thousand microseconds, or 300 million to 400 million nanoseconds to refresh themselves. Now, look, a good MPI implementation will deliver at least 1.5 million to 2 million messages while you blink, and that's in one direction only; ours will deliver twice to three times as many.

If you want to learn how to make the MPI standard, the system software, and the underlying hardware interplay to deliver this remarkable result, you need to know how all of them work and what makes them fly. This is why, starting this journey, we will first revisit the basics: the parallel hardware, the MPI and the related standards, the MPI subsetting, as well as the time-proven HPC software development principles.

1.1 Parallel Computer

Every serious computer these days is parallel; HPC computers even more so (see Figure 1.1).

Going top to bottom (and counterclockwise), a parallel computer is normally composed of several computer nodes connected with each other over a network or two. Each node contains several processors with connections between each other, as well as to the main memory and the peripheral devices over a processor interconnect or bus. Each processor in turn consists of several processor cores, possibly running more than one hardware thread, each using a set of execution units. However, this is not the end of it. Let's go bottom up now to see how it all adds up and how all these components can be used to create a faster communication library.

DOI 10.1515/9781501506871-001

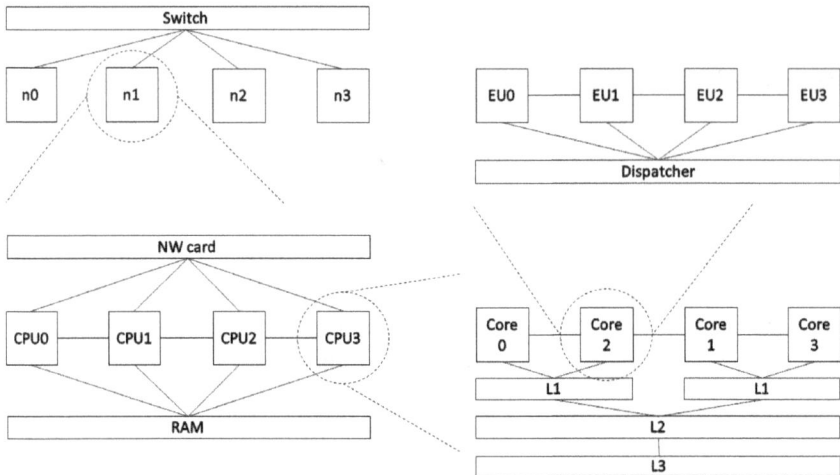

Figure 1.1: Hierarchy of a typical HPC system

1.1.1 Intraprocessor Parallelism

A typical *processor core* has many *execution units*, each dedicated to a certain set of commands, be that logic/arithmetic, floating point, memory manipulation operations, and so on. Commands are fetched from the *main memory*, often in batches of more than one, decoded and scheduled for execution on all those aforementioned units. They are normally executed *out of order* and chiefly more than one at a time, although less sophisticated processors execute them *in order* and one by one. In the end, the memory is modified in a way that corresponds to the in-order, one-by-one command execution. A good communication library will use all available data transfer execution units to the maximum while trying not to interfere with the application program activities.

All commands and values processed by the program make their way up and down the *memory hierarchy* that contains the *main memory* and several (typically two or three) cache levels. Each higher cache level is smaller and faster than the lower cache levels or the main memory, with the top *cache level 1* (L1) being the smallest and the fastest of them all. Data prefetching is employed automatically or explicitly on program request to preload data most likely to be required in the nearest future. The calculation itself is normally performed on the registers that are renamed on the fly in order to lower the amount and intensity of the data dependencies and collisions between the commands. Keeping the most frequently used data in cache is a well-known technique. A good communication library will, if possible, fit its control structures onto the cache line (normally 64 bytes long) and distribute them across the memory address space in order to avoid cache address aliasing and other conflicts. At the same time, this library will try to assure that the data passing through the cache does not spoil the activities of the application program, but rather helps them by moving the data to be processed quickly up the cache hierarchy. Note that in some cases, when the data is not going to fit into the cache or is not going to be used immediately after the transfer, it may make sense to bypass the cache, and there are ways to achieve this in the more advanced processors.

Speculative execution may be used to process conditional statements before the corresponding condition can be evaluated. Both branches may be followed at the same time, but once the condition value is known, the wrong branch is discarded. Prediction based on past decisions helps to increase the probability of following the right branch, especially if speculative processing follows only one branch that is presumed to be correct. Even though prediction works generally rather well, and the penalty for discarding the wrong branch, if at all incurred, is relatively low, a good communication library will try to avoid conditional statements in the code hot path, or at least try to avoid conditions causing the behavior to be less predictable. What is better or less predictable depends on the processor architecture, so it makes sense to have a look into the respective processor programming manual. Normally, always-true, always-false, and now-true-then-false are combinations that are tracked and predicted quite adequately by the modern processors.

A processor core typically supports several *hardware threads*, each of which acts basically as a virtually independent core. The level of resource sharing between the threads may vary, from totally independent to fully shared. However, totally independent resources are more typical of actual cores than hardware threads, and total resource sharing is rare. One thing to keep in mind is that the higher the level of sharing, the more likely are the conflicts between the threads. However, sharing of the cache between the threads may lead to lower latency when the data has to cross the thread boundary. So, a good communication library will exploit both resource sharing and resource independence to the advantage of the application program, winning better latency but avoiding unnecessary thread conflicts.

Modern processors normally have more than one core that can be thought of as basically independent processors. However, different cores normally sit on the same internal processor bus and share some cache levels, typically starting with *cache level 2* (L2). Again, this may have both negative and positive consequences, increasing the chances of the core conflict on one hand, and decreasing the latency of the data transfer once the data residing in the shared cache has to cross the core boundary. A good communication library will be tuning itself on the fly to provide the application program with the best available latency and/or bandwidth depending on the actual requirements of the program rather than some predefined strategy. At the very least, a good communication library will let the user select the right predefined set of defaults that are deemed most appropriate for the task at hand.

Very close consideration should be given to the number of cores in the processor. If there are relatively few of them, say, up to four or eight, rather simple algorithms and dense data structures are likely to perform better than complicated algorithms based on sparse data structures. If, however, the number of cores exceeds 16, both algorithms and data structures used by a good communication library will pay attention to avoiding unnecessary waste of both space and time for attending to inactive or less frequently used connections.

1.1.2 Interprocessor Parallelism

The level of integration of the modern processors continues to grow. Most of them include the *memory management unit* (MMU) and a *graphics accelerator*, while *network cards* are finding their way into the processor internals as we speak. If a certain component, be that a graphics accelerator or a network card, does not live in the processor itself, it connects to the processor either directly via a *bus* or an *interconnect*, or indirectly via one of the intermediaries (sometimes called *bridges*) that drive a standard external bus like PCI Express (PCI Special Interest Group, 2017). It is important to understand that depending on the nature of the peripheral device, it may or may not be capable of acting without a processor's participation. If this capability is available, the process of entrusting the respective peripheral unit with a certain asynchronous activity, be that computation or data transfer, is referred to as *offloading*. Offloading further increases the level of parallelism achievable by the system. A good communication library will know how to use all offload capabilities available in the system.

External peripherals may be and normally are shared by neighbor processors. It is common, for example, that a computer node contains only one graphics accelerator and one network card, although this may vary depending on the system architecture. Sharing again brings both a blessing and a curse: it drives the cost of the system down but increases the level of conflict possible due to the simultaneous processor access requests reaching the respective peripheral unit. A good communication library will use one of several techniques to avoid or at least lower the level of conflict induced by this kind of resource sharing.

Processors in a node may and usually do share access to the common main memory. Often this access goes via the shared *cache level 3* (L3), although, as with any cache level, it can be bypassed if the program wants to do so. An independent memory access path normally increases the bandwidth available to the respective processing unit, but it may also increase the latency of accessing data being used by another processing unit. In the extreme, called *Nonuniform Memory Access* (*NUMA*), access to one's "own" memory is noticeably faster than access to "somebody else's." This is typically a result of the extra complexity of the interprocessor protocol or even the number of "hops" the data has to go through along the interconnect before reaching its respective destination. A good communication library will try to use its "own" data internally as much as possible and let the application program decide how to exploit NUMA to its advantage.

Moreover, it is also possible that certain processor units may have their very own memory distinct from other processor memories, and that this memory may even be of a special type that has, for example, higher bandwidth than usual. In this case, what used to be *SPMD* (Single Program, Multiple Data) execution slowly becomes *MPMD* (Multiple Programs, Multiple Data) execution, which reaches its local pinnacle as soon as the data crosses the node boundary, since the use of a Local Area Network (*LAN*) enforces message passing in some form, either transparent or apparent to the user. Of course, networks can also connect completely different machines, in which case they are normally called Wide Area Networks (*WAN*). This is the area of so-called *metacomputing*, *grid*, or *cloud* which basically all denote the same thing in its successive historical incarnations. The use of cloud in particular implies to many people the use of a Virtual Machine (*VM*), although at least in the case of HPC this may not be necessary (US Patent No. 8,725,875, 2014). If more than one connection medium is available, a good communication library will opt for the faster one, and possibly multiplex longer messages across all available connection media or links to increase the overall bandwidth.

Finally, if the target platform provides any additional hardware-supported or other specific features relevant to MPI programming, be that additional synchronization mechanisms, special communication channels, vendor-specific system libraries or accelerators, a good quality implementation will take all that into account and use it to achieve the maximum possible performance while staying easily usable and user-friendly.

1.1.3 Exercises

Exercise 1.1 Examine your computer system and clarify how many nodes it has, what node types are available, and by what network(s) they are connected. If in doubt, look up Appendix A for an example.

Exercise 1.2 Examine your computing nodes and find out what kind of processor(s) they exhibit, how many cores and threads they run, what memory type and what caches they use, and what networking and graphics capabilities they have.

1.2 MPI Standard

The MPI standard (Message Passing Interface Forum, 2017) has come to dominate HPC message passing programming, just like its younger cousin OpenMP (OpenMP Consortium, 2017) has come to dominate HPC shared memory programming. This is a natural course of events, where the most successful interface basically takes it all, leaving only small and specialized niches to its less fortunate competitors.

1.2.1 MPI History

The MPI movement started back in 1992, when the emerging field of HPC was teaming with various mutually incompatible and yet functionally roughly equivalent message passing interfaces, both vendor-specific and portable, intended for a veritable explosion of the so-called massively parallel computing (see Figure 1.2).

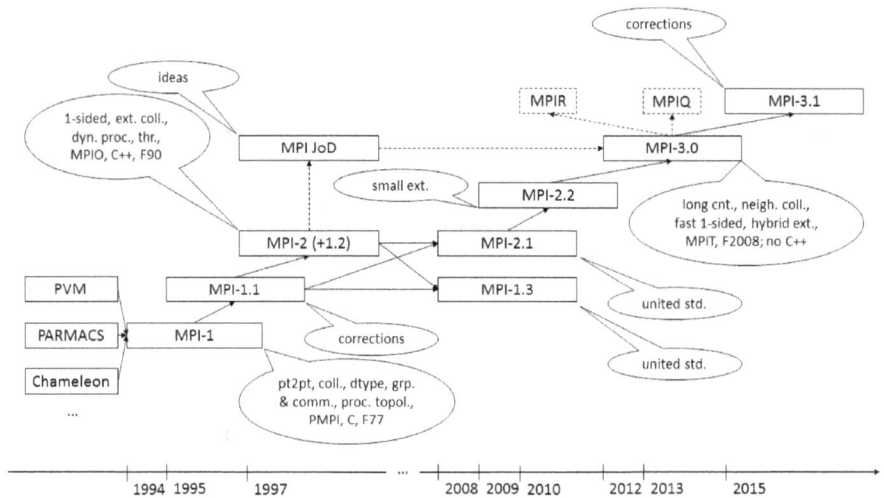

Figure 1.2: MPI history

At that time, I was involved in design and development of one of the more portable interfaces—the *PARMACS* interface version 6.0 for Intel's iPSC/2 and iPSC/860 computers (Hempel, Hoppe, & Supalov, 1992). Until version 6.0, the PARMACS (PARallel MACroS) were just a set of m4 preprocessor macros that were expanded into the respective vendor-specific call sequences right inside the Fortran application programs. The main goals of the new version were reformulation of the PARMACS as a full-blown software library, definition of the C interface, and the addition of certain new features. A notable part of that interface, namely, the process topologies, would eventually find their way into *MPI-1*, albeit reformulated inside out as virtual process topologies (Message Passing Interface Forum, 1994).

   The undisputed leader of the field however was the *Parallel Virtual Machine* interface (PVM: Parallel Virtual Machine, 2017). It had a number of sound features (like opaque task ids and dynamic process enrollment), but it also had, at least until the later version 3.0, a basic performance-relevant flaw: all the data to be transferred had to be packed into a buffer that was sent around using the XDR data representation (Eisler, 2006). The process of packing and optional data conversion (which was necessary, e.g., for the little-endian machines sitting on the traditionally big-endian networks) severely limited PVM performance as soon as the underlying network was fast enough to make those extra latencies matter for typical message lengths. Additional memory space to be allocated for an internal transfer buffer was another negative point in those times when the memory sizes were measured in mega- or even kilo-bytes. And since this buffer was global, the overall design did not sit well with the emerging threading libraries.

There existed many other interfaces, of which none survived the turn of the century. One of those interfaces—Chameleon portability system—developed by the Argonne National Laboratory, played a decisive role in the rise to prominence of the first widely available MPI implementation that was aptly called *MPIch* (MPI over CHameleon) (Argonne National Laboratory, 2017). This ever-growing package was used by its authors as a proof of concept of the MPI standard, and it closely tracked and sometimes guided the standard development by trying to implement it as it evolved.

Another implementation emerged later as a formidable competitor to the MPICH that was principally suited for tightly coupled parallel machines. This was *LAM/MPI* (Local Area Multicomputer MPI) that, as follows from its name, was specifically targeted for connecting the then-popular workstations over LAN—normally, the *Ethernet* (IEEE 802.3 ETHERNET WORKING GROUP, 2017). However, performance of the LAM/MPI on so-called tightly coupled machines would rarely match that of the MPICH, so both of them stayed in business, each taking a niche to itself. Whatever the implementation, of which more and more would be created going forward, they provided the basic functionality defined by the MPI-1 standard: the point-to-point and collective communication, datatypes, group and communicator management, virtual process topologies, profiling interface, as well as the C and Fortran 77 language bindings.

Soon after the emergence of MPI-1 it became apparent that this standard had a number of errata and was too limited in scope to compete head to head with the PVM. A corrected *MPI-1.1* (Message Passing Interface Forum, 1995), and then *MPI-2* (with *MPI-1.2* as its part) (Message Passing Interface Forum, 1997a) appeared in quick succession. As a by-product of the MPI Forum discussions, the *MPI Journal of Development* (Message Passing Interface Forum, 1997b) and the *MPI Real Time Interface* (Kanevsky, Skjellum, & Rounbehler, 1997) were created. The former would be partially reused, if only as a source of inspiration, in the further standard development. The latter died out. Major extensions introduced by the MPI-2 standard included dynamic process management, one-sided communication, extended collective operations, thread support, file I/O (aka MPIO), as well as C++ and Fortran 90 language bindings. However, MPI-2 did not become as widely popular as MPI-1.1, and the MPI-2 and then *Interoperable MPI* (IMPI) standards (George, Hagedorn, & Devaney, 2000) that my friends and I implemented for Pallas GmbH in the year 2000 remained relatively unused.

After that there followed a rather long lull. The MPI Forum decided to reconvene and revisit the MPI standard in the light of new developments only in 2005, i.e., about a decade after the publishing of the original MPI-1 standard. By then, the glorious times of the tightly coupled, custom built, massively parallel computers were coming to an end under the increasing pressure exerted by the so-called Beowulf clusters (Beowulf.org, 2002) built from off-the-shelf components and commodity networks like Ethernet and the nascent *InfiniBand* (InfiniBand Trade Association, 2017). Processor, memory, network, and packaging technologies all came a long way, new HPC applications emerged and grew in complexity and scale, while the related standards (like OpenMP) and the programming languages and compilers for them all evolved to reach new levels of expressive power and speed. Thus, the HPC field was ripe for another great leap.

However, before starting anything really new, the MPI Forum soundly decided that first they had to have a common document representing the whole of the MPI standard, rather than two separate standards MPI-1.1 and MPI-2 (with MPI-1.2 curiously as part of it). So *MPI-2.1*, uniting the whole of the MPI-1.1 and MPI-2 documents, came to be (Message Passing Interface Forum, 2008a). The complementary *MPI-1.3* standard (Message Passing Interface Forum, 2008b) contained only the MPI-1.1 and MPI-1.2 related parts of the MPI-2.1 standard compiled into one book. It was meant as a reference for those vendors and users who (for whatever reason) would not want to go for the full MPI-2 standard functionality. Both standards included the respective standard errata documents published by the Forum earlier. They also corrected other glaring errors in the original standards and clarified certain semantic matters while generally (but not quite) maintaining backward compatibility with the original standards.

The second necessary step was taken with *MPI-2.2* (Message Passing Interface Forum, 2009) that was designed to correct all known errors of the past MPI standards and add several new small features while still maintaining the overall backward compatibility. The full list of changes is long and rather technical—you can look it up in the Change-Log chapter there. What was essential to the MPI implementor, apart from the C++ binding deprecation, was the addition of the MPI_IN_PLACE buffer designation to all collective routines, the removal of the restriction of the accumulation operations to the elementary datatypes, the addition of a whole slate of new datatypes, and the addition of the distributed graph routines as well as several other routines of lesser import, accompanied by a number of smaller and yet essential semantic clarifications of the existing entities and their behavior. The minor standard version MPI_SUBVERSION was incremented to let application programs differentiate between subversions of the MPI-2 standard.

The real major step came with *MPI-3* (Message Passing Interface Forum, 2012). It removed the C++ bindings (a bold move), added support for long counts, neighbor collective operations, fast one-sided operations, the MPIT Tool Information Interface, and hybrid programming extensions, as well as modern Fortran 2008 bindings. Some errata of the MPI-2.2 standard were corrected as well. In addition to this, the already existing *MPIR Process Acquisition Interface* (DelSignore, 2010) and the *MPI Message Queue Dumping Interface* (Vo, 2013) were formalized to promote unification between different MPI implementations on the one hand, and various MPI tools (predominantly parallel debuggers) on the other. These documents were endorsed by the MPI Forum but they were not published as part of the MPI standard per se.

Although the goal of maintaining backward compatibility was not explicitly pursued by the Forum during the development of the MPI-3 standard in particular, due attention was paid at all times to the introduction of the minimum required changes. Thanks to that and a bit of luck, I managed to achieve backward compatibility between MPI-2.2 and MPI-3 later on by inventing the common MPICH Application Binary Interface (Supalov & Yalozo, 20 Years of the MPI Standard: Now with a Common Application Binary Interface, 2014) that was accidentally identical to the Intel MPI Library ABI of the time. With the advent of that ABI, the MPI world finally took optimal shape where two major implementations—MPICH and by now *Open MPI* (Open MPI Consortium, 2017) of LAM/MPI descent—started to compete for the crown on fairly level terms. This provided an enormous boost to the quality of both implementations because this ended the fragmentation of the MPICH part of the world, and now both MPICH and Open MPI extended development teams achieved the critical mass that allowed them to stay on par with the competition at all times.

MPI-3 proved too great a leap, however, so that *MPI-3.1* had to be published some three years later (Message Passing Interface Forum, 2015). It corrected the Fortran 2008 binding as well as certain other issues of the MPI-3 standard. In comparison to the volume of corrections, the number of actual new features was rather low, as suits a minor standard version increment. Again, the Change-Log chapter there is your friend.

The MPI Forum is currently busy working out the next MPI standard, to be named *MPI-4*. The biggest and most elusive element that has successfully avoided standardization so far that should be addressed is the fault tolerance support. Further extension of the hybrid programming models, one-sided operations, as well as addition of performance assertions and persistent collectives are also all on the table. We will see what comes out of this in due time.

1.2.2 Related Standards

The MPI standard does not stand alone. It is dependent upon and has influenced a whole set of related standards in return. Of those, the most important are the OpenMP specification (OpenMP Consortium, 2017), the programming languages C and Fortran, as well as the networking technologies such as Ethernet (IEEE 802.3 ETHERNET WORKING GROUP, 2017) and InfiniBand (InfiniBand Trade Association, 2017), and the related application programming interfaces (API) like POSIX sockets (Wikipedia.org, 2017), uDAPL (Openfabrics.org, 2007), the emerging OpenFabrics libfabric (OFI Working Group, 2017), and several other portable or vendor-specific interfaces of various degrees of power and generality. Historically, many more networking technologies and interfaces were relevant, but none of them is visible at large today. Another notable connection is the venerable IEEE 754 standard (Goldberg, 1991) that rules the floating-point operation domain in modern computers. Also of substantial importance are the operating systems, with Linux in its various variants enjoying a huge lead over MS Windows, MacOS, and those remaining vendor-specific operating systems like AIX that have survived the grand OS crunch at the beginning of the twenty-first century.

OpenMP is a standard that defines how threads can be programmed in a more or less system- and language-independent fashion. It is available in the C and Fortran variants, thus covering the two most popular programming languages in the HPC. This is not the only standard in this area, with the POSIX threads (Wikipedia.org, 2018b) in particular being a viable competitor for those C and C++ programmers who do not want to rely on the somewhat more formalized and less flexible expressive means provided by the OpenMP. Since the MPI standard wades ever deeper into the shared memory domain that is tightly coupled to thread programming, there is a lovely interplay between several aspects of both standards, with the MPI trying not to replace the OpenMP but rather to complement it and provide the necessary support for the respective runtimes. Since both MPI and OpenMP want to exercise control over the system resources, with the processor cores and memory being the most important areas of the intersection, one can imagine that both runtimes have to be mutually aware of each other in order not to step on each other's considerable toes. Process allocation and binding in particular have to be synchronized, ideally by the runtimes, practically by the application programmer.

In my view, the C and Fortran programming languages remain mainstays of hardcore scientific programming. Programmers lauding Python, R, MATLAB, Go, and other languages may all make their case, but for a completely different type of application. Ritchie's wunderkind C has achieved something of a maturity after it became ANSI C, which in my opinion also made the language a little less comprehensible and manageable than it was originally intended to be. Fortran in particular suffered a lot from the insatiable drive toward "progress." A simple and straightforward scientific language of yore that once had the now removed three-branch arithmetic if statement (for positive, zero, and negative outcomes) that directly matched one of the commands of the underlying historical IBM systems, has become a veritable monster in its 77, 90, 2000, and now 2008 incarnations. The latter in particular influenced the MPI-3 standard that added a Fortran 2008 language binding which was designed to resolve the venerable choice argument issue. This move finally succeeded in MPI-3.1, and the respective extension to the Fortran language was discussed by both the MPI Forum and Fortran Standardization Committee members, some of whom sat in both bodies. C++, being in my eyes an object-oriented assembler, has come and gone for good as far as the MPI standard is concerned.

In the area of networking, one can trace a perpetual tendency toward the creation of more capable, faster, more intelligent networks that strive to take as much load as possible off the shoulders of the CPU, with some notable exceptions. Likewise, the interfaces provided by the respective network vendors and their trade associations keep growing in expressive power and the level of abstraction that they provide to the MPI implementor. In particular, the handling of tag matching, which is an essential part of the MPI standard, is now almost universally done at the level of the network interfaces rather than by the MPI itself. Directly connected to this is the tendency of going from rather simple packet or Remote Direct Memory Access (RDMA) interfaces like OFA verbs (Openfabrics.org, 2017) or uDAPL (Openfabrics.org, 2007) to the more advanced, message-based interfaces—although the ways of expressing the message passing may vary greatly, from the most direct in the emerging OFI libfabric interface (OFI Working Group, 2017) that closely resembles MPI itself, to the most intricated in the Portals interface (Sandia National Laboratories, 2017) that works in terms of hardware supported message queues. On a level with the message passing, many networking interfaces retain the lower level interfaces mentioned above, so that an MPI implementor is confronted with the need to find out what particular part of the provided interface works best on the particular target platform.

The IEEE 754 standard defines the set of floating point datatypes covered by the MPI interface. Even though in the past the well-known C and Fortran types like double and DOUBLE PRECISION might have had nonstandard lengths and representations, by now one can safely bet that a computer that can do floating point computations will at least have an IEEE 754 execution mode, most likely as a default. Of course, the MPI standard defines, and an MPI library may provide support for, vendor- or machine-specific data types like REAL*2 and the like. However, using them in an application defeats the purpose of portability that the MPI standard was conceived for in the first place.

Finally, the OS revolution and advent of Linux dominance came as a blessing and a curse combined. On one hand, Linux is a very reasonable operating system originally designed, as was its grandma Unix, with the task of programming in mind. On the other hand, the popularity fragmented the Linux market segment, with Red Hat and SUSE in particular vying for the commercial crown, and many smaller more or less standard distributions like Debian, Ubuntu, and others filling up the less stringent fringes and targeting more or less professionally astute user groups. Slight differences between Linux distributions create additional programming and testing burdens, as do rather frequent and sometimes backward incompatible updates to the OS kernel and especially of the programming tools. In any case, the by now mature support for threading, process binding, memory locality control, interprocess communication, networking, and file I/O provided by all modern Linux distributions have all evolved with time under the constant pressure of the IIPC interfaces like the MPI standard.

1.2.3 Exercises

Exercise 1.3 Read the latest MPI standard. I mean it: you will need this to proceed sensibly beyond this point.

Exercise 1.4 Explain the difference between the blocking and nonblocking communication. Is blocking local or non-local?

Exercise 1.5 Find out what the standard point-to-point communication mode is comparable to in your favorite MPI implementation. In particular, find out whether it is buffered or not, and if so, to what degree.

Exercise 1.6 Clarify whether your favorite MPI implementation provides nonblocking collective operations and how they compare in performance to the blocking ones, and whether they support any computation/communication overlap.

Exercise 1.7 Examine corresponding MPI documentation and source code to see whether one-sided communication is mapped upon the point-to-point communication or supported natively, and if so, over what communication fabrics.

Exercise 1.8 If you are unable to give articulate answers to any of the above questions, read the answers in the Appendices, then go back to Exercise 1.3 and repeat until done.

Exercise 1.9 Find out what operating system is installed on your computer, whether it is up-to-date, and what support for the threading, process binding, memory locality control, interprocess communication, and networking it offers.

1.3 MPI Subsetting

There are basically two ways of developing programs. One of them is to precisely define the overall architecture, write detailed specifications, carefully distribute the work between the subordinates, monitor it diligently, and ... find out in the end that nothing works as expected. The other is to start small, integrate continuously, test permanently, keep adding features on user feedback, and ... end up with an unmanageable monster whose architecture resembles the house that Jack built. Reasons? Only hindsight is 50–50, and most users should be told what they really want. In other words, trying to map out the complete product architecture from the start is probably going to fail without due revision, while shaping the product in response to the user feedback alone is simply futile.

My experience shows that the golden middle lies in between. You have to have a pretty good idea of both the product architecture and the user needs up front, but the precision of both may and should grow over time. Staging the work so that you end up with a working product (rather than a prototype) every now and then is another recipe for success. I call this approach *pragmatic subsetting*. It is subsetting because the product goes through a series of steps that extend its functionality in a sensible manner while maintaining or even improving performance, if that is possible. It is pragmatic because it is driven by the ever current and changing understanding of the priorities.

In application to the MPI development, pragmatic subsetting calls for an overall design that will in the end allow a full implementation of the MPI standard. This is only possible if the architecture is kept sound and extensible from day one. On the other hand, thinking big requires acting small, so that every implemented subset should not only work and be fast, it should provide a sensible set of features that correspond to a realistic and practically relevant set of targeted MPI applications and MPI users who write them.

This book sets out to show how this approach can be applied to implementing a very fast, reasonably capable, and extremely compact MPI library in a few relatively simple steps.

1.3.1 Motivation

The idea of subsetting is based in part on my personal experience with a certain message passing library that I proposed, designed and developed back in the 1990s. I had the joy then of starting from scratch and having full creative freedom except for the predefined set of functions and certain semantics. I started small, making the library pass just one byte from one process to the other over shared memory. I got latency under one microsecond, which was really fantastic back then, when CPU frequencies were measured in mega- rather than gigahertz. Then I added longer messages and more processes. Latency grew to about 1.5 microseconds. Then I added acknowledged delivery as an option and got latency of about 2 microseconds overall, also in the original path without acknowledgements. This was the cost of the extended protocol and extra conditionals required to implement it. By the time I was done with all the necessary semantics, latency went up to about 4 microseconds, which was still well within the design target. However, this crawling growth of latency that was roughly proportional to the number of features added made me think whether MPI could be made faster if its semantics were sensibly restricted.

Indeed, MPI is a very long and rather complicated standard. This is explained in part by the style selected long ago for presenting it. Instead of formal definitions known in some other standards, the MPI standard is a fluent narration that does contain all the necessary information but spreads it across the text. In this form, it trades compactness for readability—big time. For example, the "big bang" `MPI_Init()` function is mentioned in at least 36 pages, all of which need to be looked up for the correct implementation. The same is true of the "big crunch" `MPI_Finalize()` function that is mentioned in at least 23 pages, and so on (see Function Index chapter of the MPI-3.1 standard).

As a result, to fully comprehend the MPI standard (and very few people in this world can honestly claim that), you have to read and reread the entire text several times, spend a lot of effort implementing it in all its completeness and intricacy, while staying in constant touch with other professionals to permanently refine your knowledge and understanding—only to find out almost by chance how far away from the truth you could stray now and then.

At the same time, MPI complexity is objectively present due to the sheer number of entities the standard deals with. In order to illustrate the growth of the MPI standard complexity over time, I counted the number of pages and the number of MPI entities (constants, predefined handles, type definitions, callback prototypes and functions, etc.) in the standard documents available so far (see Table 1.1).

Table 1.1: MPI standard complexity growth

Standard Version	Page Count	Entity Count	Of them: Functions[1]
MPI-1	235	244	128
MPI-1.1	238	249	128
MPI-1.3	245	255	129
MPI-2	374	504[2]	216
MPI-2.1	608	665[3]	332
MPI-2.2	647	867	332
MPI-3	852	841	440
MPI-3.1	868	855	450

Eyeballing it, there is about one MPI entity per standard page (with the exception of the MPI-2 standard, due to the C++ binding) and about one function per two pages on average (same exception). Thus, about half of the entities defined by the standard are functions (again, except for the MPI-2). And if extrapolation is any guide, the coming MPI-4 is going to surpass a thousand pages, and hence a thousand entities and a

---

1 Language independent definitions. Multiply by the number of language bindings (3 for MPI-2*, 2 otherwise) to get the approximate real number.
2 Including separate C++ entities, excluding info keys and values.
3 Excluding some C++ entities thanks to reformatting (MPI-2.1) and the following deprecation (MPI-2.2).

half of a thousand functions. This complexity measure does not take into account the interrelation of the entities, so the overall complexity is actually much higher.

In order to manage this complexity, a full-scale MPI implementation resembles a pyramid that has been turned upside down, so that it rests on its tip (see Figure 1.3).

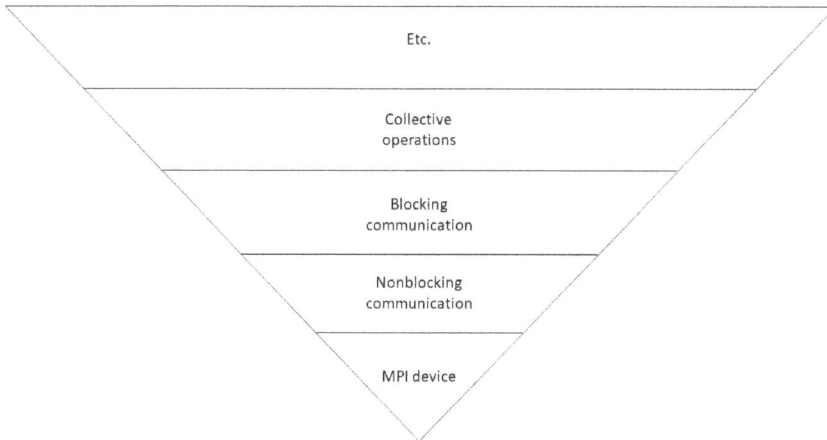

Etc.

Collective
operations

Blocking
communication

Nonblocking
communication

MPI device

Figure 1.3: MPI internals

The tip is referred to as the *MPI device*. This part of the MPI implementation is responsible for the process start-up, connection establishment, and subsequent interaction with the underlying data transfer fabric(s). On top of this lower part, the rest of the MPI is built in layers. Typically, the first layer includes the blocking or the nonblocking communication calls like MPI_Send() and MPI_Recv() or MPI_Isend() and MPI_Irecv(), respectively. With those bits in place, higher level capabilities are added one by one, typically starting with the collective operations. All these layers can either be configured, built, and bound together during the MPI build procedure (which is typical of the MPICH, for one), or can sort of find each other and start co-operating at run time (which is the more modern and modular approach used by the Open MPI).

It is intuitively clear that the larger the pyramid, the stronger its tip needs to be. Indeed, if you intend to implement the whole of the MPI-3, you have to take into account a veritable multitude of features that will most likely never be used in their completeness by any one given application. Still, the MPI device has to support all those features, and this leads to the increase in both the code complexity and in the performance overhead (compare this to my experience with the communication library mentioned above). Logically, if the structure is large and heavy, and its tip is strong, the rest of the structure has to be made strong as well. Consequently, if the structure is made smaller, both its tip and the structure itself can be made lighter.

Just ask yourself and honestly answer the following two questions:
- Is all MPI power needed by an average MPI user all the time? Definitely not.
- Is MPI complexity hurting its performance? Probably yes.

These two answers lead to the idea of the MPI subsetting. An *MPI subset* is understood as a smaller set of MPI functionality that would be sufficient for a certain class of applications. Thus, at least a part of the existing application code base is reused, and yet you can cut very substantial corners when implementing the respective MPI subset. Among other things, this book will show that, skillfully played, this bet indeed pays off both in speed and code size.

1.3.2 Sensible MPI Subsets

The simplest sensible subset can be seen in the very first MPI "Hello world" program shown in the MPI standard. It includes six essential MPI calls (MPI_Send(), MPI_Recv(), MPI_Init(), MPI_Finalize(), MPI_Comm_size(), and MPI_Comm_rank()) as well as one MPI communicator (MPI_COMM_WORLD), one predefined elementary datatype (MPI_CHAR), and one predefined semi-opaque MPI type (MPI_Status). Indeed, if you replace the MPI_CHAR with the MPI_BYTE, you can write a lot of sensible MPI programs for a homogeneous parallel system and may never even want to learn more of the MPI.

Another well-known example of an MPI subset is... the MPI-1 itself. As mentioned in Section 1.2.1, for a very long time indeed this was the highest common denominator between many different MPI implementations. Some of the corner features of it might or might not be implemented in all their completeness, but the basic functionality defined by the standard was generally provided, give or take a quirk or two. Even now, quietly, some of the existing implementations provide only a subset of the functionality defined by the current MPI standard because implementing and testing all of it may not be sensible from the point of view of their respective customer base.

However, between these two extremes (the infamous six-call-MPI mentioned above and the MPI-1.3 or, for that matter, the MPI-3.1 standard as a whole), there exist a few other meaningful intermediate MPI subsets. They can be defined in terms of the provided entry points, associated MPI objects, and their features. As a matter of fact, when a new communication library like an MPI implementation is being developed, it is quite reasonable to go through these subsets one by one (see Figure 1.4).

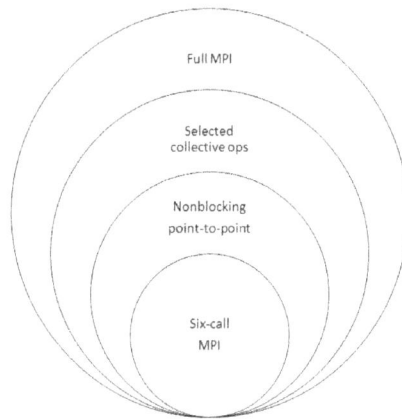

Figure 1.4: Sensible MPI subsets

The first sensible subset beyond the six-call-MPI includes several popular collective operations (MPI_Bcast(), MPI_Barrier(), and MPI_Allreduce()) that can be implemented first using the blocking point-to-point calls, or even directly if the communication fabric allows this. Others can be added as the need arises, if it ever will.

Another reasonable subset includes nonblocking point-to-point communication. Only the standard communication mode (i.e., MPI_Isend() and MPI_Irecv()) can be provided for starters. Note that as soon as the MPI_Isend() and MPI_Irecv() functions are available, you can also formulate a reasonable MPI_Sendrecv() function in terms of them, which may come in handy during further collective operation development. What to do first—the simple collectives mentioned above or the nonblocking point-to-point communication—is a matter of your priorities.

Beyond this there are a few inflection points that an implementation may need to go through in a certain sequence. Adding data heterogeneity, thread support beyond MPI_THREAD_FUNNELED, dynamic process management, one-sided communication, and nonblocking collectives may require big changes to the implementation reaching pretty deep into the dark matter, unless these extensions have been thought of in advance. The remaining MPI features—even file I/O and communicator management—can be thought of as an extra set of balconies, towers, bridges, and such—they won't change the main structure of the building, just add extra beauty and complexity on top of it. In other words, the last reasonable step is a full MPI implementation.

It is a big question whether this kind of (over)complication makes sense in certain cases when it is known *a priori* that some more advanced features will never be called upon. Thus, subsetting borders on the 10/90 principle, according to which maximum effort should be invested into the first 10% of work that gives 90% of the required functionality, while the remaining 90% of work asymptotically fills out the remaining 10% of the functionality and should be expended only if really necessary. We will be using subsetting actively throughout this book with this pragmatic consideration in mind. Indeed, chasing the last ounce of functionality sometimes (er, almost definitely) comes at the cost of performance lost elsewhere.

1.3.3 Exercises

Exercise 1.10 Define a sensible MPI subset a step beyond one that includes collective operations and nonblocking point-to-point communication.

1.4 Fine Points

Well, that was a lot of stuff for an overview. However, we are not done yet, not by a long shot. Knowing the standard and having a rough development plan is good, but this will not lead us anywhere unless some technology is added to the mix. Of course, what I say below is my point of view, rooted in my nature and nurture. Your mileage may vary. However, with my peak performance of about a thousand lines of debugged code per day, and the sustained output of about 75 lines of debugged code per day in the long haul, which is about 20 times higher than the normal 3–4 lines per day, I thought you might be interested in what I have to say on this highly polemical topic.

1.4.1 Programming

This is not an abstract book on program architecture and development. So, you will find no patterns, templates, and fancy interfaces like OLE, DCOM, or CORBA here. They all are too general, too slow, too clumsy for what we have to do: program hardware to the utmost of its innate performance.

Still, some advanced methods do deserve requisite attention. No one is going to argue against structured programming with all that it entails, including a mostly harmless goto statement here and there. Object orientation may help as a general approach, even though my language of choice—C—does not readily provide support for objects per se, just allowing a comparable method to be used by hand. Patterns are useful as additional knowledge, while templates and other fancy features can be very inspiring if enjoyed with due caution.

There are many points of view on source code formatting and commenting. Mine is that the fewer symbols there are, the more comprehensible the program is going to be. For example, calling a loop variable i is perfectly OK as long as it is clear from the context what it will be used for. Moreover, having been trained in the noble art of programming in those times when King Fortran and Queen Algol ruled the world in perfect harmony, I became accustomed to using letters i, j, k, l, m, n for integer variables without much ado. Likewise, I skip pointless blanks, use a tab to indent code when I can, and tend to put very few comments in, and those just to clarify design choices rather than to describe what exactly the program is going to do. After all, a properly written program itself is its best description.

And (predictably) I do not like nor do I use any Integrated Development Environments (IDE) if I can help it. First, I like my system native editor most. Second, I do not really need any syntax highlighting. Third, I hate changing my habits when someone out there decides that a feature well known for decades should work differently starting tomorrow. And fourth, and most important: I want to stay in complete control of the things that I create. This includes all aspects of the process, from design through development, code management, building, debugging, and benchmarking, to optimization and delivery.

1.4.2 Code Management

There are several mature and popular code management systems—cvs(1), svn(1), git(1), to name only a few—as well as several online portals that will be glad to offer you services of code management, version control, and even limitless backup. It's up to you what you use in your work on the exercises. I personally can use any of those systems mentioned above, but in this book, I chose not to do this. Every piece of software we are going to create will be rather small in size and limited in purpose. So, it is just easier to keep, say, a particular MPI subset implementation and all its source and service files in a separate subdirectory.

In fact, all that I wrote for this book is contained in a rather straightforward archive that can be downloaded from the accompanying website, https://www.degruyter.com/view/product/488201 under Supplementary Information. There, every chapter has a major directory, and every program from that chapter is contained in a corresponding subdirectory. If the program evolves with the text, I create special directories for different versions of the program. The program names are mentioned in the text and, most often, in the code listing captions. To complete the picture, these subdirectories contain README.txt files outlining what lives where.

1.4.3 Building

Here, again, I personally prefer to go on foot. There are wonderful tools like autoconf(1), automake(1) and so on that will take the burden of creating the standard configure and Makefile off your shoulders. In exchange, you will get code that is nearly unmanageable without those tools, and once in a while does weird things you will need an expert's advice on.

In this book, things are rather simple. So, in accordance with the 10/90 principle I write all `Makefiles` by hand, avoid using `configure`, and refrain from any help in choosing what will be started, when, and how. The standard GNU tool chain, including the GNU C compiler of the day and the associated build tool chain are used throughout. If you happen to have something else that you are more comfortable and familiar with, just go ahead and use that instead. Note that the closer your compiler syntax is to that supported by the GNU compilers, including both the language definitions and the compiler options, the easier it is going to be in the long run. However, the ease of use should be put against the performance, and here one does detect enormous differences that we will look into when the opportunity presents itself.

The same considerations are true of the operating system. In this book, I use the free Linux OS like Ubuntu 16.04 LTS and Debian 8.4. Commercial distributions like Red Hat Enterprise Linux (RHEL) and SUSE Enterprise Server (SLES) or their respective open source pendants Fedora Core and OpenSUSE may be pulled in once in a while for comparison, to the extent the license terms and availability allow. When it comes to supporting new hardware, you are likely to fare better with the commercial Linux distributions, at least early on in adoption. However, I do feel great respect for the developers of a truly open OS like Debian, and I use this OS or its derivatives like Ubuntu predominantly in this book.

This has a practical aspect, too: if your software is free, who is going to shell out very considerable license fees for a commercial OS to run it? This does not exclude the necessity of being compatible with the latter, of course, but puts this wish into proper perspective.

1.4.4 Testing and Debugging

Before a program is optimized, it must be correct. As my teacher once put it: "*Any correct program is infinitely faster than an incorrect one.*" Indeed, who needs a lightning fast program that produces rubbish on output, and who cares how fast it does that? Now, there are many ways to create correct programs, some of them very nice and scientific, most rather boring, but none of them misses the testing and debugging pit in practice. Since this is unavoidable, it makes sense to come prepared.

Tests should be created alongside the program. This is time consuming, yes, but this is also absolutely essential. In addition to providing immediate feedback after program changes, and indeed verifying new changes that may lead to the creation of new tests, those tests that output some data may also be used for regression testing, especially if run automatically and regularly. A test may also check its output itself, which is a bit of debugging—this is why I decided to put these two topics together in one section. In fact, testing and debugging are two sides of the same coin.

Programs in general, and parallel programs in particular, behave in some quite counterintuitive ways. In my experience, nothing helps to understand their behavior better than a full debugging output produced by the tried-and-true printf(3) statements put in at strategically important points in advance. Of course, printing is a system level activity that spoils any performance, and so it must be eliminated in a debugged program. Conditional compilation controlled by a macro like DEBUG is perfect for this. Why is this better than going through a program in an interactive debugger step by step? It is better for me but may not be better for you. I personally like knowing all about the program execution, as this helps to find issues that could otherwise go unnoticed. This does not exclude the usage of a good parallel debugger like DDT or TotalView, if you can afford one and cannot do without. I can, apart from looking into an occasional core file, to which end the gdb(1) works just fine.

Another conventional macro—NDEBUG—can be used to tell the assert(3) macros to fade into the background. These assertions are very useful in checking implicit assumptions and argument sanity, but they will rarely make sense if left active in the production code.

Finally, programs should check their own health. It does not make sense, normally, to print out the contents of the data buffers, especially if there are smarter methods to check that the program behaves in the expected way. Reducing a whole array of data to one number is a good first step. Checking the buffer contents against the expected template is another. Putting in a macro like CHECK to control the respective conditional compilation statements crowns this approach.

1.4.5 Benchmarking

Working with a product like MPI means fighting for every nanosecond that can be squeezed out of the critical code path, within reason. Benchmarking is an essential instrument that is used for finding out what runs well and what could be improved further still. There are just a few principles that need to be kept in mind.

Benchmarking should measure target performance in a *realistic setting*. If you want your program to work well under Linux OS, you should also benchmark it under the OS that works in the usual mode. Some manuals will tell you that you should disconnect your networking, disable all nonessential systems services, and even go down to privileged single user execution mode. Well, if your program will eventually work under these particular conditions, do this. If it does not, what's the point? Who other than the marketing department will ever enjoy your *"peak performance"*?

Benchmarking should measure *out-of-the-box performance*. Again, some manuals will tell you that you need to tune your program settings to achieve the maximum performance possible. Well, few users will ever open your program manual, and even fewer will be brave enough to start flipping the controls. Even more important, almost none will have the time for that. The only exception is when you measure processes that are influenced by something that you want to investigate, be that process binding or the influence of the aforementioned systems services. In that case alone does it make sense to do more than a normal resource-stripped, deadline-driven user would.

Benchmarking should be *reproducible*, that is, properly set up and well documented. The OS version and update level, the compiler version used and the flags employed, the system library versions, the BIOS (or EFI) version and settings relevant to performance, the network card driver version and other specifics, the time of day, the date, the benchmarking source code used, and any other extra data pertinent to the reproducibility should be noted and kept safe and sound for future reference. It is outright incredible how efficiently our mind filters out everything that we do not need to survive. You may have to recheck your prior findings when software changes, or you will discover a new effect that was not taken into account before, or your results will be challenged by a colleague or a customer. The possibilities are infinite. The answer is however unique: proper setup and documentation.

Benchmarking results should be *statistically sound*. This means that you should never trust a single measurement, unless this was a counter example, of which, of course, one instance is enough. Practically speaking, running a benchmark under well-controlled conditions three times (at most five) will normally show whether there is anything fishy with the method or the setup. When measuring a very fine-grained, particularly unpredictable, or known erratic behavior, more care and thus more attempts and more mathematical rigor will be required. Ask your mathematician friend to help you out in case of doubt or master the statistics yourself and save on the beer promised.

This said, nothing can replace your own *common sense*. Ask yourself what you are going to measure, how, why, and answer yourself honestly. This will help.

1.4.6 Presentation

Next, there is always a question of what to show and how to show it. Indeed, until data is processed and presented in a sensible fashion, it is difficult to comprehend it, to say nothing about actually using it to drive any optimization.

I personally prefer graphical representation. In the case of latency, a general overview of the whole graph, mapping latency against message length, as well as a close-up of the smaller messages (up to 1 KiB or 100 KiB) in the linear and, most often, logarithmic scaling of the message size axis, has proved to be most valuable. Indeed, by looking at the latency as a whole using linear scaling of the message size axis (see, e.g., Figure 3.5b), you learn what tendency you end up with, which allows direct comparison of different communication methods. By looking at the latency for short messages using logarithmic scaling of the message size axis in particular (see, e.g., Figure 3.5a), you learn in utmost detail what method gives you better results.

For bandwidth, a general overview in linear format as well as, most often, in logarithmic format of the message size axis provides the necessary overview (see, e.g., Figure 3.6a for the latter). It also hides unnecessary small message details better observed using the respective latency graph. Remember: to get latency right, you need to do something brilliant; to get bandwidth wrong, you have to do something terrible.

Finally, once in a while latencies and bandwidths of different fabrics will be so different as to force us to scale the value axis logarithmically as well (see, e.g., Figures 3.5b and 3.6b). Beware: fancy scaling has been shown to make a pig out of an elephant, graphically speaking, so use it with care! Due to this, we will use logarithmic scaling of the message size axis, and linear scaling of the value axis by default in this book, and explicitly highlight any deviation from this rule.

However, despite all the beauty and visual power of the graphical method, once the number of data series and factors to consider surpasses a very modest number of four to six, graphical representation loses its appeal, and one has to look into details using the tabular form (see, e.g., Table 2.4).

As for the values to be used for building those graphs and tables: some people prefer average values, some swear by geomean, some favor median, some select the best result available and proudly present it, some honestly show everything they have observed. None of these methods is flawless. Average values, either mean or geomean, do not show data observed in practice, only some approximation of it. None is good if the distribution is skewed and adding sigma does not help much. Median is not bad, unless the series selected accidentally represents an ugly outlier between two sound subgroups. The best result is not really scientific, especially if it is glued piecewise (as sometimes happens) from different series. And a sea of data is evidence of a poor understanding of the underlying data set.

In this book, we will use all of the above representations depending on the issue at hand, with preference given to actual observed median results, backed if necessary by statistical investigation performed in the background.

1.4.7 Optimization

MPI is all about performance. Apart from the sound development and benchmarking practices outlined above, we will have to deal with optimization. In fact, my friends and I have written a whole book on this topic alone (Supalov, Semin, Klemm, & Dahnken, 2014).

In short, here is how to optimize programs the right way. We called this approach top-down, closed-loop optimization. It is called *top-down because it addresses the levels of parallelism in the presumed order of potential effectiveness, namely:*
–   *Algorithms.* You can do whatever you want with the bubble sort, but it will be beaten hands down by any better sorting algorithm in most, if not all, situations.

Selecting a proper algorithm is normally the most efficient but also the most time- and skill-demanding optimization method. Think ahead.

- *Distributed memory*. If the target program uses distributed memory parallelism, as the programs in this book always will, the next step is to see if the data the program processes has been distributed sensibly and properly, that there is no load imbalance, and that the communication patterns and operations used by the selected algorithms are suitable for the purpose. This includes proper usage of the underlying communication mechanisms, including networks. In other words, this is a step where MPI usage is going to be optimized, either by tuning the MPI itself, or by changing the application so that it uses MPI better, or by improving the application to make it more amenable to distributed memory parallelization in general.

- *Shared memory*. If the target program uses threads, this usage can and should be analyzed next. If the program does not use threads, good reasons must exist for why this is not being done. In any case, unused cores or hardware threads means performance can be lost. From a technical point of view, attention is going to be paid to the data distribution, to the synchronization methods used, and to the data movement, if any is required.

- *Processor parallelism*. At last, there comes the time for the traditional intraprocessor optimization many people would tend to start with unless they were initiated to the top-down approach. Here, several sublevels exist:

    o First of all, commands and data should be coming to the execution units in a possibly uniform, uninterrupted stream at the maximum available memory or, better, cache bandwidth. So-called pipeline bubbles that happen when control or data flow is broken should be eliminated.

    o Second, the commands and data should be arranged in such a way that the aforementioned execution units are loaded with useful work as heavily and uniformly as possible: the multipliers should always be multiplying, dividers should be dividing, adders adding and subtracting, and so on.

    o Third, the results of these operations should reach the intended memory level without undue delay caused by data dependencies, cache aliasing, and still worse, paging or swapping.

You would hope that some of this petty work should be handled by compilers and even hardware. Indeed, computers have become exceedingly clever of late, so clever that they do beat us humans in chess and go. Well, to tell you the dirty little secret of the trade: computers still like simple programs most. The more elaborate your computation is, the likelier it is that something will go astray. So, keep it simple.

Thus far, we have talked about *what* should be optimized. Now, let's turn to *how* this should be done. The method we advocate is called *closed-loop* where several steps are done repetitively in a particular order leading to a predictable and systematic performance improvement:

- First, establish a performance baseline. This is a set of relevant benchmarking results showing how the target program(s) behave under well-controlled, repetitive, realistic conditions that correspond to the expected user practice.
- Next, find a bottleneck. This is a piece of the program or its execution that consumes the most time in proportion to its purpose. In a typical numerical program, this may be the inner loop, in a shared memory program this may be excessive synchronization, in a distributed memory program this may be huge data amounts being sent along the slowest communication paths, and so on. When looking for a bottleneck to address, use the top-down approach described above. If you find nothing at the current level, go down to the next lower one.
- Once you have found the reason for the performance slowdown, devise a cure and implement it and, this is very important, do this one reasonably sized and specifically targeted modification at a time. Avoid the urge to change this, this, and that all at once. Even if you feel that this is an "obvious" improvement, follow discrete steps if at all possible.
- Repeat the benchmarking and compare the results to the performance baseline established earlier. Ideally, they should show improved performance. If they don't, find out what went wrong and roll back the changes as required, or fix issues you have inadvertently introduced so that they screwed up the intended performance hike.
- Once performance does get better, repeat the process, making no assumptions as to what level your next bottleneck is going to live at. Keep in mind that benchmarking results that you have collected to prove the performance improvement are not equal to the new performance baseline that will normally be more extensive.

Of course, this is tedious. A stroke of a genius would have bypassed all these chores to create a perfect program, hopefully from scratch. Well, if you feel like it, give it a try. After all, it is never too late to scrap a program that went wrong, right? However, remember what one of the most recognized experts in the field once said about premature optimization (Knuth, 1974). And check program correctness regularly, too.

1.4.8 Issue Tracking

This is a problem you should not be overly bothered with while working with this book. I will always be around to help you out if need be. If, however, you want to make things full scale, go ahead and start an issue tracker for the programs you are going to write. I usually prefer post-it notes and well-composed README files that should document the sources of inspiration (most usefully as links), the design decisions, the program behavior and options, as well as its build procedure, running sequence, and expected results. If the program goes through several revisions, those should be documented or at least referenced as well, together with the dates of issue and the most important features added and/or changed. This very book is going to be such a README file in our case.

1.4.9 Exercises

Exercise 1.11 Examine the code management utilities offered and choose one you like best. You will probably want to use it throughout this book, so be careful when choosing between the cvs, svn, git, or something else.

Exercise 1.12 Experiment with the program building tools, including the C and the Fortran compilers, and find out what level of the respective language standard they support. If you have both GNU and other compilers, compare them using standard benchmarks like SPECint, SPECfloat, and Stream.

Exercise 1.13 Try out the MPI libraries that come with your OS distribution (if any) and make yourself comfortable with the respective MPI commands like mpicc, mpif77, and mpiexec (or mpirun).

Chapter 2
Shared Memory

In this chapter, we will create a simple MPI subset capable of basic blocking synchronous point-to-point communication over shared memory. While not being a complete MPI implementation, this subset will take a sensible first step toward the complexities expected in the following chapters, and at the same time set a useful reference point in performance, since shared memory is certainly the most sensitive and capricious of all communication fabrics we will deal with.

2.1 Subset Definition

Let's make our first MPI subset very simple. Shared memory is lightning fast, and the simpler the subset, the lower latency we will be able to achieve. To this end, we will:
- Provide the basic MPI functionality:
  - o The initialization and termination calls MPI_Init() and MPI_Finalize()
  - o The query calls MPI_Comm_size() and MPI_Comm_rank()
  - o The wall clock query function MPI_Wtime()
- Implement only rudimentary blocking communication:
  - o Standard mode blocking point-to-point calls MPI_Send() and MPI_Recv()
- Restrict the rest of the MPI functionality as follows:
  - o Assume the MPI_THREAD_SINGLE mode
  - o Support any process count
  - o Provide only the MPI_BYTE datatype
  - o Permit any nonnegative message count
  - o Allow message tags up to 32767
  - o Operate only over the MPI_COMM_WORLD communicator
  - o Assume the predefined MPI error handler MPI_ERRORS_RETURN

This basically gives us the six-call-MPI extended by the MPI_Wtime() function, as well as two MPI constants/predefined handles mentioned above plus the MPI_SUCCESS function return code, plus three type definitions (MPI_Comm, MPI_Datatype, and MPI_Status), i.e., 13 entities all in all; of them, seven are functions. Compare this number to 855 entities (of them 450 functions) in the full MPI-3.1 standard and feel the difference: we are going to implement only about 1.5% of the total MPI functionality and, in my opinion, still be able to serve about 98.5% of all MPI applications out there.

DOI 10.1515/97815015068971-002

2.1.1 Exercises

Exercise 2.1 Define a couple of MPI subsets a bit wider than the one described above but stopping short of adding nonblocking communication.

Exercise 2.2 Define a subset or two still narrower than the one described above. Think about message length, tag values, and other entities.

2.2 Communication Basics

Before we dive into the implementation of the MPI subset described above (oh, I understand that your hands are itching to get to the keyboard), let's look into the basic communication mechanisms that we will use. There will be enough programming involved, and we may learn quite a bit in the process.

2.2.1 Intraprocess Communication

Pushing bytes back and forth inside a process is normally done with the help of the memcpy(3) function in the hope that it was optimized to the maximum by the compiler runtime developers or the OS programmers. This is not necessarily always true, so you should start any implementation by verifying all assumptions you may have. It would be rather easy to find out how the memcpy(3) behaves inside a process by simply building the netpipe(1) benchmark and running it a couple of times. However, this way we would not get our hands dirty, and getting our hands dirty is what this book is about, right?

2.2.1.1   Naïve Memory Benchmark
So, let's write a naïve memory benchmark and see what it outputs (see Listing 2.1).

Listing 2.1: Naïve memory benchmark m1 (see shm/prep/m1/m1.c)

```
#include <string.h>
#include <sys/time.h>
#include <stdio.h>

#define IMAX      10000
#define LMAX      32*1024*1024

static double loh_wtime(void)
{
    struct timeval tv;
```

```
    gettimeofday(&tv,NULL);

    return tv.tv_sec + 0.000001*tv.tv_usec;
}

int main(int argc,char **argv)
{
    int i,imax = IMAX,l,lmax = LMAX;
    double t;
    char *b1,*b2;

    b1 = malloc(lmax);
    b2 = malloc(lmax);

    for (l = 0; l <= lmax; l = (l) ? l << 1 : 1) {
        t = loh_wtime();
        for (i = 0; i < imax; i++)
            memcpy(b2,b1,l);
        t = loh_wtime() - t;
        printf("bytes = %-8d\titers = %-8d\ttime = %-12.6g\t"
               "lat = %-12.6g\tbw = %-12.6g\n",l,i,t,t/i,l/t*i);
    }

    return 0;
}
```

The source code is pretty straightforward. After buffer allocation to the preset lmax size of 32 MiB performed by the malloc(3) function, the input buffer b1 is copied imax times over to the output buffer b2 to make the time interval measurable. This is done for all buffer lengths between 0 and lmax in steps that correspond to the powers of two. The time interval is measured by taking time stamps before and after copying with the help of the system timer queried by the gettimeofday(2) called by the function loh_wtime(). The "wtime" can be decoded as "wall clock time," but what does this funny prefix loh_ stand for? Sure, this only means "low hassle," in line with the 10/90 principle we discussed before, and has absolutely nothing to do with the legendary HAL 9000 (Wikipedia.org, HAL 9000, 2018c).

Listing 2.2 shows a sample output on my workstation.

Listing 2.2: Naïve memory benchmark m1: sample output

```
bytes = 0     iters = 10000   time = 0.000183821 lat = 1.83821e-08 bw = 0
bytes = 1     iters = 10000   time = 0.000192881 lat = 1.92881e-08 bw = 5.18455e+07
bytes = 2     iters = 10000   time = 0.000163078 lat = 1.63078e-08 bw = 1.2264e+08
bytes = 4     iters = 10000   time = 0.000147104 lat = 1.47104e-08 bw = 2.71916e+08
bytes = 8     iters = 10000   time = 0.0001719   lat = 1.719e-08   bw = 4.65387e+08
bytes = 16    iters = 10000   time = 0.000160933 lat = 1.60933e-08 bw = 9.94205e+08
```

```
bytes = 32        iters = 10000  time = 0.000160933 lat = 1.60933e-08 bw = 1.98841e+09
bytes = 64        iters = 10000  time = 0.000241041 lat = 2.41041e-08 bw = 2.65515e+09
bytes = 128       iters = 10000  time = 0.000327826 lat = 3.27826e-08 bw = 3.90452e+09
bytes = 256       iters = 10000  time = 0.000270128 lat = 2.70128e-08 bw = 9.47698e+09
bytes = 512       iters = 10000  time = 0.000398874 lat = 3.98874e-08 bw = 1.28361e+10
bytes = 1024      iters = 10000  time = 0.00057888  lat = 5.7888e-08  bw = 1.76893e+10
bytes = 2048      iters = 10000  time = 0.00129318  lat = 1.29318e-07 bw = 1.58369e+10
bytes = 4096      iters = 10000  time = 0.00110412  lat = 1.10412e-07 bw = 3.70975e+10
bytes = 8192      iters = 10000  time = 0.00196505  lat = 1.96505e-07 bw = 4.16886e+10
bytes = 16384     iters = 10000  time = 0.00382805  lat = 3.82805e-07 bw = 4.27999e+10
bytes = 32768     iters = 10000  time = 0.00911689  lat = 9.11689e-07 bw = 3.59421e+10
bytes = 65536     iters = 10000  time = 0.0181119   lat = 1.81119e-06 bw = 3.61839e+10
bytes = 131072    iters = 10000  time = 0.0336289   lat = 3.36289e-06 bw = 3.8976e+10
bytes = 262144    iters = 10000  time = 0.0581489   lat = 5.81489e-06 bw = 4.50815e+10
bytes = 524288    iters = 10000  time = 0.114553    lat = 1.14553e-05 bw = 4.57682e+10
bytes = 1048576   iters = 10000  time = 0.226953    lat = 2.26953e-05 bw = 4.62023e+10
bytes = 2097152   iters = 10000  time = 0.490494    lat = 4.90494e-05 bw = 4.27559e+10
bytes = 4194304   iters = 10000  time = 1.63837     lat = 0.000163837 bw = 2.56004e+10
bytes = 8388608   iters = 10000  time = 3.45719     lat = 0.000345719 bw = 2.42642e+10
bytes = 16777216  iters = 10000  time = 7.0863      lat = 0.00070863  bw = 2.36756e+10
bytes = 33554432  iters = 10000  time = 14.3233     lat = 0.00143233  bw = 2.34264e+10
```

Let's forget the statistics and other relevant stuff for a moment, just to learn what these numbers tell us when represented graphically (see Figure 2.1, note the logarithmic scale along the horizontal buffer length axis).

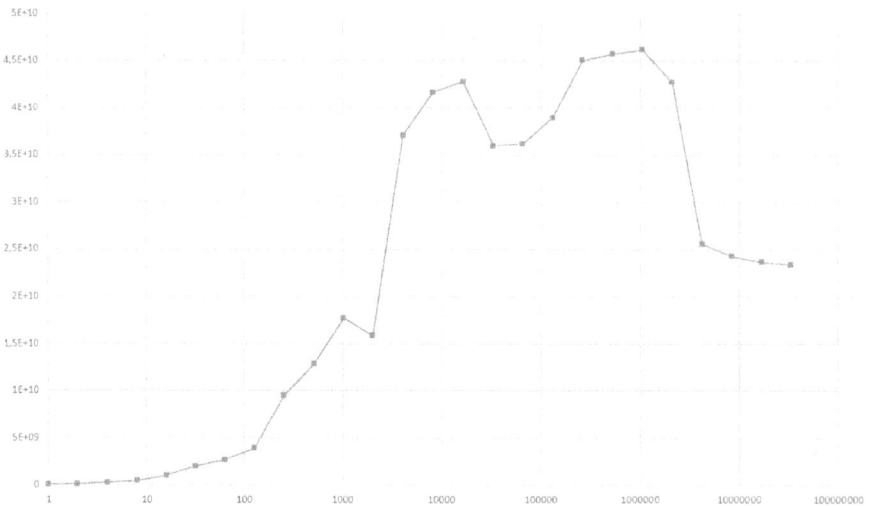

Figure 2.1: Naïve memory benchmark m1: bandwidth sample (bytes per second)

What we see is a rather typical memory copy bandwidth graph. After going through the caches, around message size of 1 megabyte, performance goes down to the band-width of the L2 cache that on this machine is 64 MiB in size. If we were to continue the measurement beyond that point, we would probably see the curve go down to the main memory bandwidth of about 25 GiB per second. Depending on the size and nature of the caches, one can normally see a couple of steps in the curve that corre-spond to the cache bandwidth going down.

If we look at the latency for a smaller buffer length range, we will see that it stays roughly unchanged for short buffers up to 128 bytes, and then grows linearly, give or take an outlier or two (see Figure 2.2; again, note the logarithmic scale along the horizontal buffer length axis; we cut the length range to 100000 bytes in order to get a meaningful picture for the lower lengths):

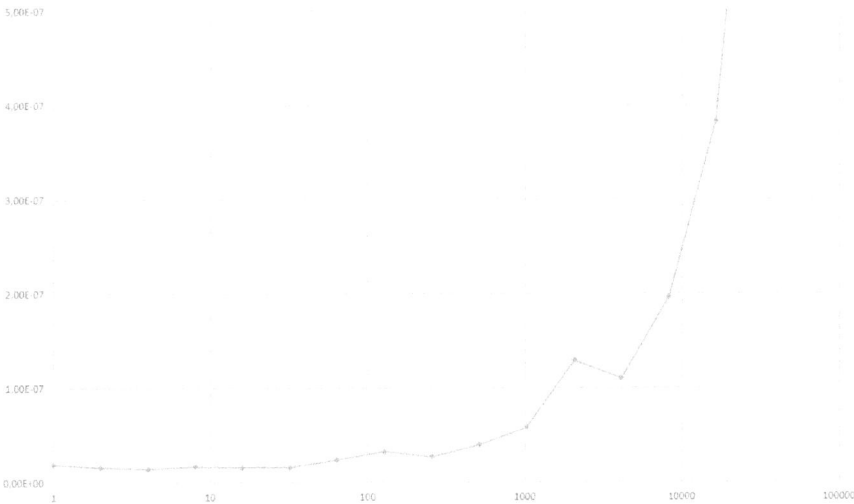

Figure 2.2: Naïve memory benchmark m1: latency sample (seconds)

So, we can sum up that as long as messages stay short enough, their latency is going to be rather low and nearly constant, while for very long messages everything will be determined by the last level cache and ultimately memory bandwidth. This is all we want to know at the moment, since this provides a most useful framework.

2.2.1.2 Ping-Pong Pattern

Let's modify this benchmark a little in order to explore a different data transfer pattern. What we have done so far was a one-way transfer, where the data was going in one direction only, from buffer b1 to buffer b2. This is a well-known data transfer pattern that is sometimes referred to as *ping* (not to be confused with the ping(1) utility typically used to crudely assess roundtrip performance of IP-based networks). When this pattern is used, what you get is how fast you can get data from point A to point B or, in other words, how many messages you can send over the link in a unit of time. There is another popular pattern, when data gets from point A to point B and then back—for example, after it has been processed, or when another data piece was generated on the other side and needs to be passed over. This exchange pattern is aptly called *ping-pong* (actually, the aforementioned ping(1) utility uses the ping-pong pattern). This is what the result looks like (see Listing 2.3):

Listing 2.3: Naïve memory benchmark m2: ping-pong pattern (see shm/m2/m2.c)

```
#include <string.h>        // memcpy(3)
#include <stdio.h>         // printf(3)

#include "loh.h"

#define IMAX      10000
#define LMAX      32*1024*1024

int main(int argc,char **argv)
{
    int i,imax = IMAX,l,lmax = LMAX;
    double t;
    char *b1,*b2;

    b1 = malloc(lmax);
    b2 = malloc(lmax);

    for (l = 0; l <= lmax; l = (l) ? l << 1 : 1) {
        t = loh_wtime();
        for (i = 0; i < imax; i++) {
            memcpy(b2,b1,l);
            memcpy(b1,b2,l);
        }
        t = loh_wtime() - t;
        printf("bytes = %-8d\titers = %-8d\ttime = %-12.6g\t"
               "lat = %-12.6g\tbw = %-12.6g\n",l,i,t,t/i/2,l/t*i*2);
    }

    return 0;
}
```

Here, we have already moved the earlier developed function loh_wtime() into a separate source file loh.c (see Listing 2.4):

Listing 2.4: Naïve memory benchmark m2: ping-pong pattern (source file `shm/prep/m2/loh.c`)

```
#include <sys/time.h>    // gettimeofday(2)
#include <stdio.h>       // NULL

#include "loh.h"

double loh_wtime(void)
{
    struct timeval tv;

    gettimeofday(&tv,NULL);

    return tv.tv_sec + 0.000001*tv.tv_usec;
}
```

The contents of this source file are described by the header `loh.h` (see Listing 2.5):

Listing 2.5: Naïve memory benchmark m2: ping-pong pattern (see `shm/prep/m2/loh.h`)

```
double loh_wtime(void);
```

Here, by the way, is a fitting `Makefile` that takes care of building the program as well as cleaning up and so on (see Listing 2.6). This is about the best way of documenting what needs to be done for the program to be used.

Listing 2.6: Naïve memory benchmark m2: ping-pong pattern (see `shm/prep/m2/Makefile`)

```
all: m2

run: all
        ./m2

m2: m2.c loh.c loh.h
        cc -O -o m2 m2.c loh.c

distclean:
        rm m2

.PHONY: all run distclean
```

Figure 2.3 depicts what the bandwidth looks like.

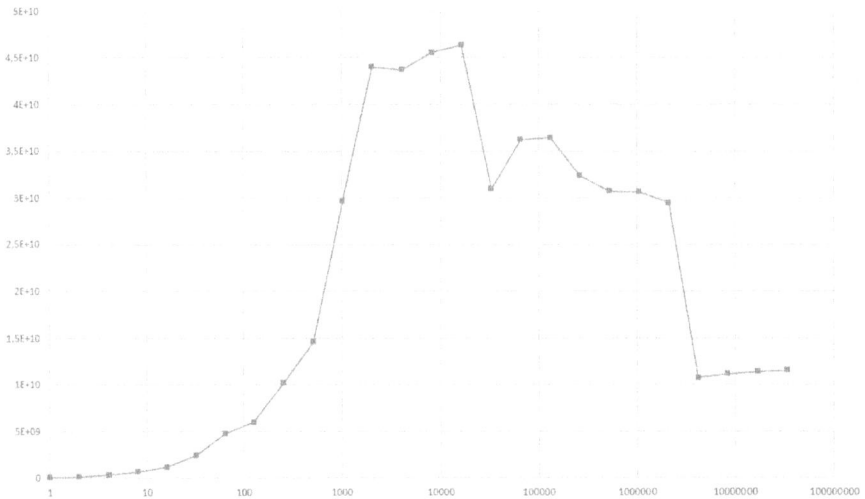

Figure 2.3: Naïve memory benchmark m2: bandwidth sample for the ping-pong pattern (bytes per second)

We see the usual step-like graph that has maxima corresponding to different cache levels and then to the last level cache or even the main memory. Note that apart from the peak, the values correspond to roughly half of those achieved by the original naïve benchmark.

2.2.1.3   Potential Improvements

Of course, both programs could be extended and improved. One could, for example, add runtime options to control the maximum buffer sizes, the maximum iteration count, and the message lengths that the process goes through. The maximum buffer size in particular could be of great interest since memcpy(3) performance may vary depending nontrivially on buffer size, with data transfers of buffer lengths of powers of two being sometimes substantially better optimized than others, and especially odd buffer lengths. However, we will not do any of this. On one hand, all this has been done many times over in benchmarks like the aforementioned netpipe(1) benchmark. On the other hand, short of replacing the compiler or even writing your own, better memcpy(3) function, there is little you can do about a memory related performance issue. Again, there are volumes dedicated to this topic alone.

Another, more pertinent consideration is that the naïve memory benchmark does not fill the buffers with any sensible information, so that they are most likely allocated in a lazy fashion and actually made available to the process only on the first access. This might spoil the first-time latency and hence influence the overall result at least for the zero-sized messages. A comparable consideration is true of the first pass of any measurement because it might eventually, depending on the relative buffer and cache sizes as well as buffer alignment, have to load at least a part of data into the higher-level cache in case prefetching has not done so already. Some kind of *warm-up*, that is, prior execution of the benchmark at hand without actually timing it, might take care of this. However, adding the warm-up phase may also spoil the following measurement, because it will then take place under more artificial conditions than in practice. In addition to this, the warm-up phase might also hide or at least soften some initial effects that could be relevant for real-life applications. This is why it may be advisable to do the first performance investigation without the warm-up and add it only if the effect of its absence appears detrimental to the message passing benchmarking. Finally, since in real life, an MPI implementation will never be sending zero bytes due to the presence of the control information in every message, some disturbance for very short messages may also safely be ignored.

In the same vein, the benchmark measures so-called *in-cache performance*, when whole buffers or at least sizeable parts of it fit into the last level cache and can be considered *warm*. Another approach would be to force an *out-of-cache* data transfer to see how *cold* buffers would work. Again, this is mostly a matter of choosing the right application-relevant scenario than anything else.

A smaller and yet notable consideration is the precision of the timer. Doing ten thousand iterations most likely levels this factor out, but for purely scientific purposes, you might want to know how low the actual timer tick is, so that the precision of the measurement be established.

Note that calling the `printf(3)` function inside the main benchmarking loop is normally considered acceptable and sort of enlivens waiting for results. However, in the case of fine measurements like those we are performing now, this may influence performance. In principle, one should collect the timing data into an array first, and only afterward print the data out as the rest of it is easily derivable.

This is also the motivation behind keeping the whole of the benchmark inside the main program and avoiding the very clearly possible introduction of at least one extra level of functions that would hide the nature of the benchmark from the main program. This is done to avoid any additional effects possibly introduced by the call and return sequences. Likewise, calling `memcpy(3)` even for the smallest messages may lead to a bit of overhead that is again undesirable. After all, passing zero-sized buffers does nothing else but calling the `memcpy(3)` that returns as soon as it finds out that there is nothing to process. Look how much this costs you! A part of that price is the payment for the lack of the warm-up mentioned above: not only user data but also the `memcpy(3)` code has to be loaded and make its way up the cache hierarchy.

Finally, the program is very careless about checking for and reacting to various error conditions. Memory allocation might conceivably fail, as might the output if it were redirected to a file residing in a nearly full file system. However, the 10/90 principle discussed earlier dictates that this kind of virtually impossible eventuality be ignored, because the program would fail either catastrophically and noticeably or quietly and gracefully, both of which are acceptable. Another seemingly weak point is the lack of proper cleanup that in this case would include the memory deallocation. However, the program termination takes care of that all right anyway.

2.2.2 Interprocess Communication

We are not going to compare all possible ways of passing data between, and synchronization of, different processes. We are writing a subset MPI implementation, so let's go to the fastest methods right away. If you want a comprehensive overview of all those mutexes, semaphores, remote procedure call protocols, and various data passing methods, you should read another book (Stevens, Unix Network Programming, Volume 2: Interprocess Communications (2nd Edition), 2012).

2.2.2.1 Naïve Shared Memory Benchmark
The easiest of the fast interprocess communication methods uses a shared memory segment to push the bytes, and polls for a flag to determine if the message has already arrived. Using the naïve memory benchmark as a source of inspiration, and knowing a couple of tricks of the trade, we can very quickly create a naïve shared memory benchmark (see Listing 2.7). Again, this is more instructive than trying to use anything else, since we will be reusing these basic mechanisms later on.

Listing 2.7: Naïve shared memory benchmark sm1 (see shm/prep/sm1/sm1.c)

```
#include <string.h>      // memcpy(3)
#include <sys/mman.h>         // mmap(2)
#include <unistd.h>      // fork(2)
#include <stdlib.h>      // exit(3)
#include <stdio.h>       // printf(3), perror(3)

#include "loh.h"

#define IMAX        10000
#define LMAX        32*1024*1024

int main(int argc,char **argv)
{
    int i,imax = IMAX,l,lmax = LMAX;
    double t;
    char *buf;

    volatile char *shm;
    pid_t pid;

    if ((shm = mmap(NULL,lmax + 1L,PROT_READ | PROT_WRITE,
                    MAP_ANONYMOUS | MAP_SHARED,-1,0L)) == MAP_FAILED) {
```

```
            perror(argv[0]);
            exit(EXIT_FAILURE);
        }

        if ((pid = fork()) == -1) {
            perror(argv[0]);
            exit(EXIT_FAILURE);
        }

        buf = malloc(lmax);

        for (l = 0; l <= lmax; l = (l) ? l << 1 : 1) {
            if (pid == 0) {      // "sender"
                t = loh_wtime();
                for (i = 0; i < imax; i++) {
                    while (*shm)
                        ;
                    memcpy((char *)shm + 1,buf,l);
                    *shm = 1;
                }
                t = loh_wtime() - t;
            }
            else {                   // "receiver"
                t = loh_wtime();
                for (i = 0; i < imax; i++) {
                    while (!*shm)
                        ;
                    memcpy(buf,(char *)shm + 1,l);
                    *shm = 0;
                }
                t = loh_wtime() - t;
            }
            printf("pid = %d\tbytes = %-8d\titers = %-8d\ttime = %-12.6g\t"
                   "lat = %-12.6g\tbw = %-12.6g\n",pid,l,i,t,t/i,l/t*i);
        }

        return 0;
}
```

The code is a little more involved, but just barely. First, we use the mmap(2) system call to allocate an anonymous shared memory segment shm to the size of the buffer lmax plus one byte to keep the transmission control flag at the head. We make it simple and rough by directly using the pointer arithmetic rather than by introducing a structure or a type definition. The hidden wisdom of this decision will become apparent later. Note that the flag is effectively declared volatile, which is important: it will be changed by the other process, so we need to make sure that those asynchronous changes become visible in our process' cache hierarchy and registers, too.

Then we clone the process by using the fork(2) system call. The shared memory segment allocated before the fork is inherited by both processes: the parent that gets the child process id (pid) as the fork(2) return value, and the child that gets zero. This time we do care if the return value -1 indicates an error, if only to show how system errors can be handled by committing a dignified *seppuku* with the help of the exit(3) function. Note that writing the last *haiku* is an integral part of the ritual, and the perror(3) function does just that.

We arbitrarily assign the child to be the sending process, and the parent to be the receiving process. Unlike the shared memory segment, the user buffer buf is allocated after the fork(2), as it is not going to be shared. We could have done this before, but this might be misleading. Since we operate in two different processes, the buffer variable may safely have the same name buf in both.

The actual exchange follows in the main loop that is already somewhat familiar to you. Since the mmap(2) system call not only allocates memory, but also initializes it to zero values, we do not need to care about redoing this in our program. Now, the sender process (child) waits for the shared buffer to become ready by polling the transfer control flag (the first byte of the shared memory segment) until it gets down to its initial value of zero. Then it fills the shared memory buffer using the contents of the user send buffer and sets the transfer control flag to one. The receiving process (parent) waits until the transfer control flag is set to a nonzero value. Then it copies over the contents of the shared buffer into the user receive buffer and sets the transfer control flag back to zero. And this goes on and on until the iteration count reaches its limit. Again, we skip all the cleanup, like buffer unmapping, parent waiting for the child, or memory deallocation and so on since the program termination will take care of this fully automatically: bang!

This is definitely not the best way of doing this transfer, and we see this confirmed (see Figure 2.4):

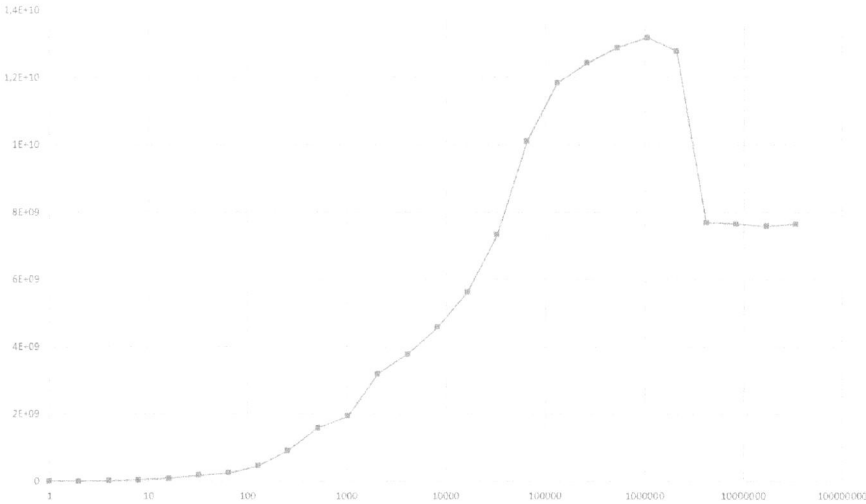

Figure 2.4: Naïve shared memory benchmark sm1: bandwidth sample (bytes per second)

2.2.2.2 Potential Improvements

Figure 2.4 shows that although the general shape resembles the naïve memory benchmark bandwidth in shape (see Figure 2.1), it runs at about one-third of its maximum.
At least the factor of two can be explained rather easily. Indeed, when the sending
process writes its data into the shared memory segment, the receiving process polls
for completion and does nothing else. Likewise, once the receiving process starts
copying the contents of the shared memory segment into the user buffer, the sending
process polls for completion and does nothing else. In other words, not only is double
the work performed, it is also fully serialized on the shared memory segment.

We will attack both the serialization and then the double copying later, since there
are well known ways of doing so: *double buffering* and *Direct Memory Access* (DMA).
Double buffering was invented very early in the history of computers to let various
parts of the system cooperate without stalling each other. In our case, it works by
letting the sending and the receiving processes operate in parallel: while the sender
writes data to one part of the shared memory segment, the receiver reads it from the
other part, and then they swap the parts and continue. With time, the sender and the
receiver automatically synchronize, so that their respective actions perfectly overlap.
Moreover, if the data to be transferred is split into smaller pieces, we can shorten
the first stage, when this exchange waits for the first data to be written. A look at the
bandwidth diagram with its characteristic step-like shape due to the memory hierarchy (see Figures 2.1 and 2.3) also suggests that by doing so we may eventually achieve
additional performance while we will be passing the data via the cache rather than
the main memory.

The DMA works by letting the process originating the operation write data directly to or read it directly from the address space of the targeted process. Of course, in order to do this, the originating process (or simply origin) has to learn the address of the buffer in the targeted process (or simply target) somehow, and, more importantly, have the means of overriding the address space protection that comes as part of the standard operating system services. It is somewhat ironic but not unusual, by the way, that extra protection offered by default requires additional hassle to do things informally. Fortunately, modern operating systems, including Linux, do offer additional services that allow just that without undue overhead. However, you must note that the use of those additional services comes at a cost: you need to have this feature available and activated, you have to set up the control structures, including learning the target buffer address somehow, and you have to pay whatever price the trap to kernel that is unavoidable in this case is going to incur. In other words, it may be less profitable to send all data this way, but it may become attractive above a certain buffer size, especially if the target address is known not to change too frequently or at all.

Using the double buffering or DMA will remove the factor of two, one way or another. For now, however, let's consider what else could be leading to the observed slowdown that amounts to the factor of three rather than two. This may help us deduce where the additional factor of 1.5 may be coming from.

First off, on allocation, the shared memory segment is probably aligned on the page (typically, 4 KiB) boundary, or at least on the word, i.e., the 4- or 8-byte boundary, depending on the 32- or 64-bit architecture, respectively. The latter is what the malloc(3) routine will normally do, and you can even enforce a particular alignment if that is necessary. At the moment, by putting the transfer control flag ahead of the shared buffer space, we effectively spoil the buffer alignment, which in turn makes the internal life of the memcpy(3) function unnecessarily complicated. Indeed, it may have to copy data from the user send buffer that is at least word-aligned, into the shared memory buffer that starts at an odd address, and then again copy data from the latter to the word aligned user receive buffer. In other words, we enforce two partially misaligned data transfers, which cannot be good.

Before we hasten to correct this apparent flaw, which can easily be done by, say, allocating at least a word at the beginning of the shared memory segment for the transfer control flag(s), let's think about another possible issue of potentially great import. Unlike the naïve memory benchmark, here we have to deal with two different processes. The operating system will normally put them onto different processor cores. What are these cores? Are they hardware threads running on the same physical core, or separate physical cores? If the latter, how do those cores relate to each other, the cache(s), and the memory interface(s)? All this is out of our control at the moment, and thus all our measurements hang in the air, being subject to forces we do not influence, although we could and should do so. Fortunately, modern operating systems provide the means for controlling the process placement that is called either *process binding* or *process pinning*.

Now let's make a plan of our further activities, spelling out what we want to do, what we want to achieve, and, most pertinently, in what order we are going to proceed:

- Use a properly aligned shared memory buffer to promote the fastest possible memory copying
- Research the influence of the process binding upon the data transfer performance
- Exploit double buffering to avoid the data transfer serialization and possibly gain additional performance
- Optionally, we may also want to:
- Add DMA capability to avoid the double copying and increase performance still further
- Find out when DMA overtakes double buffering in order to determine the switch-over point

We will do all this step by step, some in this chapter, some in later chapters, and will keep modularizing and cleaning up the code as we go. After all, we want to end up with a functional and well performing MPI subset rather than the most sophisticated shared memory benchmark on this planet, right?

2.2.2.3 Buffer Alignment
With this in mind, let's refactor the code so that it starts remotely resembling our final goal (see Listing 2.8):

Listing 2.8: Naïve shared memory benchmark sm2: refactored main program (see shm/prep/sm2/sm2.c)

```
#include <stdlib.h>    // exit(3)
#include <stdio.h>     // printf(3), perror(3)

#include "loh.h"

#define IMAX      10000
#define LMAX      32*1024*1024

int main(int argc,char **argv)
{
    int i,imax = IMAX,l,lmax = LMAX;
    double t;
    char *buf;
    pid_t pid;

    if ((pid = loh_init(lmax)) == -1) {
        perror(argv[0]);
        exit(EXIT_FAILURE);
    }
```

```
    buf = malloc(lmax);

    for (l = 0; l <= lmax; l = (l) ? l << 1 : 1) {
        if (pid == 0) {      // "sender"
            t = loh_wtime();
            for (i = 0; i < imax; i++)
                (void)loh_send(buf,l);
            t = loh_wtime() - t;
        }
        else {                    // "receiver"
            t = loh_wtime();
            for (i = 0; i < imax; i++)
                (void)loh_recv(buf,l);
            t = loh_wtime() - t;
        }
        printf("pid = %d\tbytes = %-8d\titers = %-8d\ttime = %-12.6g\t"
                "lat = %-12.6g\tbw = %-12.6g\n",pid,l,i,t,t/i,l/t*i);
    }

    return 0;
}
```

This time we moved certain bits into the already existing extra file loh.c that now looks like this (see Listing 2.9):

Listing 2.9: Naïve shared memory benchmark sm2: refactored message passing (see shm/prep/sm2/loh.c)

```
#include <sys/time.h>      // gettimeofday(2)
#include <stdio.h>         // NULL
#include <string.h>        // memcpy(3)
#include <sys/mman.h>      // mmap(2)
#include <unistd.h>        // fork(2)

#include "loh.h"

#define SOFF        1L
static volatile char *shm;

int loh_init(int len)
{
    if ((shm = mmap(NULL,len*2L,PROT_READ | PROT_WRITE,
                    MAP_ANONYMOUS | MAP_SHARED,-1,0L)) == MAP_FAILED)
        return -1;

    return (int)fork();
}

int loh_send(void *buf,int len)
```

```
{
    while (*shm)
        ;
    memcpy((char *)shm + SOFF,buf,len);
    *shm = 1;

    return 0;
}

int loh_recv(void *buf,int len)
{
    while (!*shm)
        ;
    memcpy(buf,(char *)shm + SOFF,len);
    *shm = 0;
}
```

...

This time we decided to return the process id from the initialization call loh_init() where it is readily available, and allocate the shared memory segment shm pointed to by a static variable so that it is visible to all routines in this source file. On the other hand, since both sending and receiving go in one direction for now, we refrained from overburdening the definition of the respective calls loh_send() and loh_recv() by adding an extra argument denoting the destination process. Both design decisions are questionable and temporary, but the result does its job, right? Note also that we introduced a predefined offset SOFF that will allow experimenting with the buffer alignment. Finally, we skipped the definition of the loh_wtime() function since it has not changed since the last time (see Listing 2.4).

Here is the matching loh.h header (see Listing 2.10):

Listing 2.10: Naïve shared memory benchmark sm2: refactored declarations (see shm/prep/sm2/loh.h)

```
int loh_init(int lmax);
int loh_send(void *buf,int len);
int loh_recv(void *buf,int len);
double loh_wtime(void);
```

Let's see whether changing the offset is going to influence performance, and whether we can see any effect of the refactoring on top of that (see Figure 2.5):

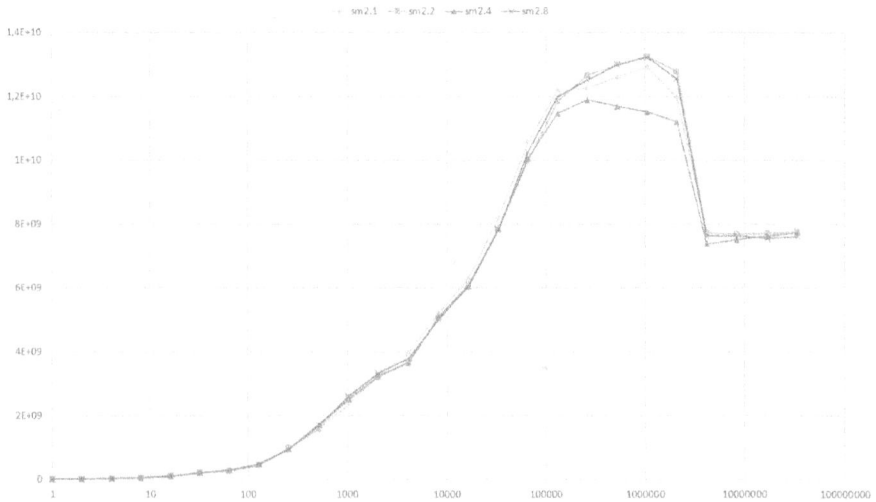

Figure 2.5: Naïve shared memory benchmark sm2: bandwidth depending on the offset value (bytes per second)

So, there seems to be no big difference between offsets of 1, 2, 4, and 8 bytes represented by the curves sm2.1, sm2.2, sm2.4, and sm2.8, respectively (with the counterintuitive exception of the lower maximum for offset 4 that needs to be explained). This means that either the memcpy(3) was written very well indeed, or that it was done very badly. In both cases there would be no difference, and we can only find out what case we have by comparing this series to another, presumably still better, memcpy(3) implementation. As to the maximum and limit numbers (again, with the exception of offset 4), this graph looks similar to the one before refactoring (see Figure 2.4). Of course, there remains the nagging suspicion that the offset should have been set to 64 bytes (cache line length).

2.2.3 Ping-Pong Pattern Revisited

Let's now make another decisive step by adding the sending and receiving capabilities to both processes. You could have used the same buffer for both purposes, but this would require additional synchronization to make sure that only one process is going to be writing data into the shared memory buffer. Of course, this would add extra overhead and thus spoil the data transfer latency. However, with memory as cheap as it is nowadays, we can bypass this nuisance by allocating two buffers: one for the parent to send data to the child, and the other one for the opposite direction. At the moment, they will be used independently, but consecutively, rather than simultaneously. Even though this looks like an outright waste, we will see why this flexibility

is going to be welcome. Moreover, we will soon drastically reduce the memory consumption, so let your love for optimization rest for a while.

One more improvement will be welcome, and that is some OS-independent way
of identifying the processes. So far, we have been using the process id, but we do not
have to do so. What if we assigned a small integer value, counting from 0 up, to the
processes we create? Then we would have a simple and sure way of addressing any of
them. This is also the basic idea behind the notion of the *MPI process rank*. Of course,
the send and receive calls have to get this extra argument, too.

With these changes, the benchmark and the accompanying source files look as
follows (see Listing 2.11):

Listing 2.11: Naïve shared memory benchmark sm3: ping-pong pattern main program (see shm/
prep/sm3/sm3.c)

```
#include <stdlib.h>        // exit(3)
#include <stdio.h>         // printf(3), perror(3)

#include "loh.h"

#define IMAX       10000
#define LMAX       32*1024*1024

int main(int argc,char **argv)
{
    int i,imax = IMAX,l,lmax = LMAX,r,o;
    double t;
    char *b1,*b2;

    if ((r = loh_init(lmax)) == -1) {
        perror(argv[0]);
        exit(EXIT_FAILURE);
    }

    b1 = malloc(lmax);
    b2 = malloc(lmax);

    o = !r;

    for (l = 0; l <= lmax; l = (l) ? l << 1 : 1) {
        if (r == 0) {           // "sender"
            t = loh_wtime();
            for (i = 0; i < imax; i++) {
                (void)loh_send(b1,l,o);
                (void)loh_recv(b2,l,o);
            }
            t = loh_wtime() - t;
        }
```

```
        else {                  // "receiver"
            t = loh_wtime();
            for (i = 0; i < imax; i++) {
                (void)loh_recv(b2,l,o);
                (void)loh_send(b1,l,o);
            }
            t = loh_wtime() - t;
        }
        printf("r = %d\tbytes = %-8d\titers = %-8d\ttime = %-12.6g\t"
               "lat = %-12.6g\tbw = %-12.6g\n",r,l,i,t,t/i/2,l/t*i*2);
    }

    return 0;
}
```

So, instead of one process id variable pid we now have two variables: this process' rank r and the other process' rank o that are computed only once for convenience. Of course, this process' rank r is also printed out when it comes to identifying the output lines.

The loh.c file changes accordingly (see Listing 2.12):

Listing 2.12: Naïve shared memory benchmark sm3: ping-pong pattern message passing (see shm/prep/sm3/loh.c)

```
#include <sys/time.h>      // gettimeofday(2)
#include <stdio.h>         // NULL
#include <string.h>        // memcpy(3)
#include <sys/mman.h>      // mmap(2)
#include <unistd.h>        // fork(2)

#include "loh.h"

#define SOFF    1L
static char *shm;
static int lshm = 0;

int loh_init(int len)
{
    int r;

    lshm = len + (int)SOFF;

    if ((shm = mmap(NULL,lshm*2L,PROT_READ | PROT_WRITE,
                    MAP_ANONYMOUS | MAP_SHARED,-1,0L)) == MAP_FAILED)
        return -1;

    return ((r = (int)fork()) > 0) ? 1 : r;
}
```

```
int loh_send(void *buf,int len,int dest)
{
    volatile char *pshm = shm + lshm*dest;

    while (*pshm)
        ;
    memcpy((char *)pshm + SOFF,buf,len);
    *pshm = 1;

    return 0;
}

int loh_recv(void *buf,int len,int src)
{
    volatile char *pshm = shm + lshm*(!src);

    while (!*pshm)
        ;
    memcpy(buf,(char *)pshm + SOFF,len);
    *pshm = 0;
}
```

...

Upon allocating double the amount of memory for the shared memory segment, and a fork(2), and some trickery with its return value in the loh_init() function, this process sends data via a portion of the shared memory segment identified by the destination rank dest by calling the loh_send() function, and receives data via the other portion identified by the negation of the source rank src by calling the loh_recv() function. This way process rank 0 uses the first portion of the shared memory segment for receiving data, and the second portion for sending it over to process rank 1. Process rank 1 in turn does the opposite, and everybody is as happy as can be, because there is no conflict or interference to resolve. This way, we get the maximum available data transfer performance by, yes, squandering some memory—at least for now.

The loh.h header pulls along to account for the extra arguments (see Listing 2.13):

Listing 2.13: Naïve shared memory benchmark sm3: ping-pong pattern declarations (see shm/prep/sm3/loh.h)

```
int loh_init(int lmax);
int loh_send(void *buf,int len,int dest);
int loh_recv(void *buf,int len,int src);
double loh_wtime(void);
```

If you look closer into the output, you will notice that the last line output by process rank 0 may get garbled by intermixing with the command prompt output by the controlling shell instance. This indicates that when process rank 1 terminates and control is returned to the shell, process rank 0 (remember, this is a child of the process that has just terminated) outlives its parent and so potentially interferes with the shell output. We could counter this by waiting on the child inside the parent, of course. For that, we would require a termination call, say, loh_finalize(), and a permission from the MPI standard to make it blocking on the termination of the child process. Another possibility is to assign rank 0 to the parent process in the hope that waiting for the last child's message, it will stay around a little longer and hence return control to the shell a little later, saving the output from garbling.

For now, this is what bandwidth looks like (see Figure 2.6):

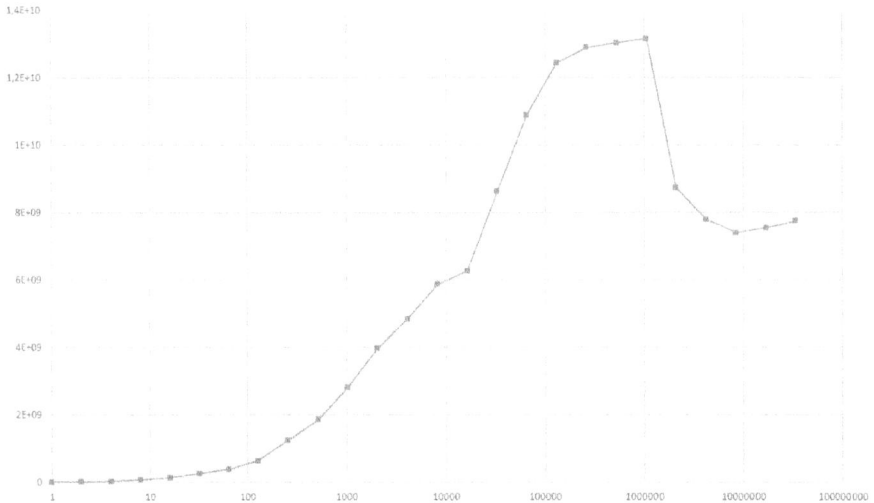

Figure 2.6: Naïve shared memory benchmark sm3: ping-pong pattern bandwidth (bytes per second)

Before we get enthusiastic and start implementing an MPI subset using the knowledge we have gained, let's compare latencies and bandwidths we have observed thus far. First, let's look at the bandwidths, using our familiar logarithmic scaling of the buffer length axis (see Figure 2.7).

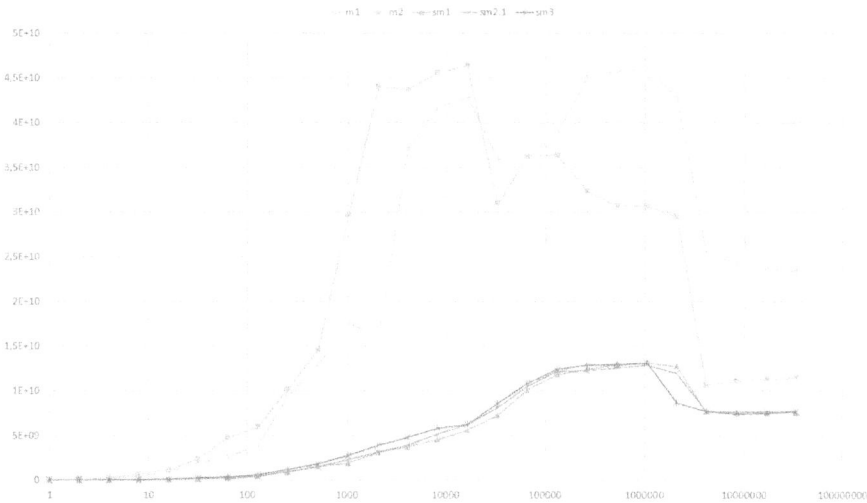

Figure 2.7: Observed memory and shared memory bandwidth summary (bytes per second)

You can see that interprocess data transfer performance represented by the sm1, sm2.1 and sm3 curves does suffer substantial disadvantage compared to the intraprocess data transfer performance represented by the m1 and m2 curves, within the many limitations of our benchmarking technology. This is why superfluous communication should be avoided, and the hassle of introducing parallelization must be well justified by the expected gain in application performance. In other words, when done improperly, MPI and its use may well hurt.

Another interesting observation is while ping or ping-pong intraprocess transfers go relatively up and down, with the ping-pong pattern being asymptotically about half as fast as the ping pattern; interprocess ping and ping-pong, apart from being substantially slower overall (apparently due to cache issues) do not differ too much, except for the area around 2 MiB.

Let's look into latency, again in logarithmic format and for lower buffer lengths (see Figure 2.8):

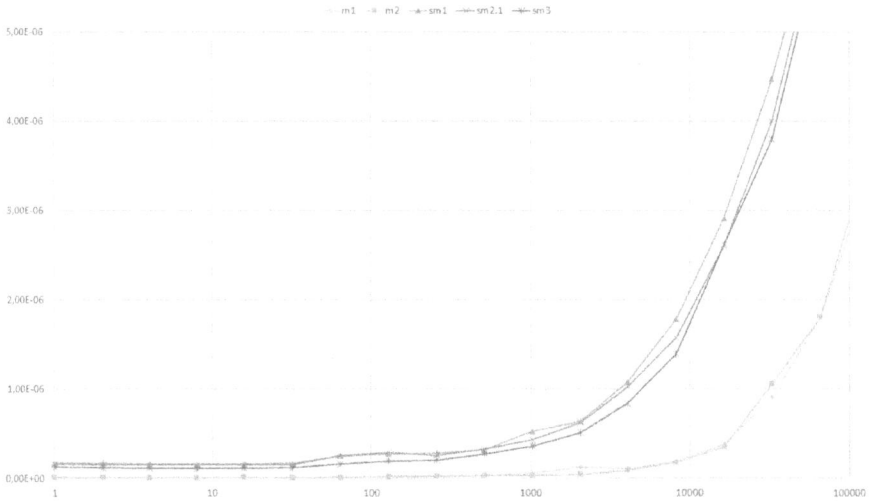

Figure 2.8: Observed memory and shared memory latency summary (seconds)

It is outright remarkable that both ping and ping-pong intraprocess transfers represented by the m1 and m2 curves have nearly the same latency until about 128 KiB. After that they split and start living completely different lives. At the same time, the interprocess transfers represented by the sm1, sm2.1, and sm3 curves stay close to each other, as already observed in the bandwidth graphs above (see Figure 2.7).

2.2.3.1 Process Binding

Now that we have a couple of benchmarks, we can deal with the process binding in a sensible fashion. There are several ways of controlling this binding. The easiest for the purposes of our benchmarking is the taskset(1) command. In order to use it, you have to understand the processor core layout.

Under Linux OS, the cat /proc/cpuinfo command provides the necessary data upon trivial postprocessing (see Listing 2.14):

Listing 2.14: Sample processor core layout

```
processor        : 0
physical id      : 0
core id          : 0

processor        : 1
physical id      : 0
core id          : 1

processor        : 2
physical id      : 0
```

```
core id         : 2

processor       : 3
physical id     : 0
core id         : 3

processor       : 4
physical id     : 0
core id         : 0

processor       : 5
physical id     : 0
core id         : 1

processor       : 6
physical id     : 0
core id         : 2

processor       : 7
physical id     : 0
core id         : 3
```

It can be seen that my laptop has only one processor with 4 cores and 8 hardware threads total (two per core). If this information is represented in tabular form, this is what we see (see Table 2.1):

Table 2.1: Sample processor core layout

Thread	Core
0	0
1	1
2	2
3	3
4	0
5	1
6	2
7	3

In other words, threads number 0 and 4 share core number 0, threads number 1 and 5 share core number 1, and so on. Let's see how our naïve shared memory benchmarks sm1 and sm3 behave depending on the core selection. It is sufficient, at least for starters, to measure how using threads 0 and 1, and then threads 0 and 4 influence performance. This way we will learn typical intracore and intercore results. First, let's look at the bandwidth (see Figure 2.9).

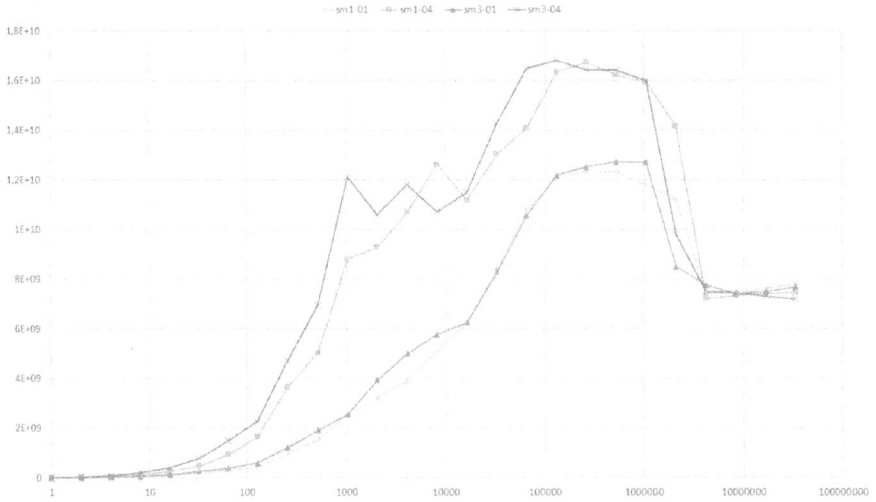

Figure 2.9: Naïve shared memory benchmarks: bandwidth depending on the process binding (bytes per second)

It can be seen that threads 0 and 4 (allocated by the command `taskset -c 0,4 ./sm3` and represented by the curves sm1-04 and sm3-04) that do *share a core*, fare noticeably better than threads 0 and 1 (allocated by the command `taskset -c 0,1 ./sm3` and represented by the curves sm1-01 and sm3-01) that sit on *adjacent cores*. Apparently, the higher level of cache sharing inside the core leads to better performance in this case. Latency observation confirms this conclusion (see Figure 2.10):

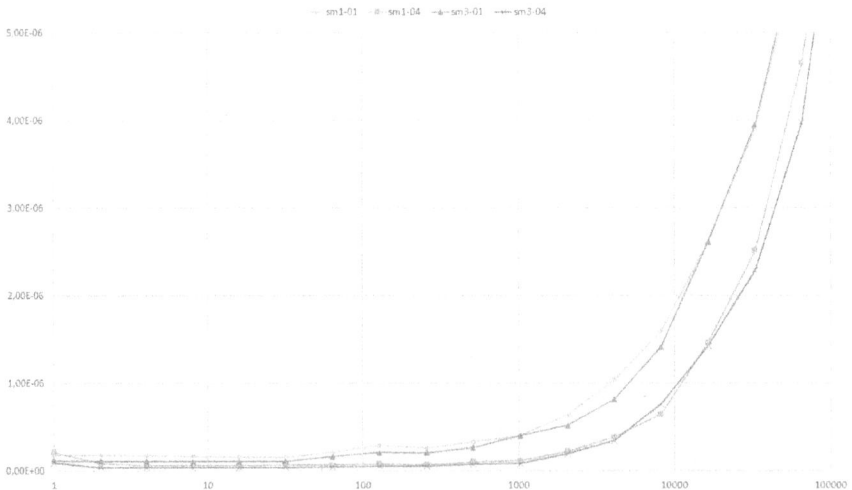

Figure 2.10: Naïve shared memory benchmarks: latency depending on the process binding (seconds)

However, both curve families still fare about three to four times worse than intraprocess data transfer performance (see Figures 2.7 and 2.8, respectively). Our task for the rest of this chapter and beyond will be to improve on this as much as possible.

2.2.4 Exercises

Exercise 2.3 Check out bigger alignment offsets (up to and including 64) in the refactored naïve shared memory benchmark. Also recheck offsets from 8 down on your system, process the data and see whether there is any difference in performance. Use the best offset detected in the rest of your exercises.

Exercise 2.4 Redo Exercise 2.3 using a different (hopefully, even better optimizing) compiler. If its memcpy(3) function performs noticeably better, use this compiler from now on. If data transfer performance demonstrates any dependency upon the alignment, adjust the offset accordingly going forward.

2.3 Subset Implementation: Blocking Communication

Now that we know the basic process creation and communication mechanisms, we have two options for how to proceed:
- Adapt an existing MPI implementation
- Create a new MPI implementation

Of course, creating a new implementation from scratch is much more fun, and this is what this book is about, after all. There are tons of literature on how to extend an existing MPI implementation, be that MPICH or Open MPI, so you will have to look it up if you need guidance on that matter.

2.3.1 Design Decisions

Before we begin, we need to make a couple of design decisions. Since we are free in our creativity, we could have done anything. However, based on my experience with the MPI implementations, we will choose one particular way that is closer to the MPICH heritage (Argonne National Laboratory, 2017).

2.3.1.1 MPI Header File
Let's formulate a brand-new header file mpi.h to suit our purpose (see Listing 2.15):

Listing 2.15: Subset MPI: declarations (see shm/mpi/v0/mpi.h)

```
typedef int MPI_Datatype;
typedef int MPI_Comm;
typedef struct {
    int MPI_SOURCE;
    int MPI_TAG;
    int MPI_ERROR;
} MPI_Status;

#define MPI_COMM_WORLD              0

#define MPI_BYTE                    0

#define MPI_SUCCESS                 0

int MPI_Init(int *pargc,char ***pargv);
int MPI_Finalize(void);

int MPI_Comm_size(MPI_Comm comm,int *psize);
int MPI_Comm_rank(MPI_Comm comm,int *prank);

int MPI_Send(void *buf,int cnt,MPI_Datatype dtype,int dest,int tag,
             MPI_Comm comm);
int MPI_Recv(void *buf,int cnt,MPI_Datatype dtype,int src,int tag,
             MPI_Comm comm,MPI_Status *pstat);

double MPI_Wtime(void);
```

There are a couple of interesting points here. First, due to the intended simplicity of the target MPI subset, we are not going to do much about communicators and datatypes just yet. It is easier to define the respective opaque objects MPI_Comm and MPI_Datatype as simple integers. Of them, we are going to use only the constants MPI_COMM_WORLD and MPI_BYTE anyway, and this design decision saves us quite a bit of unnecessary fuss. Since these are so-called opaque objects, this design decision can be revised any time.

Second, the MPI_Status object contains three standard fields that are directly accessible by the MPI user (the MPI_SOURCE, MPI_TAG, and MPI_ERROR). No other fields are defined or indeed necessary, as we do not foresee the means of setting and querying them anyway. Note however that the field MPI_ERROR is only set under very special circumstances and should be left unchanged otherwise.

The rest of the header file just follows the letter of the standard, with the added finesse of using a bit of the Hungarian notation that I personally find quite useful if administered in homeopathic doses.

2.3.1.2 Skeleton MPI Library

We can also formulate a skeleton library now, borrowing at least one function from the earlier sources (see Listing 2.16):

Listing 2.16: Subset MPI: skeleton library (see shm/mpi/v0/mpi.c)

```
#include <sys/time.h>    // gettimeofday(2)
#include <stdio.h>       // NULL

#include "mpi.h"

int MPI_Init(int *pargc,char ***pargv)
{
    return MPI_SUCCESS;
}

int MPI_Finalize(void)
{
    return MPI_SUCCESS;
}

int MPI_Comm_size(MPI_Comm comm,int *psize)
{
    return MPI_SUCCESS;
}

int MPI_Comm_rank(MPI_Comm comm,int *prank)
{
    return MPI_SUCCESS;
}

int MPI_Send(void *buf, int cnt,MPI_Datatype dtype,int dest,int tag,
             MPI_Comm comm)
{
    return MPI_SUCCESS;
}

int MPI_Recv(void *buf,int cnt,MPI_Datatype dtype,int src,int tag,
             MPI_Comm comm,MPI_Status *pstat)
{
    return MPI_SUCCESS;
}

double MPI_Wtime(void)
{
    struct timeval tv;

    gettimeofday(&tv,NULL);
```

```
    return tv.tv_sec + 0.000001*tv_usec;
}
```

2.3.1.3 Testing

Even this simple skeleton library allows for the creation and execution of the first sensible test program (see Listing 2.17):

Listing 2.17: Subset MPI: simple process creation test (see `shm/mpi/v0/t0.c`)

```c
#include <stdio.h>

#include "mpi.h"

int main(int argc,char** argv)
{
    MPI_Init(&argc,&argv);
    printf("Hello!\n");
    MPI_Finalize();
}
```

We ignore the MPI return values here. In case of an MPI error, control would normally be passed to the predefined error handler `MPI_ERRORS_ARE_FATAL` that basically aborts the process. Thus, we would never get back to the calling program. Here we will, which will lead to some interesting results later on. However, in order not to obfuscate the code, we will not be dealing with MPI errors explicitly until absolutely necessary.

A fitting `Makefile` completes the picture (see Listing 2.18):

Listing 2.18: Subset MPI: simple `Makefile` (see `shm/mpi/v0/Makefile`)

```make
all: lib

test: all run

lib: libmpi.a

libmpi.a: mpi.o
        ar -cr libmpi.a mpi.o

mpi.o: mpi.c mpi.h
        cc -c -I. -O -o mpi.o mpi.c

run: t0
        ./t0

t0: t0.c mpi.h libmpi.a
        cc -I. -O -o t0 t0.c -L. -lmpi
```

```
clean:
        -rm *.o

distclean: clean
        -rm t0 *.a

.PHONY: all test lib run clean distclean
```

This time we extend the build procedure by creating a static MPI library `libmpi.a` with the help of the `ar(1)` command, and linking a separately compiled test t0 against it. We also use the minus sign (-) in front of the `rm(1)` utility invocation so that error codes returned by it, when it misses the files to remove, do not interrupt the `make(1)` execution.

The test program output is easily predictable and thus not shown.

2.3.2 Start-up and Termination

Plenty of things have to happen for an MPI program to get started:
–   A number of processes, as requested by the user, should be created
–   Program arguments, if any, should be passed to them somehow
–   The MPI library has to be initialized

We will consider these steps below, as well as the MPI program termination and the way in which it can query the basic characteristics of its MPI universe.

2.3.2.1 Design Considerations
The only limitation we have in the already implemented process creation method (see Listing 2.12) is its ability to create no more than two processes. This is easily cured: if we keep on forking processes as if we were a living thing growing by splitting its cells, we can create as many processes as we want and do this really fast, since no other simple function is better at creating siblings than the exponent.

The last bit to contemplate is the maximum practicable size of the process universe we would ever create. You could leave this to the user who will determine how many processes they want to run. However, this would probably come at the cost of making the memory allocation a bit unpredictable. On the other hand, a typical HPC application will want to use all cores of the available processors. Starting more processes rarely makes sense, and starting fewer is perfectly fine, especially if the remaining cores are occupied by threads that, as we learned before, are much faster in exchanging data than the processes (not to mention that threads can just pass pointers to the data to each other). So, having in mind that we will hardly want to start more processes than the number of cores (or hardware threads) supported by the processor puts another reasonable cap upon the resource and especially memory consumption.

2.3.2.2 Argument Passing

Finally, how will this user-defined number of processes reach our library? The MPI standard and practice provide at least two ways to handle this:
- The mpiexec or mpirun command
- The program options

As long as we work with shared memory, it just feels more natural to use the second method. To this end, the list of options passed to the program must be parsed by the library initialization routine that should have the ability to change this list before it passes it back on to the program. Otherwise options intended for the library could possibly confuse the application and vice versa. Of course, there remains the unfortunate possibility that the library and the program may use similar looking options, which can lead to regrettable mishaps. However, since we have both under our control now, we will just make sure this never happens, right?

This will be helped by the limited number and simple syntax of the options to be supported first. Their synopsis and behavior are similar to those provided by the mpiexec command that we will learn more about later on, that is:

```
$ program [-n[p] number_of_processes] [arguments]
```

This way you can start the *number_of_processes* instances of the *program*, each supplied with the optional *arguments* to work on. Note that since we use the fork(2) system call, our processes come to be at the point of the fork(2) invocation, and so everything that has happened before this, including the results of the argument processing, memory allocation, and so on, becomes a part of their DNA, so to speak.

And what happens if no option to control the number of processes is passed to the program? In this case the MPI standard proposes that a high quality MPI implementation (and we aim at creating one) will attempt to start a so-called singleton MPI process, in which the number of processes in the MPI_COMM_WORLD communicator is equal to one. Since all our prior naïve parallel benchmarks would start two processes by default, this looks like a backward compatibility issue. However, those were not MPI programs. So, we are free and even encouraged to follow the letter of the standard in this case.

2.3.2.3 MPI Initialization

Once all these considerations have been taken into account, this is what we end up with (see Listing 2.19). Some things have been left out for clarity; we will revisit them later on:

Listing 2.19: Subset MPI: start-up (excerpt, see shm/mpi/v1/mpi.c)

```
...

#include "mpi.h"

#include <assert.h>          // assert(3)

static int _mpi_init = 0;
static int _mpi_size = 1;
static int _mpi_rank = 0;


...

int MPI_Init(int *pargc,char ***pargv)
{
#ifdef DEBUG
    fprintf(stderr,"%d/%d: MPI_Init(%p,%p)\n",_mpi_rank,_mpi_size,pargc,pargv);
#endif

    assert(_mpi_init == 0);
    assert(pargc != NULL && *pargc > 0);
    assert(pargv != NULL);

    if (_mpi_getargs(pargc,pargv) != MPI_SUCCESS ||
        _mpi_allocate() != MPI_SUCCESS   ||
        _mpi_split(0,_mpi_size) != MPI_SUCCESS) {
#ifdef DEBUG
        fprintf(stderr,"%d%d: -> failure\n",_mpi_rank,_mpi_size);
#endif
        return !MPI_SUCCESS;
    }

    _mpi_init = 1;

#ifdef DEBUG
    fprintf(stderr,"%d/%d: -> success\n",_mpi_rank,_mpi_size);
#endif
    return MPI_SUCCESS;
}
```

Let's crawl around this code snippet. First of all, we include the necessary headers, define the global data entities and assign proper default values to them. The _mpi_init tracks the state of the library (initialized or not), since almost all MPI calls should be made once the MPI_Init() had been called first. Since we only use the MPI_COMM_WORLD communicator, we can keep the process' rank and the task size in global variables _mpi_rank and _mpi_size, respectively. Alternatively, we could have defined a global struct to hold this data. Of course, in a full MPI library this would be handled this way or even linked to the MPI_COMM_WORLD. In our simple case, however, this would only confuse the code.

The MPI_Init() function gets pointers to the argument count and the argument list of the calling program. Due to this, it can not only process the argument list but also modify it once it finds options it recognizes as MPI rather than application ones. After that, we allocate the shared memory segment (we will consider this part of the program later), and finally spawn the requested number of processes in a recursive fashion. Both entry and exit of the MPI_Init() function are traced in the debugging mode that the library is compiled in once the DEBUG macro is defined, best of all, in the compiler run string spelled out in the respective Makefile. Certain prerequisites are verified with the help of the assert(3) macro that reduces to nothing as soon as the NDEBUG macro is defined. This way we can produce a streamlined or a debugging version of the library by playing with the compiler options.

2.3.2.3.1 Argument Processing

The argument processing is performed by the static function getargs() (see Listing 2.20):

Listing 2.20: Subset MPI: argument processing (excerpt, see shm/mpi/v1/mpi.c)

```
static int _mpi_getargs(int *pargc,char ***pargv)
{
    int argc = *pargc,n = _mpi_size;
    char **argv = *pargv;

    while (argv++,--argc) {
#ifdef DEBUG
        fprintf(stderr,"%d/%d: %d = \"%s\"\n",_mpi_rank,_mpi_size,
                *pargc - argc,*argv);
#endif
        if (strcmp(*argv,"-n") == 0 || strcmp(*argv,"-np") == 0) {
            if (argc > 1) {
                errno = 0;
                n = (int)strtol((const char *)*(++argv),NULL,10);
                switch (errno) {
                case EINVAL:
                case ERANGE:
                    n = _mpi_size;
                    break;
                default:
                    if (n > 0) {
                        *pargc -= 2;
                        while (argv++,--argc) {
                            (*pargv)[*pargc - argc] = *argv;
                            *argv = NULL;
                        }
```

```
#ifdef DEBUG
                        fprintf(stderr,"%d/%d: n = %d\n",_mpi_rank,_mpi_size,n);
#endif
                  }
                  else
                      n = _mpi_size;
                  break;
                }
              break;
            }
          }
      }

    _mpi_size = n;
#ifdef DEBUG
    fprintf(stderr,"%d/%d: _mpi_size = %d\n",_mpi_rank,_mpi_size,_mpi_size);
#endif

    return MPI_SUCCESS;
}
```

Note that we print out only essential data in the debugging mode. Unlike top level MPI functions of the MPI_Init() kind, the lower level function entry and exit points are not traced, for these functions only help to simplify the code rather than constitute a significant entity. If you feel a little uncomfortable with the pointer arithmetic, take it easy. What happens is that the program arguments are analyzed one by one, and if one of them matches the expected syntax, the data is interpreted and used to change the program settings, while the respective options are consumed by the function. Any interpretation error leads to the respective option being ignored, and if nothing else works out, the rank and size stay at their default values.

2.3.2.3.2 Process Creation

We will deal with allocation in the next section (see Section 2.3.3), so let's look at the process creation now (see Listing 2.21):

Listing 2.21: Subset MPI: process creation (excerpt, see shm/mpi/v1/mpi.c)

```
static int _mpi_split(int rank,int size)
{
    _mpi_rank = rank;

    if (size > 1) {
        switch (fork()) {
        case -1:
            return !MPI_SUCCESS;
        case 0:                   // child
            return _mpi_split(rank + size/2 + size%2,size/2);
```

```
            default:  // parent
                return _mpi_split(rank,size/2 + size%2);
            }
        }

#ifdef DEBUG
    fprintf(stderr,"%d/%d: _mpi_rank = %d\n",_mpi_rank,_mpi_size,_mpi_rank);
#endif
        return MPI_SUCCESS;
    }
```

Here we split ourselves recursively, leaving half of the remaining work to our firstborn child. This child process continues the spawning of its part of the job in parallel with the parent process.

2.3.2.4 MPI Termination and Query

The MPI termination and query functions are trivial compared to the process start-up ones (see Listing 2.22):

Listing 2.22: Subset MPI: termination and environment query (excerpt, see shm/mpi/v1/mpi.c)

```
int MPI_Finalize(void)
{
#ifdef DEBUG
    fprintf(stderr,"%d/%d: MPI_Finalize()\n",_mpi_rank,_mpi_size);
#endif

    assert(_mpi_init);

    _mpi_init = 0;

#ifdef DEBUG
    fprintf(stderr,"%d/%d: -> success\n",_mpi_rank,_mpi_size);
#endif
    return MPI_SUCCESS;
}

int MPI_Comm_size(MPI_Comm comm,int *psize)
{
#ifdef DEBUG
    fprintf(stderr,"%d/%d: MPI_Comm_size(%d,%p)\n",_mpi_rank,_mpi_size,comm,psize);
#endif

    assert(_mpi_init);
    assert(comm == MPI_COMM_WORLD);
    assert(psize != NULL);

    *psize = _mpi_size;
```

```
#ifdef DEBUG
    fprintf(stderr,"%d/%d: -> success, *psize = %d\n",_mpi_rank,_mpi_size,*psize);
#endif
    return MPI_SUCCESS;
}


int MPI_Comm_rank(MPI_Comm comm,int *prank)
{
#ifdef DEBUG
    fprintf(stderr,"%d/%d: MPI_Comm_rank(%d,%p)\n",_mpi_rank,_mpi_size,comm,prank);
#endif

    assert(_mpi_init);
    assert(comm == MPI_COMM_WORLD);
    assert(prank != NULL);

    *prank = _mpi_rank;

#ifdef DEBUG
    fprintf(stderr,"%d/%d: -> success, *prank = %d\n",_mpi_rank,_mpi_size,*prank);
#endif
    return MPI_SUCCESS;
}
```

Note that we do not wait for the children to die, for this would require additional effort and is not required by the MPI standard. We do not prevent the MPI library from being reinitialized, which is not supported by the standard. However, this is a dark corner anyway, and MPI-4 may yet change the situation here.

2.3.3 Blocking Point-to-Point Communication

Communication requires that the spawned processes get connected to each other. In this section, we will consider this topic and the way in which the communication can be done over shared memory.

2.3.3.1 Design Considerations

We have developed and tested our naïve bidirectional buffer system for connecting two processes earlier on. If the number of processes grows, and we keep the principle of making a pair of shared memory buffers available for any pair of communicating processes, we will end up with N^2 buffers for N processes, including N buffers for any process to communicate with itself. Yes, the latter is possible in principle in the MPI. If we make all communication blocking, however, communication with itself may become a problem that we will address in detail later on.

2.3.3.2 Memory Allocation

However, we certainly cannot afford allocating full-size data exchange buffers as we did before. Let's recall the double buffering idea mentioned earlier (see Section 2.2.2.2) and implement a relatively simple version of this mechanism here. The respective code follows (see Listing 2.23):

Listing 2.23: Subset MPI: communication setup (excerpt, see `shm/mpi/v1/mpi.c`)

```
#define BUFLEN    4096

static struct _mpi_shm_s {
    int len,tag;
    volatile char flag[2];
    char nsend,nrecv;
    char buf[2][BUFLEN];
} *_mpi_shm;

...

static int _mpi_allocate()
{
    _mpi_shm = mmap(NULL,sizeof(struct _mpi_shm_s)*_mpi_size*_mpi_size,
                 PROT_READ | PROT_WRITE,MAP_ANONYMOUS | MAP_SHARED,-1,0L);

    return (_mpi_shm != MAP_FAILED) ? MPI_SUCCESS : !MPI_SUCCESS;
}
```

Let's look at the data first. For each process pair and data exchange direction we are going to allocate a cell of the type described by the struct _mpi_shm_s. There, we will keep the message length and tag. Data will be written into and read from two buffers buf of BUFLEN bytes each. Which buffer to write to and to read from will be indicated by the fields nsend and nrecv. The sending process will use the former field, while the receiving process will use the latter field. This way we will avoid extra synchronization between the processes. They will learn about the readiness of the respective buffer part by polling and setting the corresponding fields in the flag[] array.

The actual memory allocation closely resembles the earlier examples and should be clear without extra explanation.

2.3.3.3 Data Sending

It is more interesting to look into the MPI_Send() and MPI_Recv() functions (see Listing 2.24):

Listing 2.24: Subset MPI: data sending (excerpt, see shm/mpi/v1/mpi.c)

```
int MPI_Send(void *buf,int cnt,MPI_Datatype dtype,int dest,int tag,MPI_Comm comm)
{
    struct _mpi_shm_s *pshm;

#ifdef DEBUG
    fprintf(stderr,"%d/%d: MPI_Send(%p,%d,%d,%d,%d,%d)\n",_mpi_rank,_mpi_size,
            buf,cnt,dtype,dest,tag,comm);
#endif

    assert(_mpi_init);
    assert(buf != NULL && cnt >= 0);
    assert(dtype == MPI_BYTE);
    assert(dest >= 0 && dest < _mpi_size && dest != _mpi_rank);
    assert(tag >= 0 && tag <= 32767);
    assert(comm == MPI_COMM_WORLD);

    pshm = _mpi_shm + _mpi_rank*_mpi_size + dest;
    while (pshm->flag[pshm->nsend])
        ;

    pshm->tag = tag;
    if ((pshm->len = cnt) == 0)
        pshm->flag[pshm->nsend] = 1;
    else {
        for ( ; cnt > 0; buf += BUFLEN,cnt -= BUFLEN) {
            while (pshm->flag[pshm->nsend])
                ;

            memcpy(pshm->buf[pshm->nsend],buf,(cnt < BUFLEN) ? cnt : BUFLEN);
            pshm->flag[pshm->nsend] = 1;

            pshm->nsend = !pshm->nsend;
        }

        pshm->nsend = 0;
    }

#ifdef DEBUG
    fprintf(stderr,"%d/%d: -> success\n",_mpi_rank,_mpi_size);
#endif
    return MPI_SUCCESS;
}
```

Note that in the debugging mode we check the prerequisites of our MPI subset on entry to the MPI_Send() function. Then we locate the cell responsible for sending data to process rank dest, fill out the control information, and start pushing data into the double buffers in the hope that the receiving process dest will catch up and automatically synchronize with the sending process. In this case, the sender will be writing into one buffer while the receiver will be reading from the other, and so on, until all data has been passed. We separate the zero-length message path for two reasons: first, it should be as fast as possible, and one condition is an acceptable price to pay for not calling the memcpy(3) for an empty buffer, as it would do this comparison once called anyway. Second, this approach simplifies the loop condition and processing for nonzero length buffers. By the way, here and elsewhere we do not use the modern const void * declaration for the buf argument of the MPI_Send() call just for the sake of simplicity.

2.3.3.4 Data Receiving

Now let's look into the receiving side that basically mirrors the sending side, with the important addition of setting the standard fields of the MPI_Status argument on return (see Listing 2.25):

Listing 2.25: Subset MPI: data receiving (excerpt, see shm/mpi/v1/mpi.c)

```
int MPI_Recv(void *buf,int cnt,MPI_Datatype dtype,int src,int tag,
             MPI_Comm comm,MPI_Status *pstat)
{
    struct _mpi_shm_s *pshm;

#ifdef DEBUG
    fprintf(stderr,"%d/%d: MPI_Recv(%p,%d,%d,%d,%d,%d,%p)\n",_mpi_rank,_mpi_size,
            buf,cnt,dtype,src,tag,comm,pstat);
#endif

    assert(_mpi_init);
    assert(buf != NULL && cnt >= 0);
    assert(dtype == MPI_BYTE);
    assert(src >= 0 && src < _mpi_size && src != _mpi_rank);
    assert(tag >= 0 && tag <= 32767);
    assert(comm == MPI_COMM_WORLD);
    assert(pstat != NULL);

    pshm = _mpi_shm + src*_mpi_size + _mpi_rank;
    while (!pshm->flag[pshm->nrecv])
        ;

    if (tag != pshm->tag) {
#ifdef DEBUG
        fprintf(stderr,"%d/%d: -> failure, tag %d != %d\n",_mpi_rank,_mpi_size,
                tag,pshm->tag);
#endif
        return !MPI_SUCCESS;
```

```
    }
    if (cnt < pshm->len) {
#ifdef DEBUG
        fprintf(stderr,"%d/%d: -> failure, cnt %d < %d\n",_mpi_rank,_mpi_size,
                cnt,pshm->len);
#endif
        return !MPI_SUCCESS;
    }

    pstat->MPI_SOURCE = src;
    pstat->MPI_TAG = tag;

    if ((cnt = pshm->len) == 0)
        pshm->flag[pshm->nrecv] = 0;
    else {
        for ( ; cnt > 0; buf += BUFLEN,cnt -= BUFLEN) {
            while (!pshm->flag[pshm->nrecv])
                ;

            memcpy(buf,pshm->buf[pshm->nrecv],(cnt < BUFLEN) ? cnt : BUFLEN);
            pshm->flag[pshm->nrecv] = 0;

            pshm->nrecv = !pshm->nrecv;
        }

        pshm->nrecv = 0;
    }

#ifdef DEBUG
    fprintf(stderr,"%d/%d: -> success, *pstat = { %d, %d }\n",
            _mpi_rank,_mpi_size,pstat->MPI_SOURCE,pstat->MPI_TAG);
#endif
    return MPI_SUCCESS;
}
```

Note that we check for tag match and buffer length conditions explicitly rather than using the assert(3) macro. These are true MPI error conditions that in a full-blown implementation would result in receiving a so-called unexpected message or in returning an MPI error condition MPI_ERR_TRUNCATED.

2.3.4 Testing

Of course, we want to test this library first. A simple program of the kind we have written before, but this time in MPI terms, may look as follows (see Listing 2.26):

Listing 2.26: Subset MPI: simple toast test program (see `shm/mpi/v1/t1.c`)

```c
#include <stdio.h>

#include "mpi.h"

#ifndef TOAST
#define TOAST "Cheers!"
#endif

int main(int argc,char** argv)
{
    static char sbuf[] = TOAST,rbuf[] = "Salute!";
    int rank,size;
    MPI_Status stat;

    MPI_Init(&argc,&argv);

    MPI_Comm_size(MPI_COMM_WORLD,&size);
    MPI_Comm_rank(MPI_COMM_WORLD,&rank);

    if (size > 1) {
        if (rank == 0) {
            printf("%d/%d: %s\n",rank,size,sbuf);
            MPI_Send(sbuf,sizeof(sbuf),MPI_BYTE,1,0,MPI_COMM_WORLD);
        }
        else {
            if (rank == 1)
                MPI_Recv(rbuf,sizeof(rbuf),MPI_BYTE,0,0,
                            MPI_COMM_WORLD,&stat);

            printf("%d/%d: %s\n",rank,size,rbuf);
        }
    }

    MPI_Finalize();
}
```

As you can see, MPI process rank 0 cheers everyone, while other processes cheer in return in loose chorus. However, only MPI process rank 1 is sober enough to actually hear what the other process has said and reply in kind. Since both the send buffer and the receive buffer have equal lengths, there is no issue in the possible message length mismatch. However, we can and do define other phrases in the respective Makefile, and this allows for testing different situations.

Well, this program works, even though this is not the recommended way of sending strings. For that, you would instead want to use the MPI_CHAR datatype, in case any data format conversion was necessary. This does not apply to our obviously homogeneous situation, so let's be lenient this time, even though the test program becomes non-portable due to this. Another point to take note of in this vein is the explicit passing of an extra zero byte ('\0') that conventionally marks the end of the string. Again, if our strings were Unicode, we would have to use the MPI_WCHAR datatype and send over probably more than one extra byte.

In order to do more extensive tests, especially for the border cases and error conditions, you have to write more test programs. You will find some of them in the accompanying sources (Supalov A. , 2018). Since we do not check the return code in the test program, and we assume the MPI error handler MPI_ERRORS_RETURN, the program runs through even when a typical MPI program would have aborted.

Based on my having run them, we will assume from now on that the implementation is indeed correct, and we can proceed to the benchmarking.

2.3.5 Benchmarking

In order to benchmark the resulting debugged library, a benchmarking program will have to be created as well. We are still too low on the MPI functionality to hope that one of the standard benchmarks will work out of the box, with the possible exception of netpipe(1). In addition, the benchmark we are going to create will be fully under our control, which normally creates this warm and fuzzy feeling people sometimes call "sense of ownership." It is more important, however, that this partially irrational attitude helps you do things your way.

2.3.5.1 Unified Point-to-Point Benchmark

So far, we have developed several naïve benchmarks that tested two relevant patterns, namely, the ping and the ping-pong. You will certainly recognize their traces in the source code of the following MPI benchmark. This time, however, we will combine both patterns in one program, and select the one we want to benchmark by means of conditional compilation. Indeed, it may be interesting to observe performance of either leg or none at all, so that we have built the kind of flexibility into the MPI benchmark that we will use (with several minor extensions) throughout this book (see Listing 2.27).

Listing 2.27: Unified point-to-point benchmark (see shm/mpi/v1/b.c)

```
#include <stdlib.h>        // malloc(3)
#include <stdio.h>         // printf(3)

#include "mpi.h"

#ifndef IMAX
#define IMAX       10000
```

```
#endif
#ifndef LMAX
#define LMAX      32*1024*1024
#endif

#if defined(PING) && defined(PONG)
#define DIV 2
#else
#define DIV 1
#endif

int main(int argc,char **argv)
{
    int i,imax = IMAX,l,lmax = LMAX,r,o,s;
#ifdef CHECK
    int k;
#endif
    double t;
    char *b1,*b2;
    MPI_Status st;

    MPI_Init(&argc,&argv);

    b1 = malloc(lmax);
    b2 = malloc(lmax);

#ifdef CHECK
    for (k = 0; k < lmax; k++) {
        b1[k] = (char)k;
        b2[k] = (char)(lmax - k);
    }
#endif

    MPI_Comm_size(MPI_COMM_WORLD,&s);
    if (s > 1) {
        MPI_Comm_rank(MPI_COMM_WORLD,&r);
        o = s - 1;

        for (l = 0; l <= lmax; l = (l) ? l << 1 : 1) {
            if (r == 0) {  // "sender"
                t = MPI_Wtime();
                for (i = 0; i < imax; i++) {
#ifdef PING
                    MPI_Send(b1,l,MPI_BYTE,o,0,MPI_COMM_WORLD);
#endif
#ifdef PONG
                    MPI_Recv(b2,l,MPI_BYTE,o,1,MPI_COMM_WORLD,&st);
#ifdef CHECK
                    for (k = 0; k < l; k++)
                        if (b2[k] != (char)k)
```

```
                              fprintf(stderr,"ERROR: r = %d, l = %d, i = %d, "
                                      "b2[%d] = %02x, k = %02x\n",
                                      r,l,i,k,b2[k],(char)k);
#endif
#endif
                    }
                    t = MPI_Wtime() - t;
               }
            else if (r == o) {       // "receiver"
                    t = MPI_Wtime();
                    for (i = 0; i < imax; i++) {
#ifdef PING
                         MPI_Recv(b2,l,MPI_BYTE,0,0,MPI_COMM_WORLD,&st);
#ifdef CHECK
                         for (k = 0; k < l; k++)
                             if (b2[k] != (char)k)
                                 fprintf(stderr,"ERROR: r = %d, l = %d, i = %d, "
                                         "b2[%d] = %02x, k = %02x\n",
                                         r,l,i,k,b2[k],(char)k);
#endif
#endif
#ifdef PONG
                         MPI_Send(b1,l,MPI_BYTE,0,1,MPI_COMM_WORLD);
#endif
                    }
                    t = MPI_Wtime() - t;
               }
#ifndef CHECK
            printf("r = %d\tbytes = %-8d\titers = %-8d\ttime = %-12.6g\t"
                   "lat = %-12.6g\tbw = %-12.6g\n",r,l,i,t,t/i/DIV,l/t*i*DIV);
#endif
          }
      }

    MPI_Finalize();

    return 0;
}
```

You can see that we can also check data transfer correctness (at least somewhat) if the benchmark is compiled when the macro CHECK has been defined. In fact, we do use this benchmark in this configuration as one of our test programs, having reduced the number of iterations IMAX and, if necessary, the buffer length LMAX as well. Conditional compilation helps to do this via the compiler run string rather than the more tedious source code manipulation. Since timing makes little sense when data verification interferes with the data transfer, the respective output statement is also disabled in this case. However, in order not to disturb the control flow and data layout or over-

complicate the program, we still define the timing variable t and call the MPI_Wtime()
function.

2.3.5.2 Ping-Pong Performance versus MPICH and Open MPI

Enough testing for now. This is what we get if we run this benchmark over our subset
MPI compared to MPICH 3.2 and Open MPI 2.0.2 that represented the state of the art at
the time this was written (see Figure 2.11):

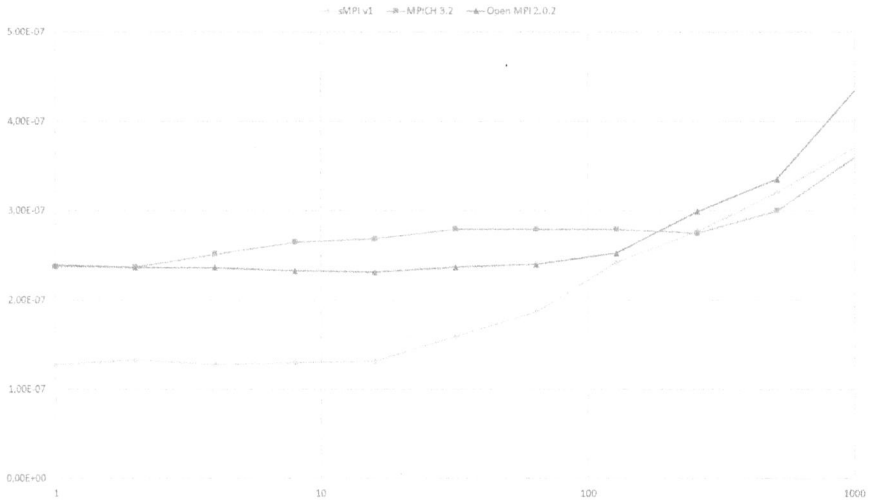

Figure 2.11: Unified point-to-point benchmark: median shared memory ping-pong latency, subset
MPI vs. MPICH 3.2 vs. Open MPI 2.0.2 (seconds)

Our subset MPI beats both MPICH and Open MPI by about two times on very small
messages! This time three runs were made over each package to see if there were any
outliers. Indeed, there were some. For our subset MPI, results fit within 9% of the
mean. For MPICH, within 11%. For Open MPI, however, one of the results differed
from the other two by about two times! We will learn what happened a little later,
although you have probably guessed it by now on the basis of the knowledge acquired
so far. For now, we show the median of the results, thus filtering out the said Open
MPI outlier.

A bandwidth graph, however, reveals another issue (see Figure 2.12):

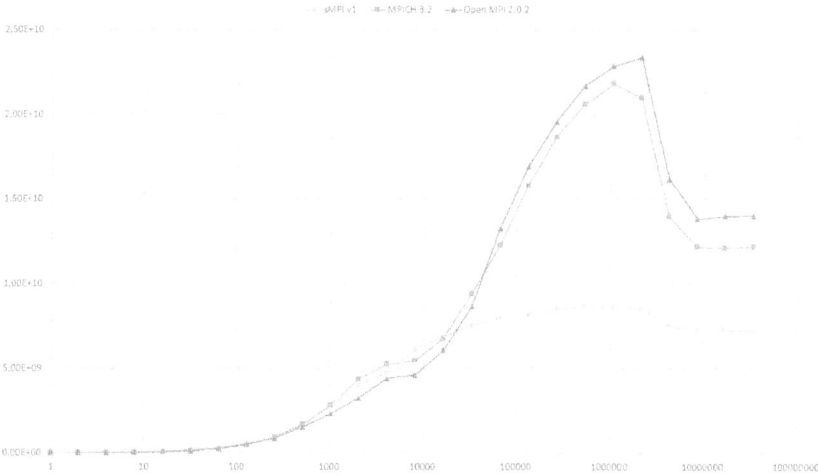

Figure 2.12: Unified point-to-point benchmark: median shared memory ping-pong bandwidth, subset MPI vs. MPICH 3.2 vs. Open MPI 2.0.2 (bytes per second)

Although our subset MPI beats or matches MPICH and Open MPI on messages up to 16 KiB, after that the competition takes off and follows a substantially more pronounced typical memory bandwidth curve than we do, ending up about twice as fast in the long run. Either they use a different mechanism or different settings for larger messages, and we need to find out what that is. However, before shooting in the dark, we need to find out what happened to that Open MPI outlier.

2.3.5.3 Influence of the Process Binding

Our primary suspect in the Open MPI matter is certainly the process binding. It may be that our subset MPI got lucky (or unlucky) and obtained a consistent binding from the operating system. MPICH was also lucky or unlucky, or used some default process binding policy to get consistent results. Open MPI, however, did not manage to build the process binding support library as of version 2.0.2 on the test machine, as documented by its output, and hence might get one binding in the first run and another binding in the following two runs. This is our first working hypothesis that is partially confirmed by the observation that once NUMA support is enabled in the operating system,[1] Open MPI successfully builds the process binding support library and shows more consistent results.

Another possible explanation of the observed anomaly is that core 0 that is involved in our measurements, is traditionally used by the operating system for I/O

---

1 In my case, running command apt-get install libnuma-dev did the trick. Your mileage may vary.

activities. It may have happened that exactly at the time of our benchmarking, this core was busy switching back and forth between the user and OS activities, which might have accounted for the observed slowdown. However, the general feeling of it is not compatible with such a situation. In case of transient I/O load we should have seen a temporary slowdown at certain message sizes rather than a general performance depression of the entire run, although that is not completely impossible either. In the future, it might make sense to avoid core 0—unless you, just like me, explicitly want to measure your performance under realistic conditions that leave core 0 wide open to the applications.

Let's look at the latency of all three packages now that we can control the binding in all of them (see Figure 2.13).

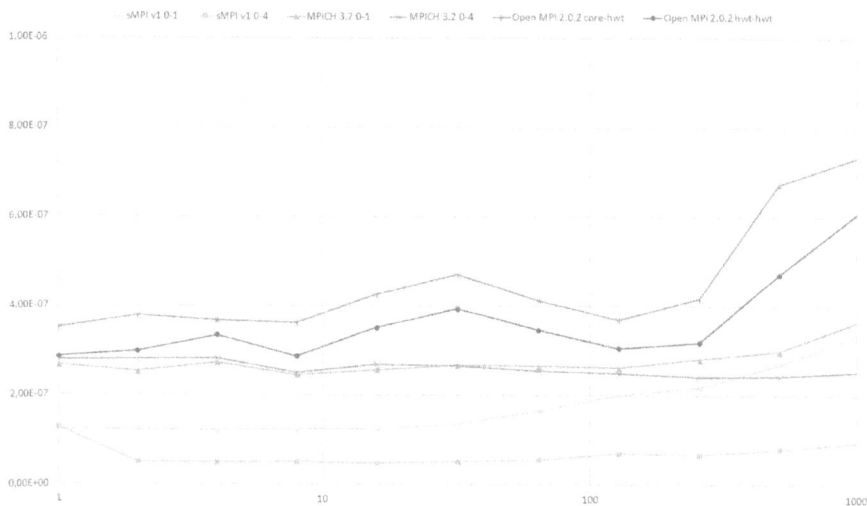

Figure 2.13: Unified point-to-point benchmark: shared memory ping-pong latency, subset MPI vs. MPICH 3.2 vs. Open MPI 2.0.2 (seconds)

Here, we present latency measured on adjacent cores (curves marked 0-1 and core-hwt) and on shared core zero (curves marked 0-4 and hwt-hwt). We can see that on small messages, unsurprisingly, our subset MPI beats MPICH and Open MPI quite handsomely—by two to four times! Comparing this graph to Figure 2.11, we can presume that while our subset MPI was placed upon adjacent cores by default, MPICH placed its processes on a shared core, and Open MPI simply relied on the OS. We also see that messages of length one (and zero, not shown due to logarithmic scaling) may suffer from the lack of the warm-up stage in the benchmark. As we have discussed before (see Section 2.2.1.3), doing the warm-up may smoothen the early results since both code and data will be pulled up the cache hierarchy. However, a more detailed review of the raw data shows that MPICH appears to be suffering less. In other words, not putting the warm-up into the program right away did help us to see something we may have missed otherwise. This was actually the primary motivation behind this seemingly counterintuitive move.

Let's look at the bandwidth now (see Figure 2.14):

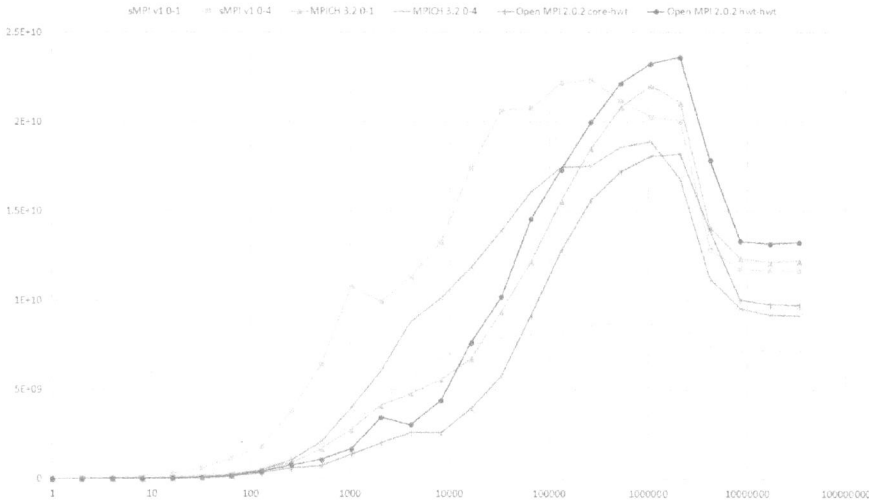

Figure 2.14: Unified point-to-point benchmark: shared memory ping-pong bandwidth, subset MPI vs. MPICH 3.2 vs. Open MPI 2.0.2 (bytes per second)

While subset MPI does pretty well against the venerable MPICH and Open MPI on a shared core where *adjacent hardware threads* share the L1 cache, double buffering apparently does not fare very well at all in the adjacent core configuration once the message length substantially exceeds the size of the shared memory buffer (twice 4KiB in the original library configuration). Moreover, when we compare the shape of the MPICH bandwidth graph to Figure 2.12, we get another confirmation that MPICH may indeed have bound its processes to the adjacent hardware threads by default or used another method to pass data between adjacent threads rather than adjacent cores. Still, on adjacent hardware threads our subset MPI happens to beat the venerable MPICH at all message ranges—up to four times on small message latency, and by about 10% to 50% on large message bandwidth! It holds out pretty well against Open MPI, too, except for the very high message sizes where Open MPI is probably using an advanced data exchange protocol.

It is possible that an increase of the shared memory buffer size would cure the performance gap versus MPICH on the adjacent cores. Should we try this right now? No. We have not exploited the entire flexibility of our benchmark yet. Maybe, when we see how our brand-new library behaves on the ping and pong patterns, we will want to change the library. In that case, any optimization performed beforehand will have to be verified again or even totally redone.

2.3.5.4 Ping Performance versus MPICH and Open MPI

Let's see how our subset MPI does versus MPICH and Open MPI on the ping pattern. It is roughly equivalent to the pong pattern, which was confirmed by additional measurements (not shown). Even though we know by now that process binding plays a decisive role in the performance, we still take the long way and check the default, out-of-the-box performance first. Who knows what we will find out?

Well, this time Open MPI has gotten particularly unlucky: out of three benchmarking runs, two were scheduled on adjacent cores, and only one got the adjacent hardware threads. We again show the median, so the best Open MPI result was pushed out as an outlier (see Figure 2.15):

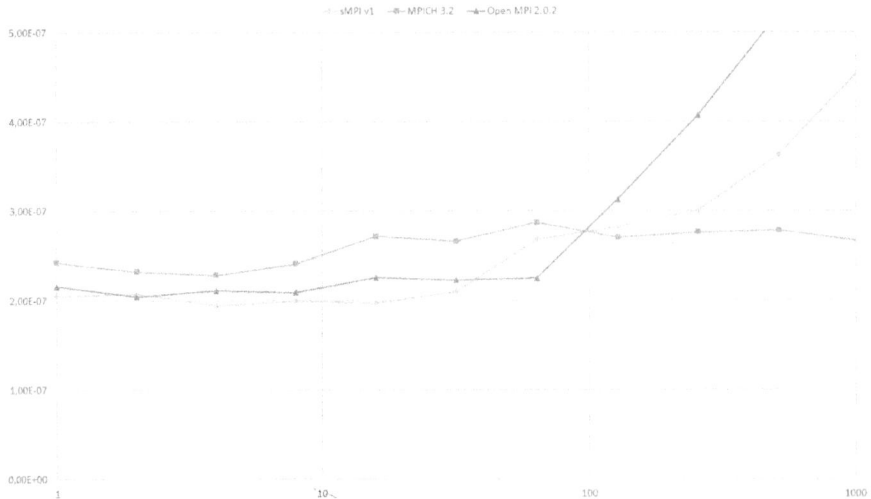

Figure 2.15: Unified point-to-point benchmark: median shared memory ping latency, subset MPI vs. MPICH 3.2 vs. Open MPI 2.0.2 (seconds)

Moreover, by the looks of it, the Open MPI curve represents a rare case of runtime process migration, where a process has been moved from the faster adjacent hardware threads that were undesirable from the OS point of view, to the slower adjacent cores arrangement. If we look at the bandwidth, we see this partially confirmed (see Figure 2.16):

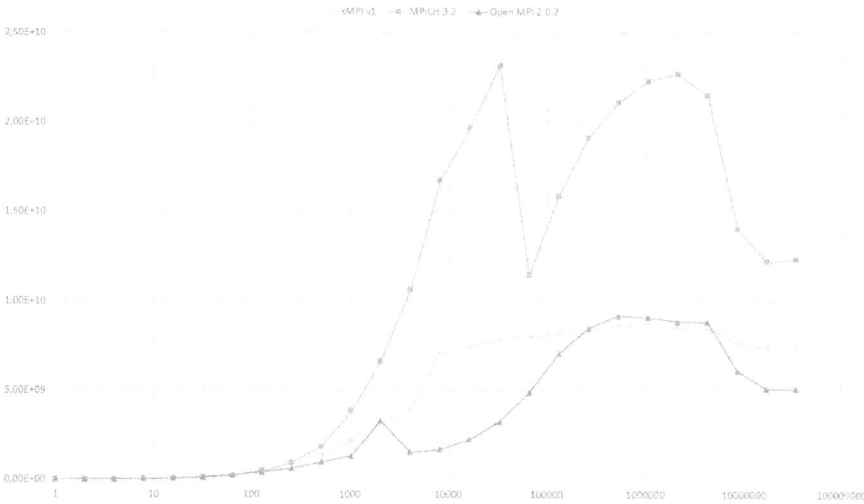

Figure 2.16: Unified point-to-point benchmark: median shared memory ping bandwidth, subset MPI vs. MPICH 3.2 vs. Open MPI 2.0.2 (bytes per second)

In fact, the funny peak around 2 KiB may even indicate that the rescheduling was done more than once. Oh well, let me report this back to the Open MPI team so that they can let me know how to fix the pinning. I also have a question to you: what happens in subset MPI when it almost doubles bandwidth between 4 KiB and 8 KiB? We will revisit this question soon. However, I also have a question to the MPICH team: the sudden drop in performance around 64 KiB despite all process binding apparently done is simply intriguing.

To get some sense out of this latter matter, let's look at the ping latency and bandwidth benchmarked with the process binding we have used before (see Figures 2.17 and 2.18):

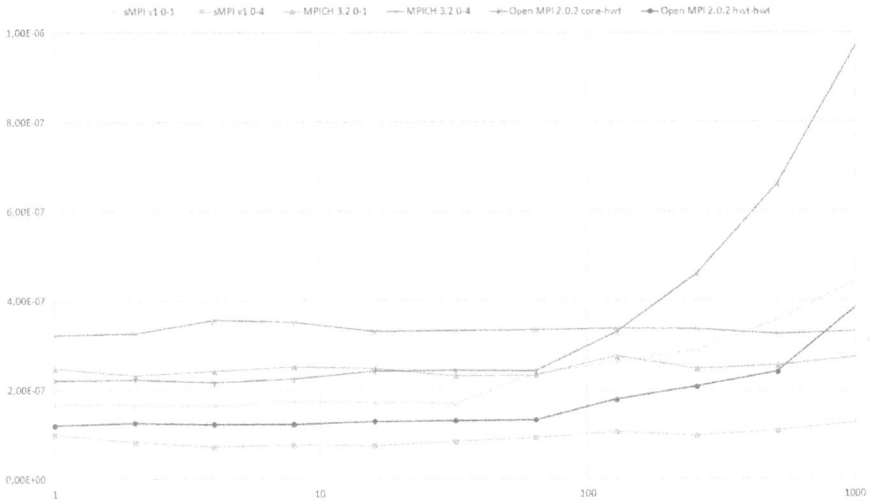

Figure 2.17: Unified point-to-point benchmark: shared memory ping latency, subset MPI vs. MPICH 3.2 vs. Open MPI 2.0.2 (seconds)

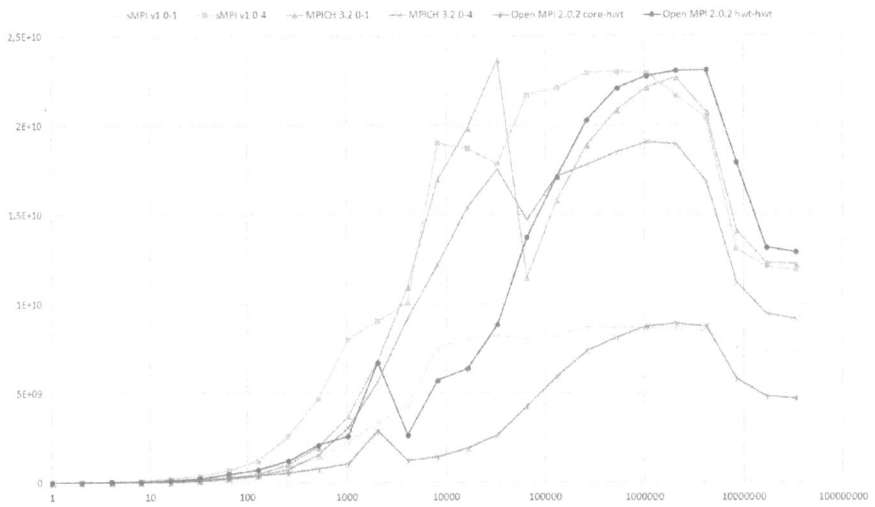

Figure 2.18: Unified point-to-point benchmark: shared memory ping bandwidth, subset MPI vs. MPICH 3.2 vs. Open MPI 2.0.2 (bytes per second)

Note that these are of course new series, and if we see some familiar features here, we can arrive at quite a few conclusions. First, MPICH must be pinning two MPI processes to the adjacent hardware threads by default. Second, Open MPI does seem to have an issue with protocol selection around 2 KiB. Third, when subset MPI is run under the same conditions, it beats or matches MPICH and Open MPI almost everywhere on the ping benchmark as well. And fourth, we again see a dramatic jump in the subset MPI bandwidth between 4 KiB and 8 KiB. What's the matter?!

2.3.6 Optimization

Gosh, this is where double buffering actually takes off! Before that, on smaller messages, if you look at the code (see Listings 2.24 and 2.25), we always start with the first half of the buffer and write data into it, and then wait for the receiver to get data out of this buffer before we may reuse it. In other words, the second half of the buffer is idling for all small messages. This is not exactly what we should be doing, right?

Another idea we have already discussed (see Section 2.2.2.2): we use just twice 4 KiB as the double buffer, hoping that this is where performance is at its best. This may or may not be true, so we need to do some more benchmarking to find out what buffer size is both bearable in terms of the memory consumption and fast in terms of the memcpy(3) performance obtained.

One more idea is to look at the memory layout, both in the shared memory segment and elsewhere. Certainly, with the double buffering improved, we could also try to rearrange the control fields so that they have a proper size and are aligned for best performance. Thinking about the register keyword, there also is possibly some scope for a hint to the compiler. Memory access is expensive, so getting some fat out of it will normally pay off.

And, by the way, while the code is fresh in your short-term memory, let's recall the old maxim stating that if there is one issue in a piece of code, there are likely to be more of them nearby. In this sense, issues are just like hobbits: you can count on making a pie once you catch at least one and don't wait for the dawn to come. With this in mind, try to ask yourself whether all statements that are seen in the code really need to be there.

Well, this is our pretty plan, or at least a draft of it. It appears that we should only attempt the BUFLEN tuning once we have attended to everything else, otherwise this tuning will have to be redone. The rest can be done in the order shown.

2.3.6.1 Improved Double Buffering

How can we use the second part of the buffer when the first half is being read by the receiving process? It would be easy to avoid resetting the nsend field to zero at the end of the MPI_Send(), and do the same about the nrecv field in the MPI_Recv(). However, who will guarantee that the values of other relevant fields, like tag or len, will not be updated while we are manipulating them on our side?

Nobody will ever guarantee that unless some mutual exclusion mechanism is employed. Or... Should we instead double those fields and let the nsend and nrecv fields point to the actual values for the MPI_Send() and the MPI_Recv() calls? We have used a comparable idea before. Maybe it will work out this time as well.

Let's start with the updated shared memory segment definition (see Listing 2.28):

Listing 2.28: Updated shared memory segment definition (excerpt, see shm/mpi/v2a/mpi.c)

```
...

static struct _mpi_shm_s {
    int len[2];
    short tag[2];
    volatile char flag[2];
    char nsend,nrecv;
    char buf[2][BUFLEN];
} *_mpi_shm;

...
```

On a level with doubling all relevant fields like len, tag, and flag, we took the liberty of changing their size and place so that longer data items come first, shorter next, and shortest last. Note that all of them fit into the first 64 bytes as does the beginning of the first half of the double buffer. Moreover, the hottest fields—flag, nsend, and nrecv—sit in one word and are word aligned at the beginning. We could have just made them straight int variables, and still keep all of them on one cache line. This is an idea we will revisit soon (see Section 2.3.6.2), but for now, let's keep these variables as specified, because this deviates only very slightly from the original layout (see Listing 2.23).

With these changes done, the rest is nearly trivial. Let's look for example at the relevant part of the MPI_Send() function (see Listing 2.29):

Listing 2.29: Updated MPI_Send() function (excerpt, see shm/mpi/v2a/mpi.c)

```
...

    pshm = _mpi_shm + _mpi_rank*_mpi_size + dest;
    while (pshm->flag[pshm->nsend])
        ;

    pshm->tag[pshm->nsend] = tag;
    if ((pshm->len[pshm->nsend] = cnt) == 0)
            pshm->flag[pshm->nsend] = 1;
    else {
        for ( ; cnt > 0; buf += BUFLEN,cnt -= BUFLEN) {
            while (pshm->flag[pshm->nsend])
                ;
```

```
                memcpy(pshm->buf[pshm->nsend],buf,(cnt < BUFLEN) ? cnt : BUFLEN);
                pshm->flag[pshm->nsend] = 1;

                pshm->nsend = !pshm->nsend;
        }
    }

...
```

The respective change to the MPI_Recv() function reflects this modification (not shown).

2.3.6.2 Memory Access Optimization

Oh, how strange the new shared memory segment looks! This will be the natural reaction of anyone who came to believe that the object orientation (OO) is the only way to do things right. Well, we can reformulate this bit to look nicer (see Listing 2.30):

Listing 2.30: Nice shared memory segment definition (excerpt, see shm/mpi/v2b/mpi.c)

```
...

#ifndef PACKED
#define PACKED     1
#endif

#ifdef PACKED
static struct _mpi_shm_s {
    int nsend,nrecv;
    struct _mpi_shm_cell_s {
        int len;
        short tag;
        volatile char flag;
        char pad;
        char buf[BUFLEN];
    } cell[2];
} *_mpi_shm;
#else
static struct _mpi_shm_s {
    int nsend,nrecv;
    struct _mpi_shm_cell_s {
        int len;
        int tag;
        volatile int flag;
        char buf[BUFLEN];
    } cell[2];
} *_mpi_shm;
#endif

...
```

Here, we split double buffers into good looking substructures and adapt both the MPI_
Send() and MPI_Recv() functions accordingly (not shown).

There is one more aspect to consider. So far, we have been packing all data nicely
into rather tight and properly aligned data structures, minding the total size of the
shared memory segment. What if we stop doing this and look into the performance
achieved when all data fields are converted into plain integers? We use a true 64-bit
platform for our development, so that access to smaller entities may not be any faster
and might actually be somewhat slower than access to the 32-bit words. And memory
is very cheap nowadays, isn't it? To reflect on this, we introduce a macro PACKED that
controls the respective definitions, making the packed version the default. Note that
in both cases, the space taken by the cell header (8 bytes in the packed case, and 12
bytes in the unpacked case, both less than the typical cache line size of 64 bytes),
combined with the size of the buffer equal to a power of two bytes in length, practically guarantee that we will suffer no cache aliasing.

Finally, remember the process binding. What is good for adjacent hardware
threads may be bad for adjacent physical cores. Thus, we get no less than 2×2×2=8
variants to consider. Comparing them all using graphs will be exceedingly messy and
doing changes one after another may very well lead us down a wrong path if we prematurely cut a branch that could prove most promising later on. So, we have no other
choice than to do a multivariate analysis. To simplify things, let's look into 4-byte
latency—far enough from zero, and yet small enough to be a very small message.

This is what we get (see Table 2.2; the best values are highlighted in **bold**):

Table 2.2: Minimum latency at 4 bytes depending on the process binding, memory layout, and
packing (seconds)

Binding/Layout	Adjacent cores		Shared core (adjacent HW threads)	
Packing	Packed	Unpacked	Packed	Unpacked
Updated layout	1.51e-07	1.45e-07	5.00e-08	3.89e-08
Nice layout	**1.37e-07**	1.39e-07	**3.75e-08**	5.25e-08

Looking at these entries, the following can be said (preliminarily and pending further,
more detailed investigation):
– For adjacent cores, the nice layout seems to work better;
– For a shared core, the picture is mixed: the nice layout behaves better for packed
   data, while the updated layout works better for unpacked data.

Still, in total it makes sense to stick to the nice, packed memory layout that showed
best latency for both process binding configurations. Of course, this needs to be verified at other message sizes.

By the way, if we use the `register` keyword on this new or the old version, things do not change either (not shown). This is relatively clear: compilers are smart enough to notice a heavily used variable and put it into a register. And, *inter alia*, how else would you dereference a frequently used pointer on this platform, huh?

2.3.6.3 Code Streamlining

Let's look very hard at the `MPI_Send()` code once again (see Listing 2.29). Why do we do the polling upfront inside the `for()` loop? We know pretty clearly that once we've got so far, the respective part of the buffer will surely be waiting for us to write into it.

Let's rewrite the main loop so that it runs a tad faster (see Listing 2.31):

Listing 2.31: Updated `MPI_Send()` function: fast main loop (excerpt, see `shm/mpi/v3/mpi.c`)

```
...

    pshm = _mpi_shm + _mpi_rank*_mpi_size + dest;
    pcell = pshm->cell + pshm->nsend;

    while (pcell->flag)
        ;

    for (pcell->tag = tag,pcell->len = cnt;
         cnt > BUFLEN; buf += BUFLEN,cnt -= BUFLEN) {
        memcpy(pcell->buf,buf,BUFLEN);
        pcell->flag = 1;

        pcell = pshm->cell + (pshm->nsend = !pshm->nsend);
        while (pcell->flag)
            ;
    }

    memcpy(pcell->buf,buf,cnt);
    pcell->flag = 1;
    pshm->nsend = !pshm->nsend;

...
```

Here, we either first do a number of iterations sending out `BUFLEN` sized chunks and then deal with the rest of the message or go directly to the latter messages shorter than or equal to `BUFLEN` in size. The receiving side is reformulated accordingly (not shown). Let's look at the numbers (see Table 2.3; best values are again highlighted in bold):

Table 2.3: Minimum latency at 4 bytes depending on the process binding, memory layout, and packing (seconds)

Binding/Loop	Adjacent cores	Shared core
Original loop	1.03e-07	3.82e-08
Fast loop	1.04e-07	3.22e-08

Note that this table is based on a different, more extensive series of measurements done for greater accuracy under another OS, hence the difference compared to Table 2.2. All in all, we can see that:

- Adjacent cores do not seem to sense the difference between the original and fast loops
- Shared core, however, does run the fast loop better by about 20%

The big difference in latency measured between Tables 2.2 and 2.3 as far as adjacent cores are concerned should be taken as a warning regarding the precision of our benchmarking technique for very small messages. Apparently, if we were to optimize this message size range more closely, we would need to refine our methodology.

2.3.6.4 Double Buffer Size Tuning

Now that we have pressed some (though still not all) fat out of the code, we are ready to perform a small manipulation with the BUFLEN value. What can one expect here? If you go down from the current BUFLEN value of 4 KiB, you can count on seeing worse bandwidth for larger messages, unless some nonlinear effect related to the cache size and sharing kicks in. An increase of the BUFLEN value might lead to an increase in bandwidth, up to a point, when the management of large intermediate buffers will offset the positive effect of bulk copying. Let's see what the measurements show (see Table 2.4; best values are traditionally highlighted in bold):

Table 2.4: Minimum latency at 4 bytes, 4 KiB, 128 KiB, and 4 MiB for adjacent cores (seconds)

Message size/ BUFLEN	4 bytes	16 KiB	128 KiB	4 MiB
1 KiB	1.15e-07	3.56e-06	2.88e-05	0.00095
2 KiB	1.10e-07	2.60e-06	2.02e-05	0.00066
4 KiB	1.17e-07	**2.15e-06**	1.59e-05	0.00052
8 KiB	1.16e-07	2.19e-06	1.46e-05	0.00048
16 KiB	**1.08e-07**	2.48e-06	0.96e-05	0.00034
32 KiB	1.12e-07	2.50e-06	**0.87e-05**	0.00027
64 KiB	1.11e-07	2.51e-06	0.98e-05	**0.00026**
128 KiB	1.20e-07	2.51e-06	1.23e-05	**0.00026**
256 KiB	1.21e-07	2.52e-06	1.23e-05	**0.00026**
512 KiB	1.14e-07	2.51e-06	1.23e-05	0.00027
1 MiB	1.13e-07	2.50e-06	1.23e-05	0.00033

Well, it looks like we should go for the BUFLEN of 16 KiB or 32 KiB. This will give us best or close to that small message latency, pretty respectable medium message latency, and twice to three times better latency for large messages (128 KiB and 4 MiB) as compared to the current BUFLEN value of 4 KiB.

Why, you may ask, do we not use any profiling tool and try to find a bottleneck as described in all the optimization handbooks? Well, first, we are just establishing the baseline, for which we need reasonable source code, and second, we have not yet looked into the alternative algorithms. We will learn more about those advanced techniques in the following sections, and only then will it make sense to go into the lower level optimizations full scale.

Also, it remains to be seen how this buffer size increase influences the shared core configuration that was doing seemingly alright on the 4 KiB double buffers.

2.3.7 Exercises

Exercise 2.5 Study the test programs provided in the source code archive and determine whether there are any corner cases and error conditions that have not been verified. Add a respective program or programs to the set in this case and extend the Makefile accordingly.

Exercise 2.6 Look more closely into the library and decide whether the current implementation is really doing only what is necessary and nothing more—even after code streamlining. If you find superfluous code segments that are executed more than once without necessity, reformulate the library code accordingly and redo the testing and benchmarking.

Exercise 2.7 Find out the optimum size of the BUFLEN constant on your system. Take care of trying suitable non-power-of-2 values as well, minding the paging and alignment requirements. Use the identified best value from now on.

Exercise 2.8 Add support for the MPI_STATUS_IGNORE status placeholder to the original MPI_Recv() implementation presented in this chapter. Use NULL as the respective constant value in the mpi.h header. Change the ping-pong benchmark to use this feature and find out whether this brings any noticeable performance improvement. Use conditional compilation throughout to make this feature selectable at build time.

Exercise 2.9 Add support for the MPI_ANY_TAG tag wildcard selector to the MPI_Recv() implementation. Use -1 as the respective constant value in the mpi.h header. Change the ping-pong benchmark to use this feature and find out whether this addition brings any noticeable performance overhead compared to the original implementation. Pay attention to running both the MPI_ANY_TAG and specific tag selection benchmarks using your new code.

Exercise 2.10 Add support for the MPI_ANY_SOURCE source process wildcard selector to the MPI_Recv() implementation. Use -1 as the respective constant value in the mpi.h header. Change the ping-pong benchmark to use this feature and find out whether this addition brings any noticeable performance overhead compared to the original implementation. Pay attention to running both the MPI_ANY_SOURCE and specific process selection benchmarks using your new code. Do not forget to try running more than two processes.

Exercise 2.11 Get rid of the double buffering altogether by restricting the maximum MPI message size to BUFLEN. Benchmark the resulting implementation to find out how much additional performance this brings in the small message area compared to the implementations created so far.

Exercise 2.12 Restrict the MPI implementation even further to pass only zero-sized messages, possibly without paying attention to the tag value. Find out how much performance this adds on top. Imagine your users will only be sending zero-sized messages (for example, for synchronization purposes). Does it make sense for them to use a full-blown MPI implementation?

Exercise 2.13 Try using clock_gettime(3) instead of gettimeofday(2) in the MPI_Wtime() function. Does this influence the timings in any way?

Exercise 2.14 Compare shared memory performance of the latest versions of MPICH and Open MPI to versions 3.2 and 2.0.2, respectively, used in this chapter. Is there any difference? Why?

2.4 Conclusions

I will be the first to point out that our MPI implementation is very limited, by the nature of the MPI subset that we set off to develop. Moreover, an MPI expert will surely point out that this subset does not really correspond to the spirit of the MPI standard and may even violate some of its basic principles. However, my conservative estimate is that about 90% of all sensible MPI applications can be represented in terms of this subset. On top of that, benchmarking shows that, at least in case of the shared memory, our subset implementation is superior in out-of-the-box performance of short messages to the popular existing implementations. This, frankly, is all that matters. And as far as the right balance of semantics versus performance is concerned, this is a question to your user as much as to you as an implementor.

Yet before we learn how to match and beat other MPI implementations on all fabrics and in all situations without sacrificing any principles, we need to learn quite a bit more about the MPI internals. Let's proceed, in the hope that you have done all the exercises provided above. For now, let's quantify complexity of the respective packages, just to highlight how much subsetting can help in reducing it. If we consider the total number of source code lines (LOC) as an approximation of the program complexity, we get the following results (see Table 2.5):

Table 2.5: T MPI subsets vs. full MPI implementations

Subset	Entities	Incl. Functions	LOC
Shared memory	13	7	343
MPI-1	244	128	~100000[2]
MPI-3.1	855	450	~350000

We've gotten substantially better results at under one thousandth of the cost. What will happen when we go for the sockets? We will see this in the next chapter.

---

2 LOC numbers for MPI-1 and MPI-3.1 packages are based on my unassisted recollection.

Chapter 3
Sockets

In this chapter, we will build upon the knowledge obtained so far to create an MPI subset capable of blocking and nonblocking point-to-point communication over Ethernet and other IP capable networks. Being about the slowest and most forgiving of all fabrics, Ethernet will allow us to go to the other extreme in the performance range and enjoy a little more freedom in the programming techniques. Again, this will not be a full MPI implementation, but it will come a step closer to the Real Thing.

3.1 Subset Definition

Let's take the next reasonable step and make our new MPI subset a bit more flexible. Sockets are normally about 10–20 times slower than the shared memory as far as latency is concerned, thus we will be able to deal with the accompanying technical challenges without undue pressure on the performance side. To this end, we will (additions compared to Chapter 2 are <u>underlined</u>):

- Provide the basic MPI functionality:
    - The initialization and termination calls `MPI_Init()` and `MPI_Finalize()`
    - The query calls `MPI_Comm_size()` and `MPI_Comm_rank()`
    - The wall clock query function `MPI_Wtime()`
- Implement extended point-to-point communication:
    - Standard mode blocking point-to-point calls `MPI_Send()` and `MPI_Recv()`
    - Standard mode nonblocking point-to-point calls <u>`MPI_Isend()` and `MPI_Irecv()`</u>
    - Completion calls <u>`MPI_Test()` and `MPI_Wait()` as well as the `MPI_Request` object</u>
- Restrict the rest of the MPI functionality as follows:
    - Implement only the `MPI_THREAD_SINGLE` mode
    - Support any process count
    - Provide only the `MPI_BYTE` datatype
    - Permit any nonnegative message count
    - Allow message tags up to 32767
    - Operate only over the `MPI_COMM_WORLD` communicator
    - Assume the predefined MPI error handler `MPI_ERRORS_RETURN`

This basically extends the earlier subset with the nonblocking communication, i.e., 18 entities all in all, including 11 functions. However, what a world of difference this will make to the internals! Still, compared to 855 entities (of them, 450 are functions) in the full MPI-3.1 standard, this will be a breeze.

DOI 10.1515/9781501506871-003

3.1.1 Exercises

Exercise 3.1 Define a couple of MPI subsets a bit wider than the one described above.

Exercise 3.2 Define a subset or two that are still narrower than the one described above. Think about message length, tag values, and other entities.

3.2 Communication Basics

Sockets operate over connection-oriented protocols like TCP/IP and connectionless protocols like UDP. In the first case, they are called *connected* or *stream sockets*. In the second case, they are named *connectionless* or *datagram sockets*. All of them split data into chunks, called packets, which travel along the wire more or less independently and may even take different routes in transit. Thus, the messages need to be reassembled on the other end. Connected protocols typically offer ordered, reliable service that automatically restores user data on the receiving end. Connectionless protocols, to the contrary, are often unordered and unreliable, and it is the duty of the caller to insure data completeness and integrity.

Each of these protocols may offer specific advantages and suffer certain drawbacks depending on the context:
- Connected, reliable protocols (like TCP/IP) normally achieve better latency and bandwidth, because they can use the preset channel established at connection time. On the other hand, the need to keep the connections active means that system resources are tied up at least linearly in each of the connected processes, and at least to the square of the total number of processes for the whole system.
- Connectionless, unreliable protocols (like UDP) excel at scalability, but they cannot match the competition on latency and bandwidth, or on ease of use. Surprisingly enough, even in a tightly coupled system datagrams do get lost once in a while. Since MPI must provide reliable, ordered delivery to the applications, it has to compensate for the datagrams' drawbacks somehow, which leads to some overhead.

Certainly, a mixture of both connected and connectionless protocols is also possible. In this case, certain process pairs are normally connected along the hottest paths, while the rest is served by datagrams in the hope that the package loss will not grind the whole to a standstill. However, management of such a mixture is a bit beyond our capabilities at the moment: we will revisit this matter later on (see Chapter 5).

There may also be unreliable connected and reliable datagram protocols, but they are used far less often. Whatever the protocol, it can be mapped upon nearly any underlying data transfer medium, be that shared memory (aka loopback interface),

Ethernet, or InfiniBand. We will look into several protocol families in this and further chapters, namely:

- Unix (or local) protocol family (AF_UNIX aka AF_LOCAL)
- Internet protocol families (AF_INET and AF_INET6)

Sockets came to life as part of the BSD operating system (Wikipedia.org, 2018a). Their original design was rather elegant and minimalistic, but since then it has grown to include so many extra features that it became very complicated indeed. In order not to get lost in this complexity, we will have to exercise substantial restraint in our treatise. If you want to learn more about sockets, read another book (Stevens, Fenner, & Rudoff, *Unix Network Programming, Volume 1: The Sockets Networking API* (3rd Edition), 2003).

3.2.1 Intranode Communication

Sockets can be used to pass data within one process or between two processes separated by any distance, both intra- and internode. Passing data within one process does not really belong to the MPI domain, even though MPI implementations based on threads are becoming more popular of late, and threads are slowly becoming first class citizens in the latest versions of the MPI standard. The reason for this is that one can pass a pointer to the data or just use the shared memory explicitly, which will be much faster than any library calls. Due to this, we will only consider interprocess communication in this chapter, and start with the intranode data transfer.

3.2.1.1 Local Sockets

These sockets belong to the AF_UNIX protocol family, also known as AF_LOCAL. Both connected and datagram protocol types are offered by the Linux OS. System call socketpair(2) establishes a couple of preconnected sockets that can be used for communication right away. Since they basically provide two sides of a bidirectional channel, it makes sense to use them in a process that is later forked in two entities that want to communicate with each other.

3.2.1.1.1 Local Stream Sockets

Let's do some programming and see how this looks in practice. We will focus on the stream sockets for starters (see Listing 3.1):

Listing 3.1: Naïve local stream socket benchmark si1 (see sock/prep/si1/si1.c)

```
#include <string.h>          // memset(3)
#include <unistd.h>          // fork(2), read(2), write(2)
```

```
#include <sys/types.h>                    //
#include <sys/socket.h>                   // socketpair(2), etc.
#include <stdlib.h>                        // malloc(3), exit(3)
#include <stdio.h>                         // NULL, printf(3), perror(3)

#include "loh.h"

#define IMAX        10000
#define LMAX        32*1024*1024

static int loh_read(int fd,char *buf,int l)
{
    int m,n;

    for (n = l; n > 0; buf += m,n -= m)
        if ((m = read(fd,buf,n)) == -1)
            return -1;

    return 1;
}

static int loh_write(int fd,char *buf,int l)
{
    int m,n;

    for (n = l; n > 0; buf += m,n -= m)
        if ((m = write(fd,buf,n)) == -1)
            return -1;

    return 1;
}

int main(int argc,char **argv)
{
    int i,imax = IMAX,l,lmax = LMAX;
    double t;
    char *buf;

    int sv[2],fd;
    pid_t pid;

    if (socketpair(AF_LOCAL,SOCK_STREAM,0,sv) == -1) {
        perror(argv[0]);
        exit(EXIT_FAILURE);
    }

    if ((pid = fork()) == -1) {
        perror(argv[0]);
        exit(EXIT_FAILURE);
    }
```

```
buf = malloc(lmax);
fd = sv[(pid) ? 1 : 0];

for (l = 0; l <= lmax; l = (l) ? l << 1 : 1) {
    if (pid == 0) {      // "sender"
        t = loh_wtime();
        for (i = 0; i < imax; i++)
            if (loh_write(fd,buf,l) == -1) {
                perror(argv[0]);
                exit(EXIT_FAILURE);
            }
        t = loh_wtime() - t;
    }
    else {               // "receiver"
        t = loh_wtime();
        for (i = 0; i < imax; i++)
            if (loh_read(fd,buf,l) == -1) {
                perror(argv[0]);
                exit(EXIT_FAILURE);
            }
        t = loh_wtime() - t;
    }
    printf("pid = %d\tbytes = %-8d\titers = %-8d\ttime = %-12.6g\t"
           "lat = %-12.6g\tbw = %-12.6g\n",pid,l,i,t,t/i,l/t*i);
}

return 0;
}
```

Here, we reuse the infrastructure created earlier for the naïve shared memory bench-mark sm1 (see Section 2.2.2.1). Upon creating a socket pair via the socketpair(2) system call, the process is split in two, and each part that inherited both sockets is arbitrarily assigned its own end of the resulting channel. Then, in line with the earlier program, we make the child process send, and the parent process receive data. This time we use the write(2) and read(2) system calls, respectively. Since both of them may and really do perform data delivery in chunks rather than in full, these calls need to be invoked repeatedly for larger messages in order to put or get the whole of the message body. This is why it is necessary to advance the buffer pointers and to decrement the counts, which is best done in the static functions loh_write() and loh_read(), respectively. Unlike the shared memory case, we are using rather slow system calls in the inner-most loop, so that an extra function call here or there does not matter.

Note that zero-sized messages are not passed across, since the write and read of the zero-byte buffer is not affecting the state of the communication channel. However, as you may remember, in MPI we will never send exactly zero bytes, because we will have to deliver some extra data comprising the message envelope. Thus, this little flaw of the naïve local socket benchmark is of no consequence to the later discussion. At the same time, we can see how much an iteration of the respective empty loop costs. Being two orders of magnitude smaller than the typical message latency, this overhead is negligible (unlike the shared memory case).

This is what we get once the data is processed (see Figure 3.1):

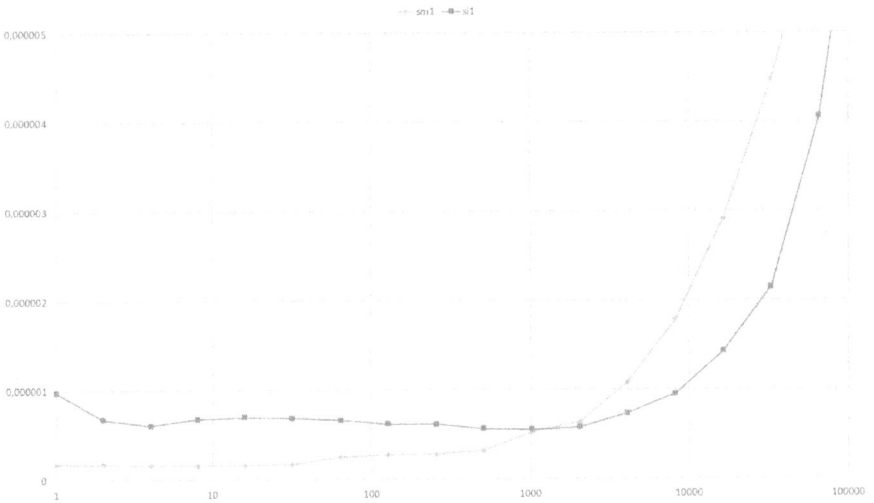

Figure 3.1: Naïve shared memory benchmark sm1 vs. naïve local stream socket benchmark si1: intranode latency (seconds)

We have mapped the sm1 curve onto the same graph (see Figure 2.8) to highlight the difference in performance observed between the shared memory and the local sockets. It should not surprise you that shared memory wins hands down on very small messages: shared memory works directly on the memory interface, while sockets trap into the kernel for each I/O operation, costing a lot of time. However, the curves cross around 1 KiB message size, which indicates that at this moment, either superior socket technology (e.g., extensive kernel level buffering) or a quirk in process binding takes over and soon doubles performance compared to our naïve shared memory benchmark. This is confirmed by the bandwidth relationship (see Figure 3.2):

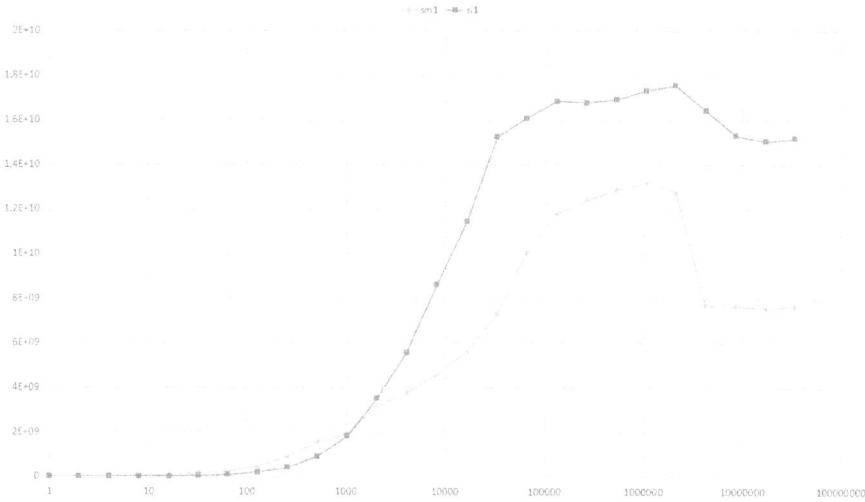

Figure 3.2: Naïve shared memory benchmark sm1 vs. naïve local stream socket benchmark si1: intranode bandwidth (bytes per second)

3.2.1.2 Internet Sockets

Unlike the essentially symmetric shared memory and local sockets we've been dealing with so far, internet sockets follow the asymmetric server-client paradigm:

- On one hand, we have a server that is there to accept and process client requests.
- One the other hand there are a number of clients, each of which transact their own business with the server.

The communication between the server and the client is bidirectional, but their roles are not equal and depend upon the socket type selected.

3.2.1.2.1 Internet Stream Sockets

Let's look at a stream socket example (see Listing 3.2):

Listing 3.2: Naïve Internet stream socket benchmark sx1 (excerpt, see sock/prep/sx1/sx1.c)

```
...

static char *host = NULL;
static char *port = "10001";

int main(int argc,char **argv)
{
    int i,imax = IMAX,l,lmax = LMAX,m,n;
    double t;
    char *buf;
```

```
    struct addrinfo ai,*pa,*pn;
    int fd,rc;
    pid_t pid = 0;

    switch (argc) {
    case 1:
        break;
    case 2:
        port = argv[1];
        break;
    case 3:
        host = argv[1];
        port = argv[2];
        break;
    default:
        fprintf(stderr,"Usage: %s [[host] port]\n",argv[0]);
        exit(EXIT_FAILURE);
    }

    memset(&ai,0,sizeof(struct addrinfo));
    ai.ai_family = AF_UNSPEC;       // IPv4 or IPv6
    ai.ai_socktype = SOCK_STREAM;
    ai.ai_protocol = 0;
#ifdef SERVER
    ai.ai_flags = AI_PASSIVE;       // for listen(2)
    ai.ai_canonname = NULL;
    ai.ai_addr = NULL;
#else
    ai.ai_flags = 0;
#endif
    ai.ai_next = NULL;

    rc = getaddrinfo(host,port,&ai,&pa);
    if (rc != 0) {
        fprintf(stderr,"%s: getaddrinfo: %s\n",argv[0],gai_strerror(rc));
        exit(EXIT_FAILURE);
    }

    for (pn = pa; pn != NULL; pn = pn->ai_next)
        if ((fd = socket(pn->ai_family,pn->ai_socktype,pn->ai_protocol)) != -1) {
#ifdef SERVER
            if (bind(fd,pn->ai_addr,pn->ai_addrlen) == 0)
                break;
#else
            if (connect(fd,pn->ai_addr,pn->ai_addrlen) == 0)
                break;
#endif
            close(fd);
        }
```

```
    if (pn == NULL) {
        perror(argv[0]);
        exit(EXIT_FAILURE);
    }

#ifdef SERVER
    if (listen(fd,1) == -1) {
        perror(argv[0]);
        exit(EXIT_FAILURE);
    }

    if ((fd = accept(fd,pn->ai_addr,&pn->ai_addrlen)) == -1) {
        perror(argv[0]);
        exit(EXIT_FAILURE);
    }

    pid = getpid();
#endif

//      freeaddrinfo(pa);

...
```

We combine the server and the client code into one program because they look nearly similar apart from the connection establishment phase. If the server side differs, it is guarded by the SERVER macro, while the client takes the opposite branch of the conditional compilation construct as necessary. The communication phase is identical to the local sockets (see Listing 3.1) and thus not shown here.

First, we parse the program arguments because both server and client need to know what address and port number to use. By default, the loopback device and port number 10001 (well beyond the reserved port range) is chosen. If only one argument is present, it is treated like a port number. If both arguments are used, the first indicates the host name or IP address, and the second argument specifies the port number.

Since we do not want to mess with the host name resolution ourselves, we use the getaddrinfo(3) function. In addition to the host name and the port number, it gets a couple of hints as to what socket family, socket type, protocol and flags are to be used, and returns a linked list of matching socket addresses. On the server side, we use the socket(2) – bind(2) – listen(2) – accept(2) system call sequence that creates a socket, binds it to a specific address, enters a listening mode, and accepts connection requests from the clients, respectively. On the client side, we use the socket(2) – connect(2) system call sequence that creates a socket and tries to connect to the server. Both bind(2) and connect(2) system calls iterate through the list of addresses returned by the getaddrinfo(3) function until success or end of list. In the latter case, the program fails.

This is what we get using the loopback device. Note that it is different from the communication path taken by the local sockets (see Figure 3.3):

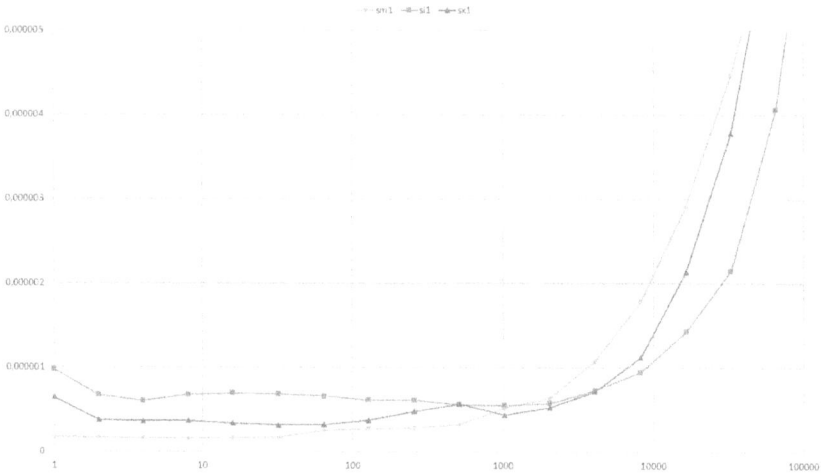

Figure 3.3: Naïve shared memory benchmark sm1 vs. naïve local stream socket benchmark si1 vs. naïve Internet stream socket benchmark sx1: intranode latency (seconds)

We see that Internet sockets behave similar to the naïve shared memory benchmark rather than naïve local stream socket benchmark. This is another confirmation of the guess that local sockets use a different mechanism. Indeed, this is again illustrated by the bandwidth graph (see Figure 3.4):

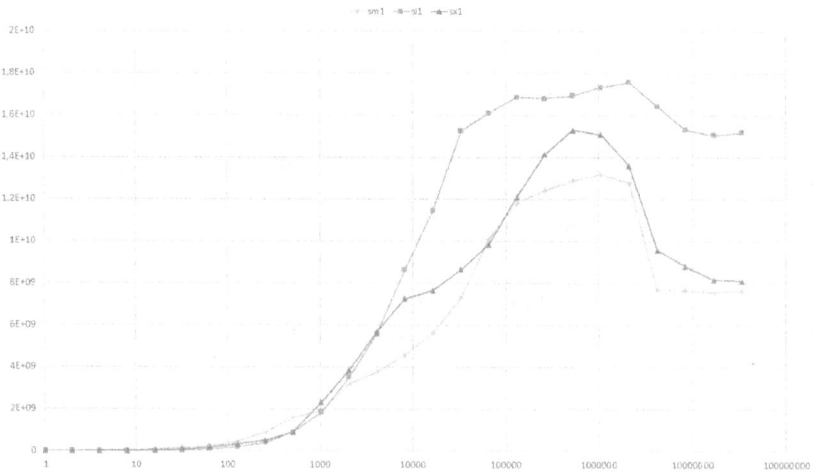

Figure 3.4: Naïve shared memory benchmark sm1 vs. naïve local stream socket benchmark si1 vs. naïve Internet stream socket benchmark sx1: intranode bandwidth (bytes per second)

3.2.1.2.2 Datagram Sockets

Datagram sockets are cumbersome when it comes to implementing the MPI with its unlimited message length, pairwise ordered delivery, and reliable data transfer. The thing is that datagrams are quite the opposite of these requirements, because:

-   The length of the datagrams is limited to a certain value, so that mapping an arbitrarily sized message may and sometimes will require a certain packaging scheme.
-   The datagrams may and in some cases do overtake each other while racing between two destinations, so that some ordering has to be imposed externally, say, by counting the instances issued and passing this ID as part of the datagram to the receiving end that should in turn take care of sorting the arriving packages into a proper sequence.
-   As has been mentioned before, even in tightly coupled systems, datagrams may and do get lost once in a while, sometimes in nontrivial quantities. This means that those who want to use datagrams for a reliable data delivery also have to take care of loss detection and retransmission, which may become especially nontrivial if losses are massive and/or occur way in the "past" as indicated by the sequence numbers.

In other words, the task of using datagrams for an MPI implementation is akin to a reimplementation of the TCP/IP protocol, this time fully in the user space. As such, this largely technical task goes beyond the scope of this book. If you fancy how datagrams can be made to work in a limited context of simple message passing with a small message size, unordered delivery and no reliability guarantee, you can look up the Linux getaddrinfo(3) man page.

3.2.1.3 Nonblocking Communication

The read(2) and write(2) system calls normally operate in the blocking mode, which would interfere with our intention to implement the blocking and nonblocking communication. Fortunately, sockets can be created in such a way that the aforementioned system calls become nonblocking.

There are several ways to make a socket nonblocking. The most appropriate for us is provided by the fcntl(2) system call. This way we can keep the whole of the connection establishment unchanged and switch the socket into the nonblocking mode just before beginning the communication. To this end, the following snippets can be added to the earlier program at the very top and very bottom of the earlier benchmarking program (see Listing 3.3, cf. Listing 3.2):

Listing 3.3: Naïve Internet stream socket benchmark sx1nb, nonblocking mode (excerpt, see sock/prep/sx1/sx1.c)

```
#ifdef NB
#include <fcntl.h>                          // fctnl(2) constants
#include <errno.h>                          // errno(3)
#endif

...

#ifdef NB
    if (fcntl(fd,F_SETFL,fcntl(fd,F_GETFL,0) | O_NONBLOCK)) {
        perror(argv[0]);
        exit(EXIT_FAILURE);
    }
#endif

...
```

Note that in our good tradition, we guard the addition by a conditional compilation construct based on the macro NB that can be set during the program build, thus giving us one source code for both blocking and nonblocking versions of the benchmark.

The data transfer calls also need to be updated a little compared to Listing 3.1. For example, for the read operation we get the following change to the innermost loop (see Listing 3.4, cf. Listing 3.1):

Listing 3.4: Naïve Internet stream socket benchmark sx1nb, data transfer calls, nonblocking mode (excerpt, see sock/prep/sx1/sx1.c)

```
...

        if ((m = read(fd,buf,n)) == -1) {
#ifdef NB
            if (errno == EAGAIN || errno == EWOULDBLOCK) {
                m = 0;
                continue;
            }
#endif
            return -1;
        }

...
```

The error codes EAGAIN and EWOULDBLOCK indicate that the respective call would have blocked if the socket were in the default, i.e., blocking mode. Measurements show no significant difference from the earlier observed performance (not shown). Note that we may be adding some extra overhead by spinning on the system calls. There are ways of avoiding this issue that we will learn later on, when they become indispensable.

3.2.2 Internode Communication

The beauty and power of sockets come to fruition when one goes for internode communication. The point is that in the programmer's view, nothing changes. What does change is performance. Since we do not need to learn too many new things here, we can simply reuse the naïve Internet stream socket benchmark (see Listing 3.2) on a cluster that has Ethernet or InfiniBand, or, for that matter, any other network that has a socket library implemented for it, which means basically all of them.

In particular, for a popular and widely used 100 Mbit Ethernet, we get the following results (see Figures 3.5a, 3.5b, 3.6a, and 3.6b):

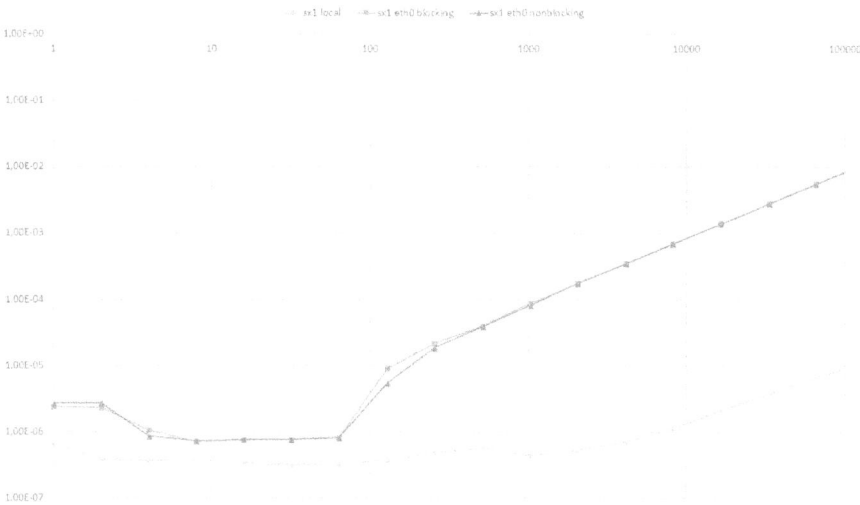

Figure 3.5a: Naïve Internet stream socket benchmark SX1: internode vs. intranode latency (seconds)

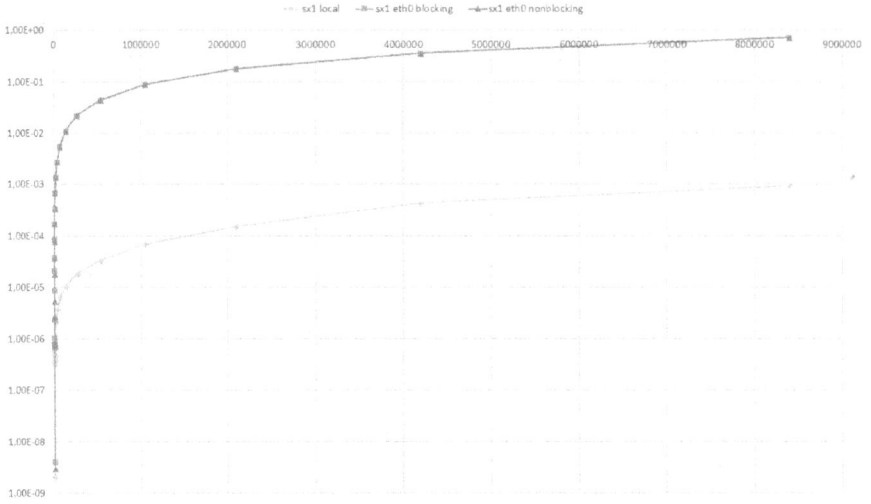

Figure 3.5b: Naïve Internet stream socket benchmark SX1: internode vs. intranode latency (seconds, linear scale of the message size axis, logarithmic scale of the value axis)

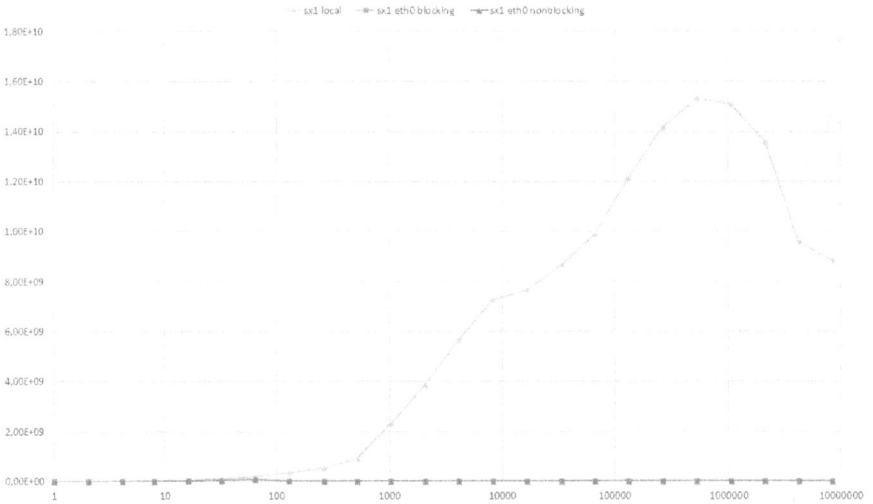

Figure 3.6a: Naïve Internet stream socket benchmark SX1: internode vs intranode bandwidth (bytes per second)

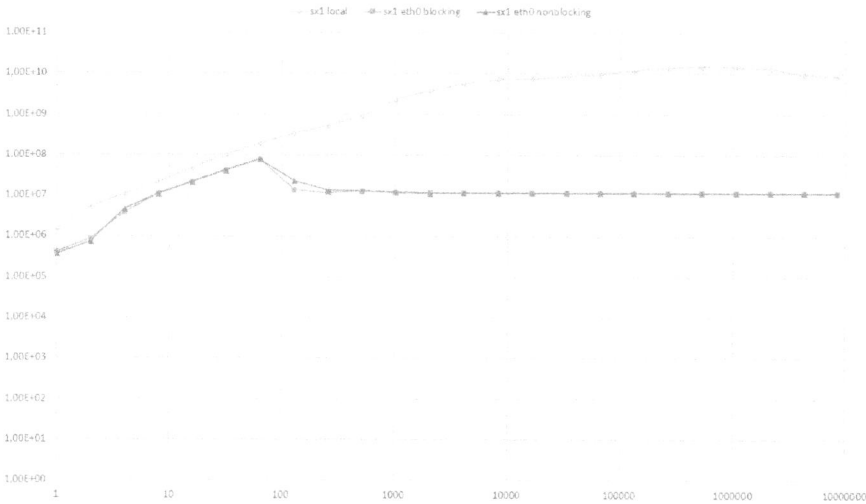

Figure 3.6b: Naïve Internet stream socket benchmark sx1: internode vs intranode bandwidth (bytes per second, logarithmic scale of the value axis)

Compare Figures 3.5a and 3.6a to their intranode pendants Figures 3.3 and 3.4, respectively, to see just how lucky we've been so far. Putting these two kinds of curves into the same graph hardly makes any sense, so dramatic is the advantage of the intranode communication. This is why we should rather use logarithmic scaling of the value axis in both graphs in order to understand what is happening here (see Figures 3.5b and 3.6b). Remember: to get latency right, you need to do something brilliant; to get bandwidth wrong, you need to do something terrible. We are apparently helped out by the operating system as far as latency of small messages is concerned: Linux is brilliant. We get roughly 95% of the theoretical 100 Mbit Ethernet bandwidth on large messages: we have not done anything terrible, since 5% is the usual toll taken by the Carrier Sense Multiple Access/Collision Detection protocol used by Ethernet under the hood.

3.2.3 Exercises

Exercise 3.3 Redo the intranode benchmarking done so far over sockets on your system. How does it fare versus my test laptop?

Exercise 3.4 If you have access to an Ethernet-based cluster, do the measurements over this network as well as intranode, and compare your results to mine. Do they essentially match?

Exercise 3.5 If you have access to cluster with InfiniBand or another network, find out how sockets are implemented over that network, and redo the measurements done above and compare the performance results with the intra- and internode data obtained so far.

3.3 Initial Subset Implementation: Start-up and Blocking Communication

It would have been very easy indeed for you to implement the simple subset that we defined for shared memory in Chapter 2. However, here we have to make a big stride forward since we want to provide both blocking and nonblocking communication. To make this stride more manageable, we will split it into two smaller steps, by using underlying blocking communication first (in this section) and underlying nonblocking communication next (in Section 3.4).

3.3.1 Design Decisions

Let's keep enhancing the existing MPI subset implementation we did before as much as we can. In this manner, we will be able to see what small changes to the subset definition can do to the implementation. Again, we generally follow the lead of the MPICH below.

3.3.1.1 MPI Header File
In line with the paradigm described above, we can extend the existing MPI header file (see Listing 2.15) by adding those extra entities that we defined as part of the subset (see Section 3.1 and Listing 3.5):

Listing 3.5: Subset MPI: declarations (excerpt, see sock/mpi/v0/mpi.h)

```
...

typedef MPI_Status *MPI_Request;

int MPI_Isend(void *buf,int cnt,MPI_Datatype dtype,int dest,int tag,
              MPI_Comm comm,MPI_Request *preq);
int MPI_Irecv(void *buf,int cnt,MPI_Datatype dtype,int src,int tag,
              MPI_Comm comm,MPI_Request *preq);

int MPI_Test(MPI_Request *preq,int *pflag,MPI_Status *pstat);
int MPI_Wait(MPI_Request *preq,MPI_Status *pstat);
```

The beginning of the header file stays unchanged and thus is not shown. The new opaque object MPI_Request is a pointer rather than an integer, because we will have to use it very often indeed, and it makes sense to avoid any conversion here. For now, the underlying opaque object contains only the MPI_Status object, as this is all we will need in our first naïve implementations.

Two calls—MPI_Isend() and MPI_Irecv()—initiate the respective nonblocking data transfer operations and return a pointer to the MPI_Request object that is managed by the MPI library. Two other calls—the nonblocking MPI_Test() and the blocking MPI_Wait()— (try to) complete the data transfer operation defined by the previously obtained pointer to the MPI_Request object. The MPI_Test() function reflects its result in the *pflag value. Both may modify the MPI_Status object that we know from the past.

3.3.1.2 Skeleton MPI Library
Likewise, we can extend the skeleton MPI library (see Listing 2.16) so that we have something to build and test with (see Listing 3.6):

Listing 3.6: Subset MPI: skeleton library (excerpt, see sock/mpi/v0/mpi.c)

```
...

int MPI_Isend(void *buf,int cnt,MPI_Datatype dtype,int dest,int tag,
              MPI_Comm comm,MPI_Request *preq)
{
    return MPI_SUCCESS;
}

int MPI_Irecv(void *buf,int cnt,MPI_Datatype dtype,int src,int tag,
              MPI_Comm comm,MPI_Request *preq)
{
    return MPI_SUCCESS;
}

int MPI_Test(MPI_Request *preq,int *pflag,MPI_Status *pstat)
{
    return MPI_SUCCESS;
}

int MPI_Wait(MPI_Request *preq,MPI_Status *pstat)
{
    return MPI_SUCCESS;
}
```

3.3.1.3 Testing
The test program t0.c (see Listing 2.17) and the Makefile (see Listing 2.18) defined earlier can be reused unchanged. Of course, we make a big assumption in the test program that the standard output stream of all programs will be redirected to our terminal, but this is what normally happens under Linux.

The output of the test program is trivial and thus is not shown.

3.3.2 Start-up and Termination

Start-up in the socket case is radically different from the shared memory scenario we considered before. There we could spawn as many processes as we wanted using the fork(2) system call. All of them shared the same shared memory segment by design, and thus could communicate with each other right away. In the Internet sockets case, which is the only practicable option when you want to start more than two intercommunicating processes potentially spanning more than one machine, both process start-up and connection establishment become nontrivial tasks on their own.

3.3.2.1 Design Considerations
It is better to delegate process start-up to some external agent instead of doing this from inside the library. The MPI standard defines basic features and the interface of such an agent, an external program called mpiexec. This external program has to pass to the MPI library information about the MPI universe—the number of processes, the rank of the current process, as well as details necessary for establishing connections between individual processes.

3.3.2.2 Process Start-up
The start-up is relatively easy. We can use one of the standard system utilities, such as the ssh(1) or the unjustly deprecated rsh(1). Both services will use their respective daemons running in the background on every reasonably configured Linux machine, so we do not have to bother with this part of the job. Of course, we will need to make sure that we can start processes this way without having to type in the passwords and such, but this is a question answered many times over by the user community (Stackoverflow.com, 2013).

3.3.2.2.1 The mpiexec Script
In order to direct said system services, we will create a small script called mpiexec and make it recognize the same options that we used before, namely:
```
$ mpiexec -n[p] number_of_processes program [arguments]
```

Doing this instead of entrusting the process start-up to the MPI library itself is quite handy, as it moves the dirty details of the process creation out of its scope into the realm of interpreters such as bash(1) that were specifically designed to handle tasks of this kind.

The next question we need to answer is how the mpiexec script will know what hosts to use for the job start-up. There are many ways to achieve this, but we will implement just two:

– By default, all processes will be created on the localhost, i.e., the same machine where the mpiexec has been started by the user. To this end we do not even need to involve the ssh(1) or another system service, which will result in better performance and less fuss with the setup, simultaneously providing an independent path for testing the more complicated second way.

– If a file named mpi.hosts is present in the current directory, however, the lines in this file will name the hosts to use, with the added finesse of the last line being taken as the name of all other hosts to use if the number of processes requested exceeds the number of the host names in the file.

A purist would say that this is a very limited approach, for example, by pointing out that we may want to use more than one host list in different situations. Well, in this case we can create files named, say, mpi.hosts.1, mpi.hosts.2, and so on, and then hard or soft link the file name mpi.hosts with the one we need at the moment. So, here again we achieve 90% of the desired result by spending only 10% of the effort required for a complete (and rarely needed) solution.

Fine. How will the processes started by an external facility learn where they belong in the MPI job? In order to make sense of their universe, they need to know the job size, their rank in it, and—for the connection establishment—the names and the port numbers of all other processes. Again, there is more than one way to cut this cake. The easiest and most flexible is to define a couple of specific environment variables, say, MPI_SIZE, MPI_RANK, MPI_HOSTS, and MPI_PORTS, that will be set by the mpiexec script and then passed as part of the environment to every MPI process. This is preferable to introducing special MPI options or configuration files, to name just a few other possibilities.

The host list is easy: we will get it from the mpi.hosts file. What about the port numbers? To be on the safe side, each process needs to be identified by a unique pair of host name and port number, since the host names may, and indeed often will be, the same for at least some of the MPI processes in a job. The method we will use is based on one port value set out of the reserved port number range, say, 10000, that will be incremented for each new process that needs to be started. This is not the ideal solution, but since we do not want to mess with the process start-up more than necessary, this is about all we can do for now.

All that said, this is what we end up with as far as the mpiexec script is concerned (see Listing 3.7):

Listing 3.7: **S**ubset MPI: the mpiexec script (see sock/mpi/v1/mpiexec)

```
#!/bin/sh
hostfile="mpi.hosts"
portbase=10000

launchLocal () {
#    echo "launchLocal $*"
    if [ $1 -lt $2 ]
    then
        local rank=$1
        shift
        local size=$1
        shift

        launchLocal 'expr $rank + 1' $size $*

        if [ $rank -eq 0 ]
        then
#            echo $rank/$size: "$*" "[wait]"
            (export MPI_RANK=$rank; export MPI_SIZE=$size; export MPI_HOSTS="$hosts";
export MPI_PORTS="$ports"; $*)
        else
#            echo $rank/$size: "$*" "[nowait]"
            (export MPI_RANK=$rank; export MPI_SIZE=$size; export MPI_HOSTS="$hosts";
export MPI_PORTS="$ports"; $*) &
        fi
    fi
}

launchRemote () {
#    echo "launchRemote $*"
    if [ $1 -lt $2 ]
    then
        local rank=$1
        shift
        local size=$1
        shift
        local host
        if ! read host
        then
            host="$last"
        fi

        launchRemote 'expr $rank + 1' $size $*
```

```
        if [ $rank -eq 0 ]
        then
#             echo $rank/$size: ssh $host "$*" "[wait]"
             ssh $host "export MPI_RANK=$rank; export MPI_SIZE=$size; export
MPI_HOSTS=\"$hosts\"; export MPI_PORTS=\"$ports\"; $*"
        else
#             echo $rank/$size: ssh $host "$*" "[nowait]"
             ssh $host "export MPI_RANK=$rank; export MPI_SIZE=$size; export
MPI_HOSTS=\"$hosts\"; export MPI_PORTS=\"$ports\"; $*" &
        fi
    fi
}

if [ $# -gt 0 ]
then
#    echo "$*"

    if [ "$1" = "-n" -o "$1" = "-np" ]
    then
        shift
#        echo "$*"

        if [ $# -gt 0 ]
        then
            size=$1
            shift
#            echo "$size x $*"

            if [ $# -gt 0 -a $size -gt 0 ]
            then
                if [ -r $hostfile ]
                then
                    last="localhost"
                    rank=0
                    while [ $rank -lt $size ]
                    do
                        if read host
                        then
                            last="$host"
                        fi
                        hosts="$hosts $last"
                        ports="$ports 'expr $portbase + $rank'"
                        rank='expr $rank + 1'
                    done < $hostfile

                    launchRemote 0 $size $* < $hostfile
                else
                    rank=0
                    while [ $rank -lt $size ]
                    do
```

```
                        hosts="$hosts localhost"
                        ports="$ports 'expr $portbase + $rank'"
                        rank='expr $rank + 1'
                done

                launchLocal 0 $size $*
            fi

            exit $?
        fi
    fi
  fi
fi

echo "Usage: mpiexec -n[p] number_of_processes program [program_arguments]"
exit 1
```

We keep the script as simple as possible, especially by processing a very limited set of options compared to the ones described by the MPI standard. A fine point that needs mentioning now is that we do take care of the order in which the processes are started. Namely, we put MP process rank 0 upon the first host mentioned in the mpi.hosts file, if any, and so on. This helps the user by making the process placement linear and predictable, keeping in mind the principle of least surprise. However, we also take care of waiting on the MPI process rank 0, as this is the only process required by the MPI standard to return from the MPI_Finalize() call. The exit code of this process becomes the exit code of the mpiexec script and thus of the MPI job as a whole. Hence, although we order the processes by the host name list, we start them in reverse temporal order, with the MPI process rank 0 starting last. If we ever decide to do funny stuff like exit code propagation from all other MPI processes, we will have to take care of that in the MPI library proper.

3.3.2.2.2 The MPI Library
The matching part of the MPI library looks like this (see Listing 3.8):

Listing 3.8: Subset MPI: MPI initialization (excerpt, see sock/mpi/v1/mpi.c)

```
...

static int _mpi_init = 0;
static int _mpi_size = 1;
static int _mpi_rank = 0;
static char **_mpi_hosts,**_mpi_ports;

...


int MPI_Init(int *pargc,char ***pargv)
```

```
{
#ifdef DEBUG
    fprintf(stderr,"%d/%d: MPI_Init(%p,%p)\n",_mpi_rank,_mpi_size,pargc,pargv);
#endif

    assert(_mpi_init == 0);
    assert(pargc != NULL && *pargc > 0);
    assert(pargv != NULL);

    _mpi_size = _mpi_getint("MPI_SIZE",_mpi_size);
    assert(_mpi_size > 0);
    _mpi_rank = _mpi_getint("MPI_RANK",_mpi_rank);
    assert(_mpi_rank >= 0 && _mpi_rank < _mpi_size);

    _mpi_hosts = _mpi_getarr("MPI_HOSTS",_mpi_size);
    assert(_mpi_hosts != (char **)NULL);
    _mpi_ports = _mpi_getarr("MPI_PORTS",_mpi_size);
    assert(_mpi_ports != (char **)NULL);

    if (_mpi_connect_all() != MPI_SUCCESS) {
#ifdef DEBUG
        fprintf(stderr,"%d/%d: -> failure\n",_mpi_rank,_mpi_size);
#endif
        return !MPI_SUCCESS;
    }

    _mpi_init = 1;

#ifdef DEBUG
    fprintf(stderr,"%d/%d: -> success\n",_mpi_rank,_mpi_size);
#endif
    return MPI_SUCCESS;
}
```

Since we have to get the process identity and other information from the environment, and this environment has at least two obligatory integer elements (the MPI_RANK and the MPI_SIZE), as well as two comma separated lists (the MPI_HOSTS and MPI_PORTS) we define a couple of auxiliary routines that take care of querying the environment variables and putting the data contained in them in the expected internal format (see Listing 3.9):

Listing 3.9: Subset MPI: environment variable query (excerpt, see sock/mpi/v1/mpi.c)

```
static int _mpi_getint(char *var,int val)
{
    if ((var = getenv(var)) != NULL)
    {
        int newval;
```

```
        errno = 0;
        newval = (int)strtol((const char *)var,NULL,10);
        switch (errno) {
            case EINVAL:
            case ERANGE:
                break;
            default:
                val = newval;
                break;
        }
    }

    return val;
}

static char **_mpi_getarr(char *var,int len)
{
    char **arr = (char **)NULL;

    if ((var = getenv(var)) != NULL)
    {
        char **newarr = malloc(sizeof(char *)*len);

        if (newarr != (char **)NULL)
        {
            char *newvar = malloc(sizeof(char)*strlen(var));

            if (newvar != NULL) {
                if ((newarr[0] = strtok(strcpy(newvar,var + 1)," ")) != NULL)
                {
                    int i;

#ifdef DEBUG
                    fprintf(stderr,"%d/%d: %s\n",_mpi_rank,_mpi_size,newarr[0]);
#endif
                    for (i = 1; i < len; i++) {
                        if ((newarr[i] = strtok(NULL," ")) == NULL) {
                            free(newarr);
                            free(newvar);
                            goto end;
                        }
#ifdef DEBUG
                    fprintf(stderr,"%d/%d: %s\n",_mpi_rank,_mpi_size,newarr[i]);
#endif
                    }

                    arr = newarr;
                }
                else {
```

```
                    free(newarr);
                    free(newvar);
                }
            }
            else
                free(newarr);
        }
    }

end:return arr;
}
```

The _mpi_getint() function resembles closely the parsing of the integer option we implemented earlier (see Listing 2.20). The _mpi_getarr() function is a little trickier. In the mpiexec script (see Listing 3.5), we start with a list represented by an empty environment variable. Hence, the first list element is prepended by a comma. In order to trim it out, we increment the string pointer by one, and hence get to the first element immediately. Then the strtok(3) takes over, and we get the remaining elements, if any, up to the value of len. Since the strtok(3) may, and normally will, change the string it is operating on, we make a copy of the value of the respective environment variable obtained earlier using the getenv(3) function. The length of the resulting string accounts for the closing null byte automatically due to the increment mentioned above. Of course, one hand washes the other here, and someone would certainly say that this way of programming is unsafe and even dangerous. To this I answer: hey, this is my program. I know how the parts fit together, and I do not see the point in hiding this from myself and complicating things beyond the call of duty.

3.3.2.3 Connection Establishment

Now that all processes have been started and fed the necessary data, they need to connect to each other somehow. Instead of delaying this in a lazy fashion to the moment when a process has to send something out or intends to receive something, we will again establish all connections upfront. Since the order matters, we will make every process with rank P:

- Open a listening socket for incoming connections under the host name and port number corresponding to its rank
- Accept all connections from processes with ranks from 0 to P-1
- Connect to all processes with ranks from P+1 to N-1, where N is the number of all MPI processes in this job

This way all processes will proceed in proper order and there will be no deadlocks. Something comparable is done in the start-up phase of the IMPI standard (George, Hagedorn, & Devaney, 2000). At the same time, due to our starting the processes from rank N-1 down to rank 0, we will be relatively certain to have a listening socket already waiting when a matching connect call is issued. As a consequence of this temporal ordering, process rank 0 will be only connecting, and process rank N-1 only accepting the incoming connections. This is not what people normally do, for rank 0 is often considered a server and hence ought to be accepting connections. We will see soon enough whether this is a good idea.

This is what the connection establishment looks like (see Listing 3.10):

Listing 3.10: Subset MPI: connection establishment (excerpt, see `sock/mpi/v1/mpi.c`)

```
...

static int *_mpi_fds;

...

static int _mpi_connect_all(void)
{
    int rc = MPI_SUCCESS;

    if ((_mpi_fds = malloc(sizeof(int)*_mpi_size)) == (int *)NULL) {
#ifdef DEBUG
        fprintf(stderr,"%d/%d: malloc: %s\n",_mpi_rank,_mpi_size,strerror(errno));
#endif
        return !MPI_SUCCESS;
    }

    if (_mpi_rank > 0)
        rc = _mpi_accept_lower();

    if (rc == MPI_SUCCESS && _mpi_rank < _mpi_size - 1)
        rc = _mpi_connect_higher();

    return rc;
}
```

We keep all socket file ids in one array that simultaneously works as the connection table. For any process P, the file id number P holds the file id of the listening socket, in case we ever need to refer to it (say, for an orderly closing). All other elements keep the socket file ids of the respective MPI process ranks.

In accordance with the earlier description of the temporal order of the process start-up, we accept connection of all ranks lower than ours (see Listing 3.11):

Listing 3.11: Subset MPI: accepting connections (excerpt, see `sock/mpi/v1/mpi.c`)

```
static int _mpi_accept(int fd,struct addrinfo *pn)
{
    int i;

    if ((fd = accept(fd,pn->ai_addr,&pn->ai_addrlen)) == -1) {
#ifdef DEBUG
        fprintf(stderr,"%d/%d: accept: %s\n",_mpi_rank,_mpi_size,strerror(errno));
#endif
        return !MPI_SUCCESS;
    }

    if (read(fd,&i,sizeof(i)) != sizeof(i)) {
#ifdef DEBUG
        fprintf(stderr,"%d/%d: read: %s\n",_mpi_rank,_mpi_size,strerror(errno));
#endif
        return !MPI_SUCCESS;
    }

    _mpi_fds[i] = fd;

    return MPI_SUCCESS;
}

static int _mpi_accept_lower(void)
{
    struct addrinfo ai,*pa,*pn;
    int i,fd,rc;

    memset(&ai,0,sizeof(struct addrinfo));
    ai.ai_family = AF_UNSPEC;       // IPv4 or IPv6
    ai.ai_socktype = SOCK_STREAM;
    ai.ai_protocol = 0;
    ai.ai_flags = AI_PASSIVE;       // for listen(2)
    ai.ai_canonname = NULL;
    ai.ai_addr = NULL;
    ai.ai_next = NULL;

    rc = getaddrinfo(_mpi_hosts[_mpi_rank],_mpi_ports[_mpi_rank],&ai,&pa);
    if (rc != 0) {
#ifdef DEBUG
        fprintf(stderr,"%d/%d: getaddrinfo: %s\n",_mpi_rank,_mpi_size,gai_
strerror(rc));
#endif
        return !MPI_SUCCESS;
    }

    for (pn = pa; pn != NULL; pn = pn->ai_next)
```

```
        if ((fd = socket(pn->ai_family,pn->ai_socktype,pn->ai_protocol)) != -1) {
            if (bind(fd,pn->ai_addr,pn->ai_addrlen) == 0)
                break;
            close(fd);
        }

    if (pn == NULL) {
#ifdef DEBUG
        fprintf(stderr,"%d/%d: bind: %s\n",_mpi_rank,_mpi_size,strerror(errno));
#endif
        return !MPI_SUCCESS;
    }

    if (listen(fd,1) == -1) {
#ifdef DEBUG
        fprintf(stderr,"%d/%d: listen: %s\n",_mpi_rank,_mpi_size,strerror(errno));
#endif
        return !MPI_SUCCESS;
    }

    for (i = 0; i < _mpi_rank; i++)
        if (_mpi_accept(fd,pn) != MPI_SUCCESS)
            return !MPI_SUCCESS;

    freeaddrinfo(pa);

    _mpi_fds[_mpi_rank] = fd;

    return MPI_SUCCESS;
}
```

The only really interesting point here is that while accepting a connection, we have no way of safely learning which exact MPI process rank has issued the respective connect(2) call. We hope that they may come in order and do everything to facilitate that, but we certainly cannot count on that, even on one host, where some processes may be delayed, and even more so in the situation of more than one host. In order to avoid any uncertainty here, we make the connecting process send its rank as the very first integer word over the established connection (see Listing 3.12):

Listing 3.12: Subset MPI: connecting (excerpt, see sock/mpi/v1/mpi.c)

```
static int _mpi_connect(int i)
{
    struct addrinfo ai,*pa,*pn;
    int fd,rc;

    memset(&ai,0,sizeof(struct addrinfo));
    ai.ai_family = AF_UNSPEC;        // IPv4 or IPv6
    ai.ai_socktype = SOCK_STREAM;
```

```
    ai.ai_protocol = 0;
    ai.ai_flags = 0;
    ai.ai_next = NULL;

    rc = getaddrinfo(_mpi_hosts[i],_mpi_ports[i],&ai,&pa);
    if (rc != 0) {
#ifdef DEBUG
        fprintf(stderr,"%d/%d: getaddrinfo: %s\n",_mpi_rank,_mpi_size,
gai_strerror(rc));
#endif
        return !MPI_SUCCESS;
    }

    for (pn = pa; pn != NULL; pn = pn->ai_next)
        if ((fd = socket(pn->ai_family,pn->ai_socktype,pn->ai_protocol)) != -1) {
            if (connect(fd,pn->ai_addr,pn->ai_addrlen) == 0)
                break;
            close(fd);
        }

    if (pn == NULL) {
#ifdef DEBUG
        fprintf(stderr,"%d/%d: connect: %s\n",_mpi_rank,_mpi_size,strerror(errno));
#endif
        return !MPI_SUCCESS;
    }

    freeaddrinfo(pa);

    if (write(fd,&_mpi_rank,sizeof(_mpi_rank)) != sizeof(_mpi_rank)) {
#ifdef DEBUG
        fprintf(stderr,"%d/%d: write: %s\n",_mpi_rank,_mpi_size,strerror(errno));
#endif
        return !MPI_SUCCESS;
    }

    _mpi_fds[i] = fd;

    return MPI_SUCCESS;
}

static int _mpi_connect_higher(void)
{
    int i;

    for (i = _mpi_size - 1; i > _mpi_rank; i--)
        if (_mpi_connect(i) != MPI_SUCCESS)
            return !MPI_SUCCESS;

    return MPI_SUCCESS;
}
```

The respective accepting process reads this value in, and thus we get the MPI rank of the process we have just accepted. Being a bit overconfident, we skip the iteration over the length of this data item written and read, for one integer will normally go through a properly setup socket in one go. Still, we guard against this eventuality, and generally pay a bit more attention to the error checking here than before.

You have certainly noticed that the roles are split unevenly between the higher level and lower level calls in the past two listings. For example, on the accepting side, the address to use is determined once in the higher level call _mpi_accept_lower(). On the connecting side, however, the addresses are determined for each connection individually inside the respective lower level call _mpi_connect(). This is no accident, just a sensible response to the lay of the land. Try to cut it differently, and you will end up with a lot more data dependencies and thus complexity.

3.3.2.4 Process Termination

Actually, let's look into the termination in more detail. Yeah, death is a part of life, so we need to look into this matter. So far, we have not really cared about cleaning up the process state, expecting, and reasonably so, that the program termination would do this for us anyway. However, what was just fine for a naïve benchmark may not be appropriate for a library that intends to approach the highest standard of quality— that of a software product usable by anyone, anywhere, anytime in an easy, intuitive, and predictable fashion. Port numbers in particular are a precious system commodity that needs to be handled with care so that it is not exhausted or blocked. We will see how this pans out as we go.

3.3.3 Point-to-Point Communication

In this section we will build upon our experience with shared memory.

3.3.3.1 Design Considerations

We are going to take quite a shortcut in this intermediate version of the MPI library. Namely, we will map the nonblocking communication upon the blocking one. Yes, it is possible, within certain limits. We will improve on this later in this chapter, so keep reading.

3.3.3.2 Blocking Communication

The only real complication here is that apart from the user buffer itself, we have to pass on the *message envelope*. In our case, when there is only one communicator and no other complications, the message tag and the message length in bytes should suffice.

One alternative would be to allocate another buffer capable of holding both the said data items and the buffer, and then send this compound buffer in one go. You may feel that this is overkill, but this is exactly what is done by the less sophisticated implementations of the OSI model (Wikipedia.org, OSI Model, 2018d).

Another alternative is to use the readv(2) and writev(2) system calls that can work on noncontiguous buffers. This being a useful lead, we will postpone it until the next iteration of our development, so that we do not overcomplicate things upfront.

This is what the relevant part of the MPI_Send() call looks like (see Listing 3.13):

Listing 3.13: Subset MPI: MPI_Send() implementation (excerpt, see sock/mpi/v1/mpi.c)

```
...

    if (write(_mpi_fds[dest],&tag,sizeof(tag)) != sizeof(tag)) {
#ifdef DEBUG
        fprintf(stderr,"%d/%d: write tag: %s\n",_mpi_rank,_mpi_size,strerror(errno));
#endif
        return !MPI_SUCCESS;
    }
    if (write(_mpi_fds[dest],&cnt,sizeof(cnt)) != sizeof(cnt)) {
#ifdef DEBUG
        fprintf(stderr,"%d/%d: write cnt: %s\n",_mpi_rank,_mpi_size,strerror(errno));
#endif
        return !MPI_SUCCESS;
    }

    for (; cnt > 0; buf += m,cnt -= m)
        if ((m = write(_mpi_fds[dest],buf,cnt)) == -1) {
#ifdef DEBUG
            fprintf(stderr,"%d/%d: write buf: %s\n",_mpi_rank,_mpi_size,
strerror(errno));
#endif
            return !MPI_SUCCESS;
        }

...
```

The matching part of the MPI_Recv() call is slightly more extensive (see Listing 3.14):

Listing 3.14: Subset MPI: MPI_Recv() implementation (excerpt, see sock/mpi/v1/mp1.c)

```
...

    if (read(_mpi_fds[src],&itag,sizeof(itag)) != sizeof(itag)) {
#ifdef DEBUG
        fprintf(stderr,"%d/%d: read itag: %s\n",_mpi_rank,_mpi_size,strerror(errno));
#endif
```

```
        return !MPI_SUCCESS;
    }
    if (tag != itag) {
#ifdef DEBUG
        fprintf(stderr,"%d/%d: -> failure, tag %d != %d\n",_mpi_rank,_mpi_size,
tag,itag);
#endif
        return !MPI_SUCCESS;
    }

    if (read(_mpi_fds[src],&icnt,sizeof(icnt)) != sizeof(icnt)) {
#ifdef DEBUG
        fprintf(stderr,"%d/%d: read icnt: %s\n",_mpi_rank,_mpi_size,strerror(errno));
#endif
        return !MPI_SUCCESS;
    }
    if (cnt < icnt) {
#ifdef DEBUG
        fprintf(stderr,"%d/%d: -> failure, cnt %d < %d\n",_mpi_rank,_mpi_size,
cnt,icnt);
#endif
        return !MPI_SUCCESS;
    }

    for (cnt = icnt; cnt > 0; buf += m,cnt -= m)
        if ((m = read(_mpi_fds[src],buf,cnt)) == -1) {
#ifdef DEBUG
            fprintf(stderr,"%d/%d: read buf: %s\n",_mpi_rank,_mpi_size,
strerror(errno));
#endif
            return !MPI_SUCCESS;
        }

    if (pstat != NULL) {
        pstat->MPI_SOURCE = src;
        pstat->MPI_TAG = tag;
    }

    ...
```

Just as earlier, we do guard against mismatching tags, since our implementation would simply hang had we not done so. In case of the buffer being smaller than the message to be received, we also bail out, leaving the buffer untouched.

3.3.3.3 Nonblocking Communication

The nonblocking calls may, within reason, be mapped upon the blocking ones. Of course, they will then retain all limitations of the blocking communication. However, what we gain in return are simplicity of the implementation and a better understanding of both the MPI rules and the role split between the transfer initialization and transfer completion calls.

The transfer initialization calls are relatively straightforward (see Listing 3.15):

Listing 3.15: Listing 3.15. Subset MPI: nonblocking communication (excerpt, see sock/mpi/v1/ mpi.c)

```
int MPI_Isend(void *buf,int cnt,MPI_Datatype dtype,int dest,int tag,MPI_Comm comm,
MPI_Request *preq)
{
#ifdef DEBUG
    fprintf(stderr,"%d/%d: MPI_Isend(%p,%d,%d,%d,%d,%d,%p)\n",_mpi_rank,_mpi_size,
buf,cnt,dtype,dest,tag,comm,preq);
#endif

    assert(_mpi_init);
    assert(preq != NULL);

    *preq = NULL;

    return MPI_Send(buf,cnt,dtype,dest,tag,comm);
}

int MPI_Irecv(void *buf,int cnt,MPI_Datatype dtype,int src,int tag,MPI_Comm comm,
MPI_Request *preq)
{
#ifdef DEBUG
    fprintf(stderr,"%d/%d: MPI_Irecv(%p,%d,%d,%d,%d,%d,%p)\n",_mpi_rank,_mpi_size,
buf,cnt,dtype,src,tag,comm,preq);
#endif

    assert(_mpi_init);
    assert(preq != NULL);

    return MPI_Recv(buf,cnt,dtype,src,tag,comm,(*preq = _mpi_newreq(&_mpi_rq)));
}
```

The MPI_Send() call does not set the status in our limited subset, so we simply nullify the particular request. The MPI_Recv() does set the status if available, so we have to carry this status somehow to the respective transfer completion call. For this we create a very small and limited implementation of the request handling component (see Listing 3.16):

Listing 3.16: Subset MPI: request management (excerpt, see `sock/mpi/v1/mpi.c`)

```
static struct _mpi_q_s {
    MPI_Status req[1];
    int n;
} _mpi_rq;

static MPI_Request _mpi_newreq(struct _mpi_q_s *pq)
{
    if (pq->n < 1)
        return prq->req + (pq->n++);

    return NULL;
}

static void _mpi_freereq(struct _mpi_q_s *pq,MPI_Request *preq)
{
    if (pq->n > 0) {
        *preq = NULL;
        pq->n--;
    }
}

static void _mpi_statcpy(MPI_Status *pstatout,MPI_Status *pstatin)
{
    *pstatout = *pstatin;
}
```

Here again we mercilessly exploit the knowledge of the data structures we have. Yes, an MPI_Request points to an MPI_Status. Further, since our test programs are not going to need more than one pending request per process, we do not need to hold more than one element. If we decide to expand this implementation later on, its external interfaces will stay unchanged. This settles the structure of the request queue for the time being.

The message completion calls are mapped upon each other. The MPI_Test() being nonblocking and thus more versatile, the MPI_Wait() is bound to sit on top of it (see Listing 3.17):

Listing 3.17: Subset MPI: transfer completion calls (excerpt, see `sock/mpi/v1/mpi.c`)

```
int MPI_Test(MPI_Request *preq,int *pflag,MPI_Status *pstat)
{
#ifdef DEBUG
    fprintf(stderr,"%d/%d: MPI_Test(%p,%p,%p)\n",_mpi_rank,_mpi_size,preq,pflag,
pstat);
#endif
```

```
    assert(_mpi_init);
    assert(preq != NULL);
    assert(pflag != NULL);
    assert(pstat != NULL);

    *pflag = 1;
    if (*preq != NULL) {
        _mpi_statcpy(pstat,*preq);
        _mpi_freereq(&_mpi_rq,preq);
    }

#ifdef DEBUG
    fprintf(stderr,"%d/%d: -> success, *pflag = %d, *pstat = { %d, %d }\n",_mpi_rank,
_mpi_size,*pflag,pstat->MPI_SOURCE,pstat->MPI_TAG);
#endif
    return MPI_SUCCESS;
}

int MPI_Wait(MPI_Request *preq,MPI_Status *pstat)
{
    int flag = 0;

#ifdef DEBUG
    fprintf(stderr,"%d/%d: MPI_Wait(%p,%p)\n",_mpi_rank,_mpi_size,preq,pstat);
#endif

    return MPI_Test(preq,&flag,pstat);
}
```

Since we know that our MPI_Isend() is not going to change the MPI_Status argument of the transfer completion calls, and that there cannot be more than one pending MPI_Irecv(), we can safely assume that any non-null MPI_Status belongs to the pending MPI_Irecv() operation and fill it in accordingly. Note that since the MPI_Isend() nullifies its MPI_Request, we use this as a sign of what operation the request comes from. Again, the 10/90 principle in action.

By the way, we could have mapped the MPI_Test() upon the MPI_Wait(), and nobody would care. The debugging output would look a little confusing, true. We do not make a big fuss about this or the fact that a call may have two entry printouts and only one exit printout. As long as we know what happens under the hood, it's fine.

Last but not least, the _mpi_statcpy() is certainly a good candidate for a macro or even a straight structure assignment. But I come from those old days when a structure assignment would not work. A macro is certainly appealing, but, frankly speaking, as soon as we see the latencies involved, you will not want to revisit this matter.

3.3.4 Testing

We can test this new library using the old test programs described earlier. Everything looks fine, as it should. In order to test the new nonblocking calls, we can extend the unified point-to-point benchmark by the *ping-ping pattern* as follows (see Listing 3.18):

Listing 3.18: Subset MPI: adding ping-ping pattern (excerpt, see `sock/mpi/v1/b.c`)

```
...

        for (l = 0; l <= lmax; l = (l) ? l << 1 : 1) {
            if (r == 0) {  // "sender"
                t = MPI_Wtime();
                for (i = 0; i < imax; i++) {
#ifdef PINGPING
                    MPI_Isend(b1,l,MPI_BYTE,0,0,MPI_COMM_WORLD,&rq);
                    MPI_Recv(b2,l,MPI_BYTE,0,1,MPI_COMM_WORLD,&st);
                    MPI_Wait(&rq,&st);
#endif
#ifdef PING
                    MPI_Send(b1,l,MPI_BYTE,0,0,MPI_COMM_WORLD);
#endif
#ifdef PONG
                    MPI_Recv(b2,l,MPI_BYTE,0,1,MPI_COMM_WORLD,&st);
#endif
#ifdef CHECK
                    for (k = 0; k < l; k++)
                        if (b2[k] != (char)k)
                            fprintf(stderr,"ERROR: r = %d, l = %d, i = %d, b2[%d] =
%02x, k = %02x\n",r,l,i,k,b2[k],(char)(k));
#endif
                }
                t = MPI_Wtime() - t;
            }
            else if (r == 0) {      // "receiver"
                t = MPI_Wtime();
                for (i = 0; i < imax; i++) {
#ifdef PINGPING
                    MPI_Isend(b1,l,MPI_BYTE,0,1,MPI_COMM_WORLD,&rq);
                    MPI_Recv(b2,l,MPI_BYTE,0,0,MPI_COMM_WORLD,&st);
                    MPI_Wait(&rq,&st);
#endif
#ifdef PING
                    MPI_Recv(b2,l,MPI_BYTE,0,0,MPI_COMM_WORLD,&st);
#endif
#ifdef CHECK
                    for (k = 0; k < l; k++)
```

```
                            if (b2[k] != (char)k)
                                fprintf(stderr,"ERROR: r = %d, l = %d, i = %d, b2[%d] =
%02x, k = %02x\n",r,l,i,k,b2[k],(char)(k));
#endif
#ifdef PONG
                        MPI_Send(b1,l,MPI_BYTE,0,1,MPI_COMM_WORLD);
#endif
                }
                t = MPI_Wtime() - t;
            }
#ifndef CHECK
            printf("r = %d\tbytes = %-8d\titers = %-8d\ttime = %-12.6g\t"
                    "lat = %-12.6g\tbw = %-12.6g\n",r,l,i,t,t/i/DIV,l/t*i*DIV);
#endif
        }
```

...

We incorporate the new pattern under the conditional compilation macro PINGPING. The rest of the code is refactored a little in order to make things work. However, as soon as we try to run this program, even as a test one with just one iteration, we stumble, because, at least on my computer, the program hangs after having sent and received a message of 8 MiB in size. What happened?

Apparently, we have overstretched some system threshold. Since we write to a socket, we basically put this message into a system buffer first, and start reading it only later. As long as the buffer allocated to us by the system is large enough, there is no problem. Once the message size gets bigger than that, however, the program has no chance of proceeding beyond the point of sending.

To fix this issue, we will have to review the library design and move a step closer to the Real Thing. However, let's have a look at the performance in those message size ranges that we do support first, so that we get some feeling of what it can look like.

3.3.5 Benchmarking

If you run this program yourself, you will more likely than not notice that it is terribly slow. So slow, in fact, that I had to reduce the number of iterations from 10000 to 10 in order to get anywhere. This subjective impression was confirmed by the resulting measurements (see Figure 3.7):

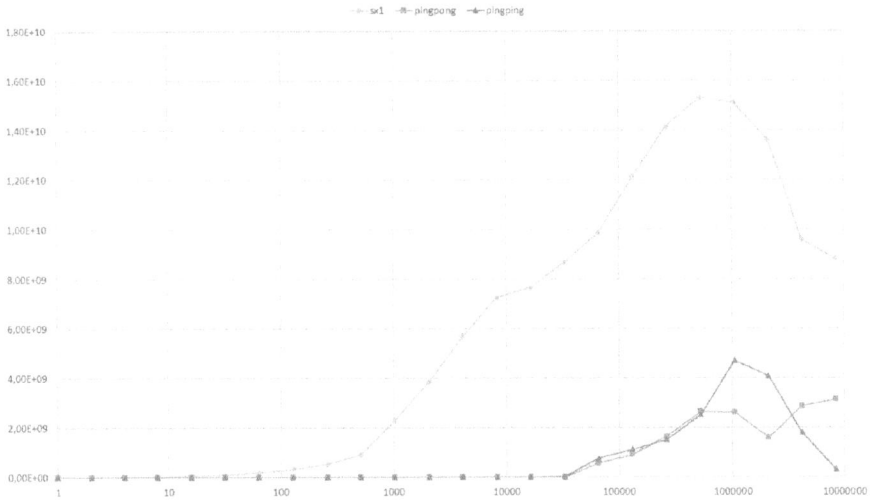

Figure 3.7: Unified point-to-point benchmark: intranode socket bandwidth (bytes per second)

Gosh, what happened here? The latency graph zooming in on the smaller messages is rather revealing (see Figure 3.8; note the logarithmic scaling of the value axis):

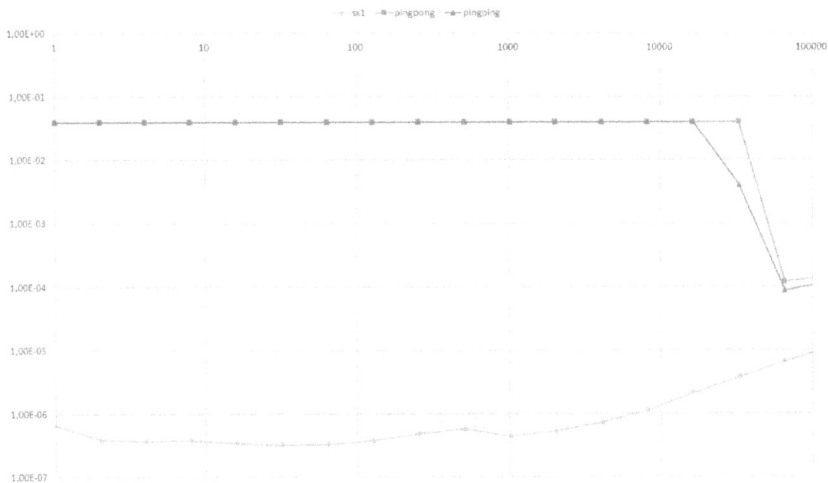

Figure 3.8: Unified point-to-point benchmark: intranode socket latency (seconds)

Apparently, we have run into some timeout that, looking at the latency graph, kicks in for messages smaller or equal to 32KiB in size. Well, this looks suspiciously like the default TCP/IP packet size for intranode messages. Indeed, setting the affected socket to the TCP_NODELAY mode using the setsockopt(2) call disables the so-called Nagle's algorithm (Nagle, 1984) and makes the system push data out at once rather than accumulate them until a packet is full. This helps with performance (see Figures 3.9 and 3.10; note that we still have to use logarithmic scaling in the latter):

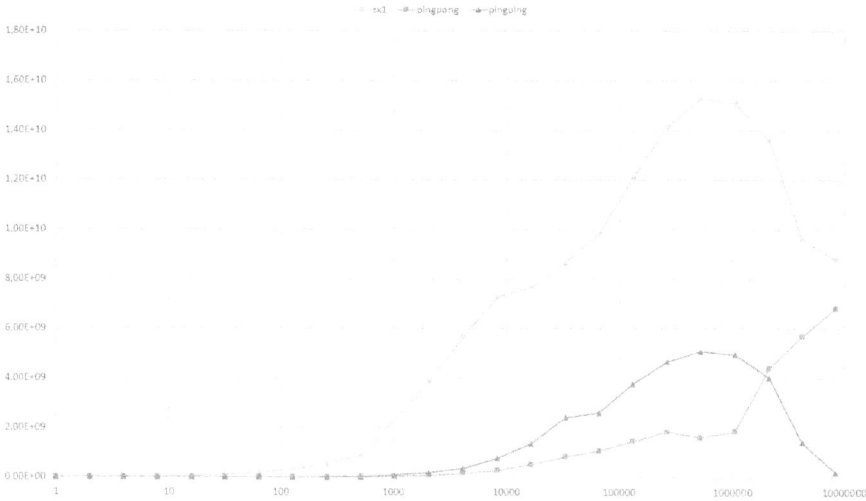

Figure 3.9: Unified point-to-point benchmark: intranode socket bandwidth w/ TCP_NODELAY on (bytes per second)

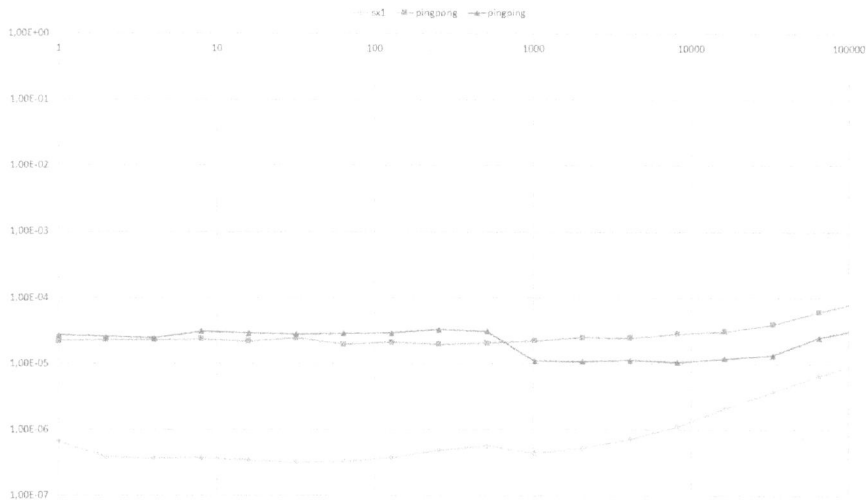

Figure 3.10: Unified point-to-point benchmark: intranode socket latency w/ TCP_NODELAY on (seconds)

Apparently, the timeout issue is gone, and even though we still have a long way to go to get closer to the underlying latency of the medium in use, we have at least got something looking like reasonable behavior for the ping-pong pattern on larger messages. The huge latency we are getting is connected to the fact that we use two or three system calls for each MPI_Send() and MPI_Recv() operation, and in doing so, we send and receive at least two or three full packets, which of course takes more time compared to the situation in which we could be sending fewer properly sized packets using fewer system calls. The ping-ping is of course looking less rosy still, but we will deal with this later on.

Here is what the required change looks like in the source code (see Listing 3.19, cf. Listing 3.8):

Listing 3.19: Subset MPI: connection management (excerpt, see `sock/mpi/v1a/mpi.c`)

```
#include <netinet/tcp.h>    // TCP_NODELAY

...

static int _mpi_nonblock(int i)
{
    int flag = 1;

    if (setsockopt(_mpi_fds[i],IPPROTO_TCP,TCP_NODELAY,&flag,sizeof(int)) == -1) {
#ifdef DEBUG
        fprintf(stderr,"%d/%d: setsockopt: %s\n",_mpi_rank,_mpi_size,strerror(errno));
#endif
        return !MPI_SUCCESS;
    }

    return MPI_SUCCESS;
}

static int _mpi_nonblock_all(void)
{
    int i;

    for (i = 0; i < _mpi_rank; i++)
        if (_mpi_nonblock(i) != MPI_SUCCESS)
            return !MPI_SUCCESS;

    for (i = _mpi_rank + 1; i < _mpi_size; i++)
        if (_mpi_nonblock(i) != MPI_SUCCESS)
            return !MPI_SUCCESS;

    return MPI_SUCCESS;
}
```

The new function _mpi_nonblock_all() can be called from the MPI_Init() in the same manner as the _mpi_connect_all() and right after it (not shown).

In addition to this, we have to take care of the occasional error "Connection refused" that can be observed when a process attempts to connect to an address that has no listening socket behind it yet. At first approximation, we can do this as follows inside the _mpi_connect() function (see Listing 3.20, cf. Listing 3.12):

Listing 3.20: Subset MPI: connection establishment (excerpt, see sock/mpi/v1a/mpi.c)

```
...

again:
    for (pn = pa; pn != NULL; pn = pn->ai_next)
        if ((fd = socket(pn->ai_family,pn->ai_socktype,pn->ai_protocol)) != -1) {
            if (connect(fd,pn->ai_addr,pn->ai_addrlen) == 0)
                break;
            close(fd);
        }

    if (pn == NULL) {
#ifdef DEBUG
        fprintf(stderr,"%d/%d: connect: %s\n",_mpi_rank,_mpi_size,strerror(errno));
#endif
        if (errno == ECONNREFUSED)
            goto again;
        return !MPI_SUCCESS;
    }

...
```

Of course, this unconditional retrial procedure is a little dangerous, since in principle we might run into a situation in which the address used is just plain wrong, and the program will hang. There are several typical ways of avoiding this issue which we will see in action later on. We do not really need any of them here.

By the way, since we had to run the benchmark more than once, yet another issue presented itself in all its ugliness: namely, the TIME_WAIT state that the TCP/IP protocol foresees. Due to this issue explained at length elsewhere (Metcalfe, Gierth, & et al., 1996), a program once in a while may refuse to start since an earlier socket that used a certain IP address/port number pair may linger in this state for a fairly long time, waiting for the eventual final extra packets that might yet come from another long dead program that used this socket once. One little modification at the beginning of the earlier mpiexec script does the trick (see Listing 3.21, cf. Listing 3.7):

Listing 3.21: Subset MPI: the mpiexec script (excerpt, see `sock/mpi/v1a/mpiexec`)

```
...

portbase=${MPI_PORTBASE:-10000}

...
```

Now, if we define the `MPI_PORTBASE` environment variable prior to the `mpiexec` invocation, the port numbers will be counted up starting from its value, otherwise, as earlier, this value will revert to its default of 10000. The `Makefile` has to be modified accordingly (not shown).

3.3.6 Exercises

Exercise 3.6 Would it not be better to put the `MPI_Send()` and `MPI_Recv()` operations into the transfer completion calls?

Exercise 3.7 How can one map the `MPI_Test()` upon the `MPI_Wait()`?

3.4 Subset Reimplementation: Nonblocking Communication

Now we have to contemplate what we have seen and generate a few directions for improvement:
- Introduce some agent that would take care of the progress even if, for example, two `MPI_Isend()` calls try to send messages both ways
- Extend the number of pending messages to more than one, ideally to an infinite number
- Map the blocking calls upon the nonblocking ones, or at least simulate this
- Resolve or mitigate the `TIME_WAIT` issue

It will take us more than this chapter, and in the case of `TIME_WAIT` even more than this book, to get through this list.

3.4.1 Redesign Directions

In an ideal situation, some agent (let's call it a *progress engine*) would take care of pushing the bytes back and forth. The MPI data transfer calls would then be reduced to creating the work requests and managing the rest of their life cycle. A work request created by a blocking call like `MPI_Send()` would be consumed inside it, giving a feeling of the blocking functionality to the MPI user. A work request created by a non-blocking call like an MPI_Isend() would be completed by the respective transfer completing calls like MPI_Wait(). The latter will have to look into the request to see what stage of processing it is at. Thus, the data pointed to by the MPI_Request object will need to include more than just the associated MPI_Status or an equivalent thereof.

Now, let's look at how we should manage those work requests. We will have send requests and receive requests posted by the MPI library. When looking for what is to be sent, we need to look into a queue containing all the send requests pending in this MPI process, and do this in order, so that we push the earliest requests out first, thus maintaining the ordering of messages between this process and its correspondents. This calls for an existence of the so-called *posted send queue*.

Likewise, we need to take care of the receive requests posted by this process. Again, the MPI standard prescribes that we work on them in the order in which they were posted, so that we process the earliest requests first. This calls for an existence of a *posted receive queue*, a precursor of which we implemented in the previous version of our library.

Is this all? No. Let's look into the brand-new ping-ping code path of the unified point-to-point benchmark (see Listing 3.18). There we have to post a send request possibly before the respective receive request has been posted by the receiving side. From the receiver point of view, the message involved is unexpected. In order to work on this, we will have to manage those unexpected receive operations, which calls for the existence of an *unexpected receive queue*. So, every time there is an MPI_Recv() or an equivalent call in the MPI program, we need to see whether a matching unexpected receive has been already posted or not, and, if so, process it instead of posting a new "expected" receive request. Only then may we proceed to the regular receive queue.

Now that we want to process as many pending nonblocking calls as we please, we do not have any excuse for ignoring the tag values. Indeed, the message ordering between any two processes is predicated by tag matching, so that message A with tag A1 sent before message B with tag B1 will be received in the reverse order if the receiving side wants to receive a message with tag B1 first. The matching is straightforward, and the wildcard MPI_ANY_TAG basically turns it off.

The same is true, sort of, for the selection of the source process: the wildcard MPI_ANY_SOURCE turns it off. Now, should we keep separate queues for each destination and source, or shared queues for all requests? This is a matter of convenience. Let's use *shared queues*: one for all send requests, one for all receive requests, and one for all unexpected messages. In principle, we could unite the latter two, but this would call for a special queue design, and it is easier to introduce two separate yet standard queues and go through them in the order mentioned above.

3.4.2 Start-up and Termination

Start-up and termination will need only a minor extension.

3.4.2.1 Design Considerations

It makes sense to keep the upfront connection establishment mechanism unchanged and switch the resulting sockets into the nonblocking mode just before the user-induced communication is to commence. This will save quite a bit of programming effort and also reduce the fuss required to deal with the nonblocking connection establishment.

3.4.2.2 MPI Header File

The only piece of the header file that needs mending is the definition of the MPI_Request layout (see Listing 3.22, cf. Listing 3.5):

Listing 3.22: Subset MPI: declarations (excerpt, see sock/mpi/v2/mpi.h)

```
...

typedef struct _mpi_request_s {
    void *buf;
    MPI_Status stat;
    int flag,tag,cnt,rank;
} *MPI_Request;

...
```

Here, the buf pointer holds the address of the buffer used in the respective operation. The stat structure holds its status, if available and necessary. The cnt, tag, and rank components describe the message envelope. The flag component indicates whether the message is either free or incomplete (0) or occupied and completed (1). Note that since we will have to access the request components, we define not only the pointer MPI_Request, but also the _mpi_request_s structure. One might argue that this is an internal detail that should better be kept in a separate, internal header file, the contents of which are not immediately visible to the user program. Well, in our case we just decide to keep it simple in the hope that no well-behaved MPI application will ever want to peek into the internals of the opaque objects in ways other than those defined by the MPI standard.

3.4.2.3 Connection Establishment

Since we have decided to change as few things as possible, the easiest way to switch sockets into the nonblocking mode is to extend the _mpi_nonblock() function introduced earlier by adding an appropriate fcntl(2) call (see Listing 3.23, cf. Listing 3.19):

Listing 3.23: Subset MPI: connection management (excerpt, see sock/mpi/v2/mpi.c)

```
#include <fcntl.h>        // fctnl(2) constants

...

    if (fcntl(_mpi_fds[i],F_SETFL,fcntl(_mpi_fds[i],F_GETFL,0) | O_NONBLOCK) == -1) {
#ifdef DEBUG
        fprintf(stderr,"%d/%d: fcntl: %s\n",_mpi_rank,_mpi_size,strerror(errno));
#endif
        return !MPI_SUCCESS;
    }
...
```

3.4.3 Point-to-Point Communication

Introduction of the progress engine will require a major rework of the code used so far.

3.4.3.1 Design Considerations

The essence of the work ahead lies in a change of paradigm. Instead of pushing the data inside the respective MPI calls, we will only work with the queues there. The actual data pushing will be delegated to the progress engine.

3.4.3.2 Blocking Communication

This time we will map the blocking communication upon the nonblocking one, as it should be done (see Listing 3.24):

Listing 3.24: Subset MPI: blocking communication (excerpt, see sock/mpi/v2/mpi.c)

```
int MPI_Send(void *buf,int cnt,MPI_Datatype dtype,int dest,int tag,MPI_Comm comm)
{
    MPI_Request req;

#ifdef DEBUG
    fprintf(stderr,"%d/%d: MPI_Send(%p,%d,%d,%d,%d,%d)\n",_mpi_rank,_mpi_size,
buf,cnt,dtype,dest,tag,comm);
#endif

    if (MPI_Isend(buf,cnt,dtype,dest,tag,comm,&req) != MPI_SUCCESS)
```

```
        return !MPI_SUCCESS;
    if (MPI_Wait(&req,NULL) != MPI_SUCCESS)
        return !MPI_SUCCESS;

    return MPI_SUCCESS;
}

int MPI_Recv(void *buf,int cnt,MPI_Datatype dtype,int src,int tag,MPI_Comm comm,
MPI_Status *pstat)
{
    MPI_Request req;

#ifdef DEBUG
    fprintf(stderr,"%d/%d: MPI_Recv(%p,%d,%d,%d,%d,%d,%p)\n",_mpi_rank,_mpi_size,
buf,cnt,dtype,src,tag,comm,pstat);
#endif

    if (MPI_Irecv(buf,cnt,dtype,src,tag,comm,&req) != MPI_SUCCESS)
        return !MPI_SUCCESS;
    if (MPI_Wait(&req,pstat) != MPI_SUCCESS)
        return !MPI_SUCCESS;

    return MPI_SUCCESS;
}
```

Each blocking call is trivially decomposable into the respective nonblocking call and the blocking completion operation MPI_Wait(). Note that in case of failure we let the failing call handle the debugging output. We will use this method throughout. Purists may object that this way the reader of the debugging output might miss the nesting of the respective function entry and exit points. Frankly, my experience tells me that seeing a number of similar failure messages coming one after another does not make for easy or pleasant reading either.

3.4.3.3 Nonblocking Communication
Gee, this bit is going to be rather interesting. This is probably the most complicated piece of code so far in this book, but's it's nature is rather simple.

3.4.3.3.1 Message Initiation
Let's look into the sending part first, leaving the receiving part and all the necessary helper functions for dessert (see Listing 3.25):

Listing 3.25: Subset MPI: nonblocking data sending (excerpt, see sock/mpi/v2/mpi.c)

```
int MPI_Isend(void *buf,int cnt,MPI_Datatype dtype,int dest,int tag,
          MPI_Comm comm,MPI_Request *preq)
```

```
{
#ifdef DEBUG
    fprintf(stderr,"%d/%d: MPI_Isend(%p,%d,%d,%d,%d,%d,%p)\n",_mpi_rank,_mpi_size,
buf,cnt,dtype,dest,tag,comm,preq);
#endif

    assert(_mpi_init);
    assert(buf != NULL && cnt >= 0);
    assert(dtype == MPI_BYTE);
    assert(dest >= 0 && dest < _mpi_size && dest != _mpi_rank);
    assert(tag >= 0 && tag <= 32767);
    assert(comm == MPI_COMM_WORLD);
    assert(preq != NULL);

    if (_mpi_progress() != MPI_SUCCESS)
        return !MPI_SUCCESS;

    if ((*preq = _mpi_newreq(&_mpi_sq)) != NULL)
    {
        MPI_Request req = *preq;

        req->buf  = buf;
        req->cnt  = cnt;
        req->rank = dest;
        req->tag  = tag;
        req->flag = 0;

#ifdef DEBUG
        fprintf(stderr,"%d/%d: -> success, *preq = %p\n",_mpi_rank,_mpi_size,*preq);
#endif
        return MPI_SUCCESS;
    }

#ifdef DEBUG
    fprintf(stderr,"%d/%d: -> failure\n",_mpi_rank,_mpi_size);
#endif
    return !MPI_SUCCESS;
}
```

There are at least two points worth noting here:

- The sending operation has been reduced to the creation of the respective request in the message send queue called _mpi_sq
- Actual message delivery is taken care of via the call to the MPI process engine that we will look into later on

With this in mind, the receiving side becomes a little less cryptic, too (see Listing 3.26):

Listing 3.26: Subset MPI: nonblocking data receiving (excerpt, see sock/mpi/v2/mpi.c)

```
int MPI_Irecv(void *buf,int cnt,MPI_Datatype dtype,int src,int tag,MPI_Comm comm,
MPI_Request *preq)
{
#ifdef DEBUG
    fprintf(stderr,"%d/%d: MPI_Irecv(%p,%d,%d,%d,%d,%d,%p)\n",_mpi_rank,_mpi_size,
buf,cnt,dtype,src,tag,comm,preq);
#endif

    assert(_mpi_init);
    assert(buf != NULL && cnt >= 0);
    assert(dtype == MPI_BYTE);
    assert(src >= 0 && src < _mpi_size && src != _mpi_rank);
    assert(tag >= 0 && tag <= 32767);
    assert(comm == MPI_COMM_WORLD);
    assert(preq != NULL);

    if (_mpi_progress() != MPI_SUCCESS)
        return !MPI_SUCCESS;

    if ((*preq = _mpi_findreq(&_mpi_xq,src,tag)) != NULL)
    {
        MPI_Request req = *preq;

        if (req->cnt > cnt) {
#ifdef DEBUG
            fprintf(stderr,"%d/%d: -> failure, cnt %d > %d\n",_mpi_rank,_mpi_size,
req->cnt,cnt);
#endif
            return !MPI_SUCCESS;
        }

        memcpy(buf,req->buf,(cnt = req->cnt));
        free(req->buf);
        req->buf   = buf;

        req->stat.MPI_SOURCE = req->rank;
        req->stat.MPI_TAG = req->tag;
        req->flag = 1;

#ifdef DEBUG
        fprintf(stderr,"%d/%d: -> unexpected success, *preq = %p\n",_mpi_rank,
_mpi_size,*preq);
#endif
        return MPI_SUCCESS;
    }
    else if ((*preq = _mpi_newreq(&_mpi_rq)) != NULL)
    {
```

```
        MPI_Request req = *preq;

        req->buf  = buf;
        req->cnt  = cnt;
        req->rank = src;
        req->tag  = tag;
        req->flag = 0;

#ifdef DEBUG
        fprintf(stderr,"%d/%d: -> expected success, *preq = %p\n",_mpi_rank,_mpi_size,
*preq);
#endif
        return MPI_SUCCESS;
    }

#ifdef DEBUG
    fprintf(stderr,"%d/%d: -> failure\n",_mpi_rank,_mpi_size);
#endif
    return !MPI_SUCCESS;
}
```

The only difference here is that inside the MPI_Irecv() we may need to deal with an unexpected message that matches the current receive request. Taking for granted that the unexpected message has already been received, all that we need to do is:

- Find out that there is indeed such a match. This we do via the _mpi_findreq() function over the unexpected message receive queue called _mpi_xq (see Listing 3.26)
- Move the data over from the internal MPI storage to the user buffer
- Do the necessary clean-up to avoid leaking the memory
- Mark the current request as completed

Note that instead of moving the requests between the queues, we just use the unexpected message queue throughout in this case. Contrary to this, expected messages (i.e., messages that already have a matching pending receive request before they arrive to the receiving process) are kept in the posted receive queue called _mpi_rq.

Here are the necessary internal declarations and new helper calls (see Listing 3.27, cf. Listing 3.16):

Listing 3.27: Subset MPI: request queue management (excerpt, see sock/mpi/v2/mpi.c)

```
static struct _mpi_q_s {
    struct _mpi_request_s req[1];
    int n;
} _mpi_rq,_mpi_xq,_mpi_sq;
```

...

```
static MPI_Request _mpi_findreq(struct _mpi_q_s *pq,int rank,int tag)
{
    if (pq->n > 0) {
        MPI_Request req = pq->req;

        if (req->rank == rank && req->tag == tag)
            return req;
    }

    return NULL;
}
```

3.4.3.3.2 Message Completion

The blocking MPI_Wait() call is naturally mapped upon the nonblocking MPI_Test() call (see Listing 3.28):

Listing 3.28: Subset MPI: message completion (excerpt, see sock/mpi/v2/mpi.c)

```
int MPI_Test(MPI_Request *preq,int *pflag,MPI_Status *pstat)
{
#ifdef DEBUG
    fprintf(stderr,"%d/%d: MPI_Test(%p,%p,%p)\n",_mpi_rank,_mpi_size,preq,pflag,
pstat);
#endif

    assert(_mpi_init);
    assert(preq != NULL);
    assert(pflag != NULL);

    if (_mpi_progress() != MPI_SUCCESS)
        return !MPI_SUCCESS;

    *pflag = 0;
    if (*preq != NULL)
    {
        MPI_Request req = *preq;

        if (req->flag) {
            *pflag = 1;
            if (pstat != NULL)
                _mpi_statcpy(pstat,&(req->stat));
            req->flag = 0;
            _mpi_freereq(_mpi_findq(req),preq);
        }
    }

#ifdef DEBUG
    if (pstat != NULL)
```

```
        fprintf(stderr,"%d/%d: -> success, *pflag = %d, *pstat = { %d, %d }\n",
_mpi_rank,_mpi_size,*pflag,pstat->MPI_SOURCE,pstat->MPI_TAG);
    else
        fprintf(stderr,"%d/%d: -> success, *pflag = %d\n",_mpi_rank,_mpi_size,*pflag);
#endif
    return MPI_SUCCESS;
}

int MPI_Wait(MPI_Request *preq,MPI_Status *pstat)
{
    int flag = 0;

#ifdef DEBUG
    fprintf(stderr,"%d/%d: MPI_Wait(%p,%p)\n",_mpi_rank,_mpi_size,preq,pstat);
#endif

    while (!flag) {
        if (MPI_Test(preq,&flag,pstat) != MPI_SUCCESS)
            return !MPI_SUCCESS;
    }

    return MPI_SUCCESS;
}
```

Since we do not know whether the request argument points to a posted send, posted receive, or unexpected receive queue, we need to find this out using the _mpi_findq() function (see Listing 3.29):

Listing 3.29: Subset MPI: request queue management (excerpt, see sock/mpi/v2/mpi.c)

```
static struct _mpi_q_s *_mpi_findq(MPI_Request req)
{
    if (req == _mpi_rq.req)
        return &_mpi_rq;
    else if (req == _mpi_xq.req)
        return &_mpi_xq;
    else if (req == _mpi_sq.req)
        return &_mpi_sq;

    return NULL;
}
```

The linear structure of our simple queues makes this a trivial task. Of course, we could have kept the request identification inside the request itself, say, in the flag component. However, this would be a clear overkill in this case.

3.4.3.3.3 Progress Engine

Now we come to the heart of every real MPI implementation—the progress engine. Our implementation will be relatively simple, in-order, blocking, and synchronous. Once we get a message to process, we process it until it is done. The engine is called from inside the main program thread, so that we need to do this relatively often—best of all out of every data transfer call like MPI_Send(), MPI_Irecv(), MPI_Wait(), and so on.

The top-level progress function _mpi_progress() is straightforward (see Listing 3.30):

Listing 3.30: Subset MPI: progress engine (excerpt, see sock/mpi/v2/mpi.c)

```
static int _mpi_progress(void)
{
    MPI_Request req;

    for (req = _mpi_firstreq(&_mpi_rq,&req); req != NULL; req =
_mpi_nextreq(&_mpi_rq,&req))
        if (_mpi_progress_recv(req) != MPI_SUCCESS)
            return !MPI_SUCCESS;

    for (req = _mpi_firstreq(&_mpi_sq,&req); req != NULL; req =
_mpi_nextreq(&_mpi_sq,&req))
        if (_mpi_progress_send(req) != MPI_SUCCESS)
            return !MPI_SUCCESS;

    return MPI_SUCCESS;
}
```

It is always an interesting question what to do first: process the send requests or the receive requests. Given that an unprocessed receive request may lead to the creation of an unexpected request, meaning an extra memory copy operation and additional management fuss, we decided here to do the receive requests first, if any.

In order to abstract out the internal structure of the message queues involved, we use a typical iterator pattern represented by the function _mpi_firstreq() and _mpi_nextreq(). Their implementation in our case is extremely simple (see Listing 3.31):

Listing 3.31: Subset MPI: request queue iterator (excerpt, see sock/mpi/v2/mpi.c)

```
static MPI_Request _mpi_firstreq(struct _mpi_q_s *pq,MPI_Request *preq)
{
    if (pq->n > 0)
        return (*preq = pq->req);

    return NULL;
}
```

```
static MPI_Request _mpi_nextreq(struct _mpi_q_s *pq,MPI_Request *preq)
{
    return (*preq = NULL);
}
```

Out of the actual progress calls, the sending part is again less involved (see Listing 3.32):

Listing 3.32: Subset MPI: progress engine, sending part (excerpt, see `sock/mpi/v2/mpi.c`)

```
static int _mpi_progress_send(MPI_Request req)
{
    char *buf;
    struct iovec iov[2];
    int m,n;

    if (req->flag)
        return MPI_SUCCESS;

    iov[0].iov_base = &(req->tag);
    iov[0].iov_len  = 2*sizeof(int);
    iov[1].iov_base = req->buf;
    iov[1].iov_len  = req->cnt;

    switch ((m = writev(_mpi_fds[req->rank],iov,2))) {
    case -1:
#ifdef DEBUG
        fprintf(stderr,"%d/%d: write tag: %s\n",_mpi_rank,_mpi_size,strerror(errno));
#endif
        if (errno == EWOULDBLOCK || errno == EAGAIN)
            return MPI_SUCCESS;
        return !MPI_SUCCESS;
    case 0:
#ifdef DEBUG
        fprintf(stderr,"%d/%d: write tag: unexpected end of file\n",_mpi_rank,
_mpi_size);
#endif
        return !MPI_SUCCESS;
    default:
        if ((m -= 2*sizeof(int)) >= 0)
            break;
#ifdef DEBUG
        fprintf(stderr,"%d/%d: write buf: partial transter, done = %d\n",_mpi_rank,
_mpi_size,m);
#endif
        return !MPI_SUCCESS;
    }

    for (buf = req->buf + m,n = req->cnt - m; n > 0; buf += m,n -= m) {
```

```
#ifdef DEBUG
            fprintf(stderr,"%d/%d: write buf: partial transter, rest = %d\n",
_mpi_rank,_mpi_size,n);
#endif
        if ((m = write(_mpi_fds[req->rank],buf,n)) == -1) {
#ifdef DEBUG
            fprintf(stderr,"%d/%d: write buf: %s\n",_mpi_rank,_mpi_size,
strerror(errno));
#endif
            if (errno == EWOULDBLOCK || errno == EAGAIN) {
                m = 0;
                continue;
            }
            return !MPI_SUCCESS;
        }
    }

    req->stat.MPI_SOURCE = req->rank;
    req->stat.MPI_TAG = req->tag;
    req->flag = 1;

    return MPI_SUCCESS;
}
```

We use the writev(2) system call to save on the relatively expensive system calls first and to make sure messages get transferred. Using several write(2) system calls in a row did not work for me anyway. However, if the first writev(2) system call cannot send the whole of the buffer, calling additional write(2) system calls for pushing out the rest of the message was all right. Go figure.

The rest of the code is rather self-explanatory. Note how debugging output not only explains what is happening, but also helps to trace and understand this at run time. In fact, quite a bit of debugging went into this piece of code—that is, a couple of hours. Without debugging output, it would be days and nights.

Finally, we get to the receiving part, complicated as usual by the need to deal with the unexpected receives (see Listing 3.33):

Listing 3.33: Subset MPI: progress engine, receiving part (excerpt, see sock/mpi/v2/mpi.c)

```
static int _mpi_progress_recv(MPI_Request req)
{
    char *buf;
    int hdr[2],m,n,unexp = 0;

    if (req->flag)
        return MPI_SUCCESS;

    switch ((m = read(_mpi_fds[req->rank],hdr,2*sizeof(int)))) {
    case -1:
```

```
#ifdef DEBUG
        fprintf(stderr,"%d/%d: read itag: %s\n",_mpi_rank,_mpi_size,strerror(errno));
#endif
        if (errno == EWOULDBLOCK || errno == EAGAIN)
            return MPI_SUCCESS;
        return !MPI_SUCCESS;
    case 0:
#ifdef DEBUG
        fprintf(stderr,"%d/%d: read itag: unexpected end of file\n",_mpi_rank,
_mpi_size);
#endif
        return !MPI_SUCCESS;
    default:
        if (m == 2*sizeof(int))
            break;
#ifdef DEBUG
        fprintf(stderr,"%d/%d: read itag: partial transfer, done = %d\n",_mpi_rank,
_mpi_size,m);
#endif
        return !MPI_SUCCESS;
    }

    if (req->tag != hdr[0])
    {
        MPI_Request reqx;

        if ((reqx = _mpi_newreq(&_mpi_xq)) != NULL) {
            reqx->rank = req->rank;
            reqx->tag = hdr[0];
            reqx->cnt = hdr[1];
            req = reqx;
            unexp = 1;
#ifdef DEBUG
            fprintf(stderr,"%d/%d: read itag: unexpected message, tag = %d, cnt =
%d\n",_mpi_rank,_mpi_size,reqx->tag,reqx->cnt);
#endif
        }
        else {
#ifdef DEBUG
            fprintf(stderr,"%d/%d: -> failure, tag %d != %d\n",_mpi_rank,_mpi_size,
req->tag,hdr[0]);
#endif
            return !MPI_SUCCESS;
        }
    }

    if (unexp) {
        if ((req->buf = malloc(req->cnt)) == NULL) {
#ifdef DEBUG
```

```
        fprintf(stderr,"%d/%d: malloc buf: %s\n",_mpi_rank,_mpi_size,
strerror(errno));
#endif
        return !MPI_SUCCESS;
      }
    }
    else if (req->cnt < hdr[1]) {
#ifdef DEBUG
        fprintf(stderr,"%d/%d: -> failure, cnt %d < %d\n",_mpi_rank,_mpi_size,
req->cnt,hdr[1]);
#endif
        return !MPI_SUCCESS;
    }

    for (buf = req->buf,n = hdr[1]; n > 0; buf += m,n -= m)
        if ((m = read(_mpi_fds[req->rank],buf,n)) == -1) {
#ifdef DEBUG
            fprintf(stderr,"%d/%d: read buf: %s, rest = %d\n",_mpi_rank,_mpi_size,
strerror(errno),n);
#endif
            if (errno == EWOULDBLOCK || errno == EAGAIN) {
                m = 0;
                continue;
            }
            return !MPI_SUCCESS;
        }

    req->stat.MPI_SOURCE = req->rank;
    req->stat.MPI_TAG = req->tag;
    req->flag = 1;

    return MPI_SUCCESS;
}
```

Here we have to do at least two read(2) system calls, first to get the message header (the tag and count values), and then to get the message body that may be nonexistent if the count is equal to zero. For simplicity, we keep the header in an integer array of two elements. A properly defined structure would do just as well. Note that on the sending side (see Listing 3.32) we do exploit the layout of the _mpi_request_s structure to skip the formation of a separate header. Here, however, we need to keep this data in a separate place, for the tag needs to be matched, and then the available buffer length checked against the actual one in order to detect a possible buffer overrun and react to it.

3.4.4 Testing

When the same set of tests is used, the results match the expectations. Of course, we have to extend the testing by at least one more exchange pattern that will be described in the following section (see Section 3.4.5).

3.4.5 Benchmarking

Since the unified point-to-point benchmark is used both for testing and for benchmarking, we will describe its extension here.

3.4.5.1 Unified Point-to-Point Benchmark Revisited

The unified point-to-point benchmarking program gets one more pattern that we call *pong-pong*, because it precedes the data sending by posting the matching receive operations, which resembles a little the second (pong) part of the ping-pong pattern (see Listing 3.34, cf. Listing 3.18):

Listing 3.34: Unified point-to-point benchmark: adding pong-pong pattern (excerpt, see sock/ mpi/v2/b.c)

```
…

#ifdef PONGPONG
                 MPI_Irecv(b2,1,MPI_BYTE,0,1,MPI_COMM_WORLD,&rq);
                 MPI_Send(b1,1,MPI_BYTE,0,0,MPI_COMM_WORLD);
                 MPI_Wait(&rq,&st);
#endif

…
```

3.4.5.2 Intranode Performance

Of course, this new pattern would never stand a chance in our previous implementation. Here, it will prevent most, if not all, unexpected receive operations, so better performance should be observed here compared to the ping-ping pattern. Indeed, when we compare intranode performance of all three two-way patterns (ping-pong, ping-ping, and pong-pong) to that of the naïve Internet socket benchmark, we see exactly what we expect (see Figures 3.11 and 3.12):

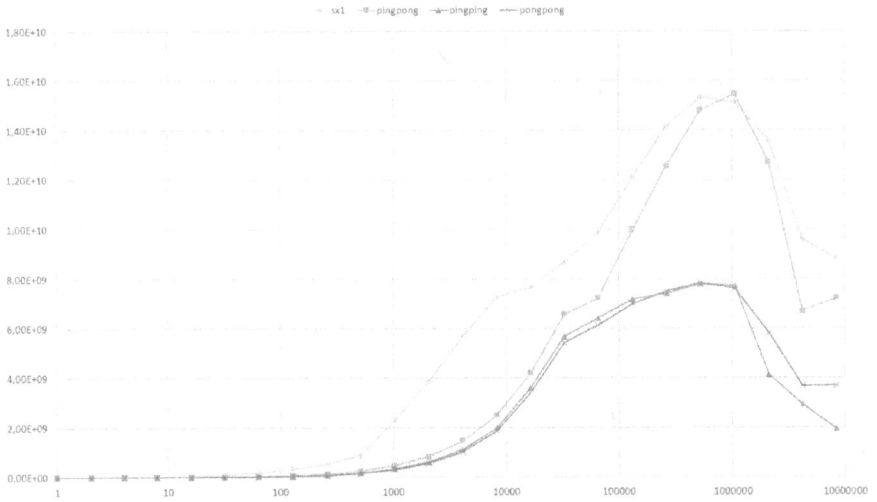

Figure 3.11: Unified point-to-point benchmark: intranode socket bandwidth (bytes per second)

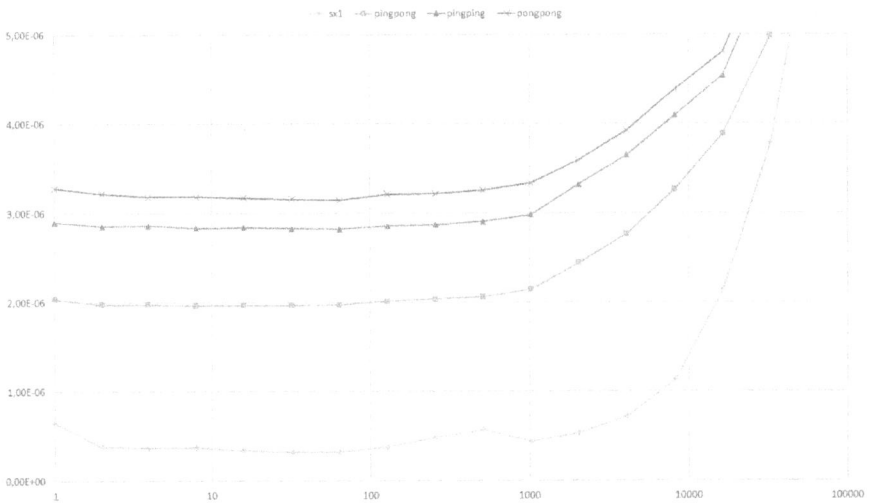

Figure 3.12: Figure 3.12. Unified point-to-point benchmark: intranode socket latency (seconds)

– Our ping-pong latency at small messages is about four to five times as big as that of the naïve socket benchmark sx1. A part of this is understandable, since we do two system calls instead of just one on the receiving side. Still, it must be possible to optimize this.

- Ping-pong bandwidth, on the contrary, is close. Remember: you need to do something terribly wrong to hurt the bandwidth, so there is apparently something to be found here, too.
- The pong-pong pattern indeed performs better than the ping-ping on long messages, sparing us the need for extra memory allocation and data copying caused when we deal with unexpected messages.
- At the same time, the pong-pong pattern runs a little slower than the ping-ping on small messages. The reason for this is yet unclear.

Note that we have to exclude the one-way patterns (ping and pong), because they break our limited message request queue implementation as soon as the receiving process fails to keep up with the sending process.

Now, let's consider competitive performance using MPICH 3.2 legacy ch3:sock device as a reference point (see Figures 3.13 and 3.14):

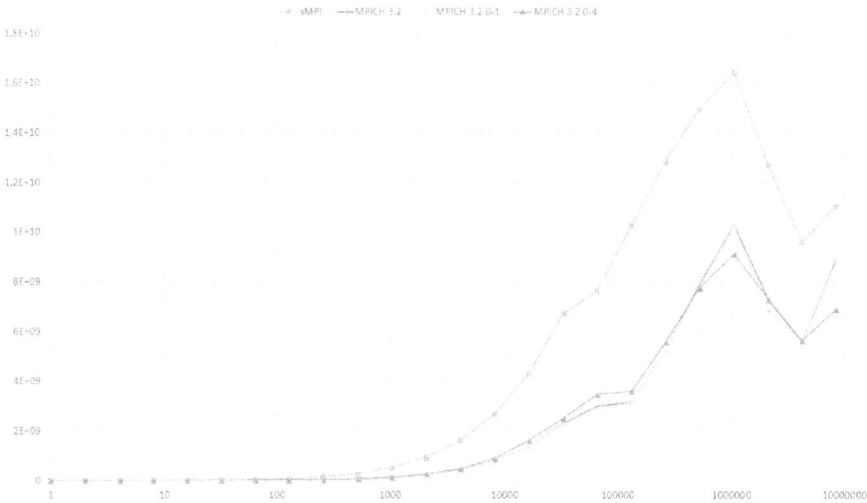

Figure 3.13: Unified point-to-point benchmark: intranode socket ping-pong bandwidth, subset MPI vs. MPICH 3.2 (bytes per second)

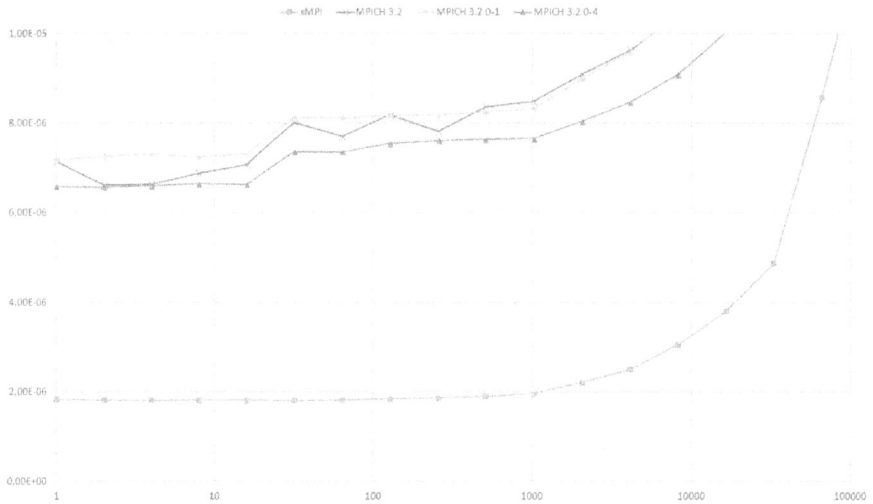

Figure 3.14: Unified point-to-point benchmark: intranode socket ping-pong latency, subset MPI vs. MPICH 3.2 (seconds)

This time, we map out subset MPI performance versus MPICH 3.2 with and without explicit pinning in order to eliminate this as an influencing factor. Whatever the settings, our subset MPI beats MPICH 3.2 by 3.5 times on small message latency, and by about 50% on large message bandwidth!

3.4.5.3 Internode Performance

Venturing internode, we first compare subset MPI performance to the naïve socket benchmark sx1. Due to dramatic difference of the value ranges, we must again use logarithmic scaling of the value axis in order to understand what is happening at lower message sizes (see Figures 3.15 and 3.16):

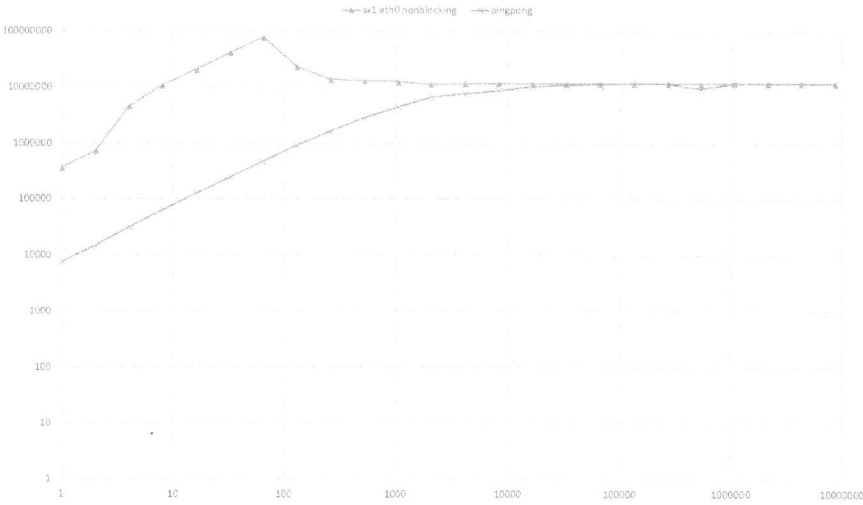

Figure 3.15: Unified point-to-point benchmark: internode socket bandwidth (bytes per second, logarithmic scale of the value axis)

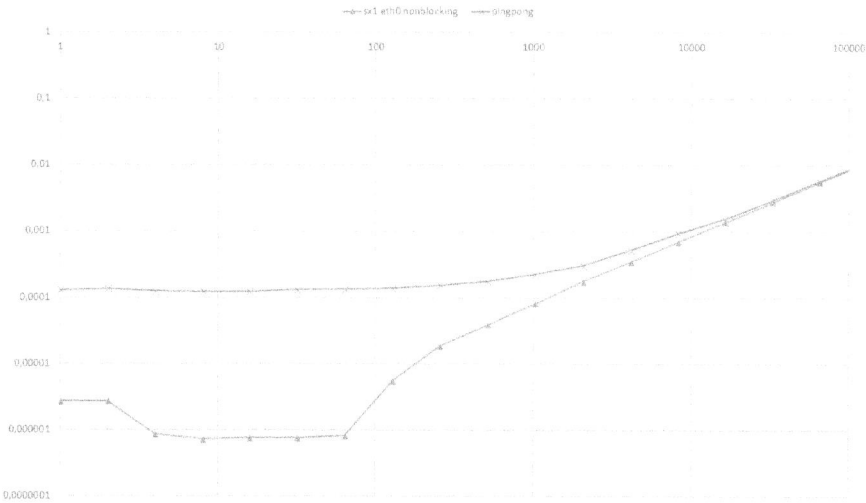

Figure 3.16: Unified point-to-point benchmark: internode socket latency (seconds, logarithmic scale of the value axis)

Well, our MPI sockets really stink on small messages compared to the available already-not-so-great raw Ethernet socket performance (cf. Figures 3.6b and 3.5b, respectively). Is this us or something else? Well, there's nothing better than to compare our own results to the competition. Alas, MPICH 3.2 would not work at all, and Open MPI 2.0.2 would segfault at about 32 or 64 KiB message size. What happened? Well, maybe because I almost accidentally used a very challenging setup for the testing: different OS versions, different word sizes, and different CPU generations on both sides of the connection, not to mention the 100-Mbit Ethernet not normally found in the current HPC installations. But, my esteemed colleagues and friends, sockets must work even if the other side is an Android smartphone handled by a Martian.[1]

Still, a bit of the Open MPI performance allows a direct comparison (see Figure 3.17):

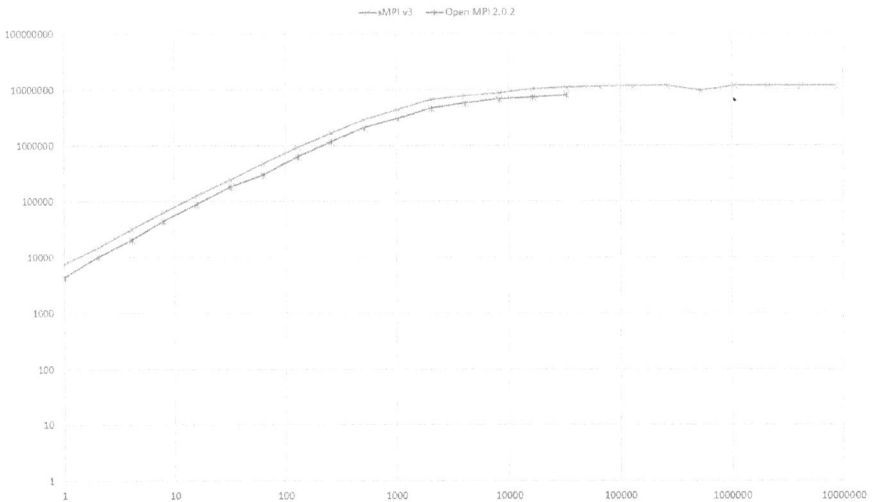

Figure 3.17: Unified point-to-point benchmark: intranode socket ping-pong bandwidth, subset MPI vs. Open MPI 2.0.2 (bytes per second, logarithmic scale of the value axis)

Minding logarithmic scaling of the value axis, we may say our subset MPI behaves noticeably better than Open MPI 2.0.2, without breaking midway through the message length range. However, Open MPI is has another handicap—it was measured using only one iteration instead of 10 used for subset MPI—so that Open MPI results could look better still, since the first iteration is normally quite expensive, serving as a sort of warm-up for the respective message length series.

It is very likely that MPICH, had it worked at all, would do about as well as Open MPI. So, it's not us who is to blame, it's life.

---

1 The respective issue was reported as fixed in MPICH 3.2.1. Open MPI team recommended upgrading to version 3.x, too.

3.4.6 Optimization

Can we do better still? Sure. If you remember, we do two system calls on the receiving side. If we were to read the message header into a temporary buffer with enough space for a short message body, we could receive short messages at the cost of just one system call and a following copying of the user data into the respective receive buffer. Since we deal with very high latencies on the networking side anyway, an extra memory copy would simply disappear in the noise. For longer messages, a second system call necessary for getting the rest of the message body will not make much difference either.

Another matter worth considering is the overwhelming frequency with which we do those system calls, mostly to learn that there is no data available to be processed. My private observations suggest that the number of these superfluous calls often stretches into hundreds before anything arrives, and I saw that in the debugging version of the library that was substantially slowed down by producing the debugging output. It is not unlikely that in the optimized library, the number of those calls goes into high hundreds or even low thousands for every MTU (Maximum Transmission Unit) worth of data served up by the network.[2] Of course, we can do better here, too, and at least spend less energy for this senseless polling.

One way is to use the select(2) or poll(2) system call. Either would lead us directly to the socket that needs attention now, and polling of the individual channels would naturally be reduced. Another idea is to back off once a few attempts to get data from the channel suggest that there is nothing there and that there will probably be nothing there for quite a while. How soon to back off and for how long, is a matter of tuning. Slower networks will tolerate bigger processing delays, faster networks will be punished by them, as will the overall performance. Yet one idea building logically on the back-off one is to fall into a so-called *wait mode* (also called *event-driven mode*) once there is nothing to be read in all channels. This means basically switching from the polling we have been practicing rather heavily so far to using the interrupt mechanism of the operating system in the way typical of the blocking I/O. This is however an advanced matter that we will revisit in Chapter 5.

3.4.7 Exercises

Exercise 3.8 Implement the buffered short message receive scheme outlined in Section 3.4.6. Do you notice any difference in latency now?

---

2 For Ethernet, normally just 1500 bytes, of which 1460 bytes (TCP MSS, or Maximum Segment Size) are available for the payload.

Exercise 3.9 Compare socket performance of the latest versions of MPICH and Open MPI to versions 3.2 and 2.0.2, respectively, used in this chapter. Is there any difference? Why?

3.5 Conclusions

Now we have seen both performance extremes that bound MPI: shared memory at its highest and 100 Mbit Ethernet at its lowest. In all cases we were able to beat or match competition, apart from those cases when it apparently used advanced protocols we will learn more about in Chapter 5.

Let us once again quantify complexity of the respective packages. We get the following results this time round (see Table 3.1, cf. Table 2.5):

Table 3.1: MPI subsets vs. full MPI implementations

Subset	Entities	Incl. Functions	LOC
Shared memory	13	7	343
Sockets	18	11	852
MPI-1	244	128	~100000
MPI-3.1	855	450	~350000

We've again gotten better performance at under one percent of the cost. Let's see if our luck lasts when we start extending subset MPI by more features and additional semantics.

Chapter 4
Extensions

In this chapter you will learn how to add essential MPI features including basic communicator management, elementary datatypes, selected collective operations, proper
error handling, and Fortran language binding.

Some of these features will be almost trivial in implementation and usage, some
will require substantial effort both in understanding and in design. Combined with
the material of Chapter 5, this MPI subset will be as close to the Real Thing as one can
get without spending years on development. Moreover, in this chapter, in accordance
with the 10/90 principle, we will focus on the most essential features of the MPI standard and tune out the remaining and less popular features (like one-sided operations,
file I/O, generalized requests, etc.).

Note that in this chapter we will deal with the features immediately visible to
the user in the form of extra calls and/or features, as well as with the necessary MPI
internals. Things hidden principally inside the MPI implementation and directed at
making it, and hence the MPI applications, run faster will be considered in the next
chapter.

4.1 Subset Definition

Let's take the sockets-based MPI subset created earlier and extend it by basic communicator management, several elementary datatypes, selected collective operations,
standard-conforming error handling, and Fortran language binding. This will make
our subset MPI capable of handling most MPI application needs. To this end, we will
(additions compared to Chapter 3 are underlined, removals struck out):
-   Provide the basic MPI functionality:
    a.   The initialization and termination calls MPI_Init() and MPI_Finalize()
    b.   The query calls MPI_Comm_size() and MPI_Comm_rank()
    c.   The wall clock query function MPI_Wtime()
-   Implement extended point-to-point communication:
    o   Standard mode blocking point-to-point calls MPI_Send() and MPI_Recv()
    o   Standard mode nonblocking point-to-point calls MPI_Isend() and
        MPI_Irecv()
    o   Completion calls MPI_Test() and MPI_Wait() as well as the MPI_Request
        object
-   Add several extensions
    o   Predefined communicators MPI_COMM_SELF and MPI_COMM_WORLD
    o   Functions MPI_Comm_dup() and MPI_Comm_split()
    o   Constants MPI_UNDEFINED and MPI_COMM_NULL for the latter call

DOI 10.1515/9781501506871-004

- o   Elementary datatypes `MPI_BYTE`, `MPI_INT`, and `MPI_DOUBLE`
- o   Functions `MPI_Type_size()` and `MPI_Get_count()`
- o   Predefined reduction operations `MPI_MAX`, `MPI_MIN`, and `MPI_SUM`
- o   Collective functions `MPI_Barrier()`, `MPI_Bcast()`, `MPI_Reduce()`, `MPI_Allreduce()`, `MPI_Gather()`, and `MPI_Allgather()`
- o   Predefined error handlers `MPI_ERRORS_ARE_FATAL` and `MPI_ERRORS_RETURN`
- o   Functions `MPI_Abort()`, `MPI_Comm_get_errhandler()`, `MPI_Comm_set_errhandler()`, `MPI_Comm_call_errhandler()`, `MPI_Error_class()`, and `MPI_Error_string()`
- o   Predefined error classes requisite to the underlying MPI subset
- o   Fortran language binding based on the `INCLUDE 'mpif.h'` method
- — Restrict the rest of the MPI functionality as follows:
  - o   Implement only the `MPI_THREAD_SINGLE` mode
  - o   Support any process count
  - o   Allow message tags up to 32767
  - o   ~~Provide only the MPI_BYTE datatype~~
  - o   Permit any nonnegative message count
  - o   Allow message tags up to 32767
  - o   ~~Operate only over the MPI_COMM_WORLD communicator~~
  - o   ~~Assume the predefined MPI error handler MPI_ERRORS_RETURN~~

This extends the earlier subset to a total of 60 entities (including types `MPI_Comm`, `MPI_Datatype`, `MPI_Status`, `MPI_Op`, and `MPI_Errhandler`, as well as `MPI_SUCCESS` plus 14 MPI error classes); of them, 27 are functions. The MPI internals will grow accordingly. Still, compared to 855 entities (of them, 450 are functions) in the full MPI-3.1 standard, it will be a relatively manageable undertaking.

4.1.1 Exercises

Exercise 4.1 Define a couple of MPI subsets a bit wider than the one described above.

Exercise 4.2 Define a subset or two that are still narrower than the one described above. Think about message length, tag values, and other entities.

4.2 Subset Basics

This time, however, we have a more complicated task at hand than the single-minded drive to make message transfer work over a certain medium. Hence, we need to change the structure of the narration. Instead of looking into the basic communication mechanisms that will remain unchanged this time, it makes sense to look over the intended extensions and decide what needs to be done, when, and why. To this end, let's look into the features we want to add one by one, clarify their purpose and import, detect their interdependencies, and make a plan.

4.2.1 Communicator Management

It is boring and sometimes even detrimental to the purpose of application programming to be limited by the full set of MPI processes described by the MPI_COMM_WORLD communicator. In certain numerical methods, for example, it may be very useful to split the MPI_COMM_WORLD into two disjunct subcommunicators containing only the odd and the even process ranks. In some computations, the set of processes involved in the computation is constantly changing, for example, in the multigrid calculations. And if you want to provide a library that uses MPI underneath, you may want to isolate whatever is happening there from the rest of the application by using a distinct communicator or two for the inner library workings. If the problem involved includes several separate subject domains, e.g., structural mechanics and fluid dynamics, it may be useful to split the application into several disjunct groups that can still communicate with each other. And if the application decides to spawn extra processes, there is again the need to represent this and ensure communication flow between the parent(s) and the children.

Whatever the application need is, we as MPI implementors must cater to it. The MPI standard provides a universal way of doing so. Processes can be combined into *groups*, groups can be bound to *communicators* that unite a process group and a *communicator context*—essentially, an extra tag that differentiates all messages sent within one particular communicator from all those in all other communicators the process figures in. Finally, the application view of processes can be changed either by renumbering and/or subsetting of the initial process set, or by defining so-called virtual process topologies that include cartesian and graph structures.

All these tricks can be reduced to one thing: a mapping of a set of processes as seen by the application to the actual set of processes as seen by the MPI implementation. An *intracommunicator* is then a combination of a group reference and a communication context. An *intercommunicator* is a combination of two group references (for the local and the remote groups) as well as of the context(s) used for local communication and for talking to the remote group. We will deal only with intracommunicators in this chapter.

Deep under the hood, whatever the application thinks of the processes is not really important. What is important is that the mapping is performed correctly and quickly, that the respective data structures do not occupy too much memory, and that the messages exchanged within different communicators do not interfere with each other.

4.2.2 Datatypes

The purpose of the MPI datatypes is twofold:
- Describe the data layout in memory, potentially facilitating hardware-supported gather/scatter operations
- Guide the data conversion, should it be necessary

Nowadays, *heterogeneous systems*, that is, systems that use components of different architecture with possibly different data representation formats, are rather unpopular once again, so that heterogeneity has lost its urgency compared to the times when the PVM ruled the world. Still, the MPI datatype subsystem is rather useful if the data to be transferred is not contained in one contiguous piece of memory. And the growing role of the accelerators may yet again push heterogeneity into the front row. All this goes in waves, as well as every time "discoveries" are made—well, they have been made many times over in the past. So, let's not forget the past and thus be better prepared for the future, that is, focus on *homogeneous systems* but keep the door open for heterogeneous ones.

Datatypes contain information about the nature of the data items, their initial offset, and the data item count. When applied recursively, with the addition of controllable *lower* and *upper bounds* that may pad the actual data on either side, the resulting *derived datatype* can describe any possible data layout. It all starts with the so-called *elementary datatypes* like MPI_INT or MPI_REAL. Additive behavior is achieved with the help of datatype constructors like MPI_Type_create_struct() that describes components of a C struct, for example. Since each of the struct components may in turn be described by a derived datatype, any struct layout accordingly gets within reach. Another set of additive constructors, like MPI_Type_contiguous() and MPI_Type_create_vector(), create derived datatypes that contain regular, repetitive patterns composed of the items described by the underlying datatype. Yet another set of additive constructors, like MPI_Type_indexed() and MPI_Type_create_indexed_block(), create derived datatypes that have an irregular pattern. Even the synopsis of the send and receive operations contains, for the sake of simplicity and usability, rudimentary means of describing an array of data items defined by their datatype and item count. The datatype here may, of course, be a derived one, too.

So much for the data layout. MPI standard prescribes usage of matching *datatype signatures* (i.e., everything but the offsets and blank space), with certain exceptions, on both sides of the transfer. This way, the MPI implementation has the freedom to decide whether any data conversion, if it is indeed necessary (and it is never necessary in a homogeneous system), should be performed by the sending, the receiving, or both sides of the transfer. There are good arguments in favor of doing the conversion once, and certain rare situations when using an intermediate universal format may be profitable as well, even though this means that the conversion may have to be done by both sides.

If you think of the data conversion, you will see that this is essentially an operation similar to the memory copy. It takes the source buffer in data format A and puts this data into the destination buffer in data format B, most likely, squeezing all unnecessary blank space out in the process. This operation is called *packing*, the reverse operation is called *unpacking*. If the byte sizes of the respective elementary datatypes in data formats A and B match, the conversion may potentially be done in place. However, with the abundance of cheap memory these days, this troublesome approach is used less and less often, and it has even been prohibited in some cases by the latest MPI standards (see the unfortunate `const void *` specifier in the `MPI_Send()` call and friends).

4.2.3 Collective Operations

Collective operations are just a convenience, with the added optimization potential thanks to sophisticated algorithms and the use of special hardware, if available. Normally, at least for starters, they are mapped upon the point-to-point communication. In fact, the use of collectives in an application is what separates beginners from veterans as far as MPI usage is concerned.

Messages sent inside the collectives must be completely independent of those sent over the same communicator using the point-to-point calls. This is where communicator context comes into play. This way or another, a very special context unique to the processes covered by the respective communicator and differing from that for the point-to-point operations, is assigned to the collective operations, and the data is then processed rather transparently.

4.2.4 Error Handling

So far, we have been pretty relaxed about error handling. You remember: whatever happened, we would simply return control back to the caller, hoping for it to handle the *seppuku* or whatever other action was most appropriate in the higher-level application context. This allowed us to learn a thing or two that would have otherwise escaped our attention, but this approach is not what the MPI standard prescribes.

The default error handler `MPI_ERRORS_ARE_FATAL` makes an MPI application terminate itself on the spot. This is the built-in self-destruct mechanism, if you wish. We however have assumed the other error handler predefined by the standard, namely, the `MPI_ERRORS_RETURN`. By not checking the return values, we permitted the program to roll over them, sometimes to a surprising and instructive effect. This was good for our naïve MPI subsets, but this is not what serious MPI applications are prepared for.

4.2.5 Language Bindings

The latest MPI standards describe two language bindings: one for C and one for Fortran. Historically, there was an extra C++ interface, but it was used so rarely that the MPI Forum decided to get rid of it, first by deprecation (MPI-2.2) and then removal (MPI-3).

However, by the nature of the Fortran language itself, the complexity of this binding more than makes up for the missing and relatively straightforward C++ binding. There are no less than three varieties of the Fortran binding:

–   Basic binding for the Fortran 77 and 90 languages (via INCLUDE 'mpi.h')—its use is "strongly discouraged" by the standard
–   Extended binding for the Fortran 90 language (via USE mpi)—its use is "not recommended"
–   Extended binding for the Fortran 2008 language with some pending extras (via USE mpi_f08)—this one is "highly recommended"

As often happens, the first binding is the easiest in implementation and the most practical in use. It works well in those cases when no tricks are done on the arrays by the application and/or the compiler. Many legacy applications still use this binding and, I bet, will be using it well into the twenty-second century, if mankind happens to last that long. The second binding is more cumbersome, and it does not solve the main problem of the so-called choice arguments completely. The last binding does that, on the second attempt, but it requires additional support from the Fortran compiler because it relies on certain extensions to the existing Fortran standard.

What is the problem of the choice arguments? Well, this is a feature of the Fortran language, in fact. In C, one can address any object using a void * pointer. In Fortran, apart from certain nonstandard extensions, there is no equivalent to the void * pointer. In addition to this, any argument in Fortran is passed by reference rather than by value. Thus, both scalar and array arguments like the buffer in the MPI_Send() call need to be handled, which leads to interface growth.

In addition to this, a Fortran compiler is allowed to make copies of certain arguments in case it decides to pass a contiguous data chunk down to the caller. This involves a so-called copy-in on the way down, and a so-called copy-out on the way up. When and how this happens is rather involved and better looked up in the MPI standard itself. As a result, nonblocking calls like MPI_Isend() and MPI_Irecv() may under certain circumstances get references to temporary arrays, so that an attempt to complete the respective request using the MPI_Test() or MPI_Wait() is certain to misfire in this case. And do not even dream of passing arrays of RECORD entities across the MPI interface boundary. The truth is that a Fortran compiler may decide to make parallel arrays out of each of the items of the corresponding STRUCTURE definition. To sum up: this is a mess. Such is the degree of complexity that it took the whole MPI Forum two attempts and heavy interaction with the Fortran standard committee itself to settle the new, correct Fortran binding. This was the primary reason for the MPI-3.1 standard appearing in close succession to the MPI-3 standard.

Another, lesser issue is the passing of the string arguments. Again, a Fortran `CHARACTER*(*)` argument is not simply a C char * argument by a long shot. In fact, there are at least three popular ways of passing string variables in Fortran as expressed by an equivalent C call. Luckily, a Fortran compiler normally uses only one of them, so that an MPI implementor needs to cover this method only if the MPI implementation needs to be portable. In addition to this, unless the compiler takes care of this somehow, Fortran strings may not be terminated by the magic '\0' value, and they will normally be trailed by blanks instead.

Finally, there are return values. All but two MPI calls are formulated as subroutines in the Fortran binding, and the remaining two functions (`MPI_Wtime()` and `MPI_Wtick()`) return a `DOUBLE PRECISION` value. The error codes come, with the same two exceptions, as the last argument of type `INTEGER` back. However, the aforementioned Fortran issues do show up for the arrays and string arguments of `INTENT(OUT)` and `INTENT(INOUT)`.

4.2.6 Interdependencies and Steps

The sequence in which all these fancy features should be added depends on how they are interrelated. Of course, we could have tried to implement everything at once, and most likely end up with a malfunctioning, poorly structured, and slow library in the end—many months later.

Instead, we want to do the work in steps, rather quickly, and testing the intermediate results so as to eliminate all errors right away. If I were not against this modern slavery, I would call this method "agile." Whatever its name, the core of the method's success lies in good planning. Also, we will not draw beautiful diagrams or argue in earnest whether clouds or tomb stones should represent the entities involved. We will just use common sense, minding, however, that at the moment of creation, any beat of a butterfly's wing can change the whole world. So, it makes sense to be very careful.

4.2.6.1 Initial Subset Implementation: Start-up, Data Transfer, and Error Handling
First off, we need to create a skeleton with all features declared: all constants, call data structures, and all function declarations. We have done this before.

Second, we need to refactor the code because we are getting now in the range of thousands of lines, and keeping that amount of code in one file gets a little stressful even for me.

Third, we need to develop a working package with the absolute minimum of functional and actually functioning features. It may even do no communication but should include all essential parts to get an MPI program up and running, at least as a "Hello world" thing. What has to be there?

We cannot do without MPI initialization and termination. We should have error handling put on the right footing immediately, otherwise we will have to redo it, and I have something against quick and dirty prototypes that land in the waste basket tomorrow. We should also have the process start-up, and if there is communication to be provided, then we need to have connection establishment, too. These are features that top the list of dependencies.

4.2.6.2. Subset Extension 1: Communicators, Datatypes, and Collectives

This done, we need to add features that cover most if not all of the intended functionality. If we choose only a subset, we may end up with a wrong design decision perpetuated by the lack of time or the sheer amount of rework needed to get things back on track. In our case, looking at the full list of features, we need to add communicator management, at least rudimentary, including their query functions. On a level with that, we need to add datatypes, and then change the point-to-point communication to handle both communicators and datatypes. With the point-to-point working again, we can implement the collectives, possibly only for a limited set of processes, just to get the feeling.

4.2.6.3 Subset Extension 2: Communicators and Collectives Revisited

Once the first extension is tested and benchmarked, we can add the remaining communicator handling and extend the collective implementation to all possible process subsets. Now we will also have both the means of creating these subsets (i.e., the MPI_Comm_split() call) and using them (point-to-point and collectives).

4.2.6.4 Subset Completion

What remains is the addition of the Fortran binding, at least for a reasonable subset of the provided MPI calls. With that, the usual testing, benchmarking, and, if necessary, mild optimization are actions that will complete the MPI subset implementation as usual.

4.2.7 Exercises

Exercise 4.3 Estimate the effort needed to implement the extensions outlined in this chapter. Once done with the implementation, compare reality to your forecast. Does it look like the "double and round up" rule works for you?

4.3 Initial Subset Implementation: Start-up, Data Transfer, and Error Handling

Due to the size of the task at hand, we will have to go through several iterations. In the first iteration, we are targeting a functional subset with a minimum of extra features necessary, namely:
- Initialization and termination
- Data transfer
- Error handling

On a level with this, we will build in enough stuff to have a good starting point for the rest of the adventure. That is, our declarations in the MPI header file mpi.h will be reasonably complete and full with respect to the MPI subset definition provided in Section 4.1, give or take a trifle or two, but the implementation of the features will definitely be incomplete at first and will mature later in stages, showing in the process, among other things, how real code evolves in real life development.

4.3.1 Design Decisions

We may want to reconsider now one general decision we made early on (see Section 2.3.1). There, if you remember, we decided to represent the MPI_COMM_WORLD communicator with an integer value. However, this is an opaque MPI object, so that we can change our mind any time and move on to a more suitable representation if we want to. There are at least two schools of thought here:
- Stay with integers as long as possible. This is what MPICH does, taking an exception only for the hotter MPI objects like MPI_Request. An attempt to access an object represented by an integer value makes it necessary to look into the special storage reserved for opaque objects. In the MPICH case, the information on what kind of object is involved and how to find it is contained in the integer value (or handle). Of course, this means that this object storage has to be managed somehow, and this creates some overhead. However, on the plus side, this representation makes it much easier to move between the C and the Fortran language bindings (see Section 4.5).
- Use pointers as much as possible. This is what Open MPI does, taking a conscious hit on the Fortran binding matter mentioned above for the sake of the simplicity and universality of the C binding and object handling. This does not preclude the use of a certain special storage, however, because in some cases opaque objects need to be allocated, retrieved, and destroyed in a very tight loop.

In our case, going to the trouble of creating a full-blown special storage—encoding and all the brouhaha associated with that—is certainly overkill. On the other hand, I am reluctant to part ways with the integers here and switch over to the pointers. So, we will again apply the 10/90 principle: stay with integers but do it the easy way for all new opaque MPI objects we need to introduce.

Another general question is how to represent null objects. For integer handles, there are again at least two schools of thought:

- Make zero a null value. This has the appeal of simplicity and symmetry with the pointer null values that are normally represented by NULL or just (void *)0. The downside of this is that you will need to reference the internal storage starting with 1. This can be alleviated by allocating a special null object at the storage area indexed by value 0.
- Move null values outside of the normal range (normally, into the negative area). Here, the null value lives outside of the normal C index value range, which on the other hand makes you check the handles for validity, something you have to do anyway.

Because of this, we follow the second approach and make null handles negative. Note that in the case of pointers representing opaque MPI objects, the choice of NULL, however natural, may not be sufficient in case there are many handles (consider, e.g., MPI_STATUS_IGNORE and MPI_STATUSES_IGNORE).

4.3.1.1 MPI Header File

Due to the number of additions, it makes sense to look at the new MPI header file mpi.h in full (see Listing 4.1):

Listing 4.1: Subset MPI: initial declarations (see ext/v0/mpi.h)

```
#define MPI_UNDEFINED           -1

typedef int MPI_Comm;

#define MPI_COMM_NULL           MPI_UNDEFINED
#define MPI_COMM_SELF           0
#define MPI_COMM_WORLD          1

typedef int MPI_Datatype;

#define MPI_BYTE                1
#define MPI_INT                 4
#define MPI_DOUBLE              8

typedef int MPI_Op;

#define MPI_MAX                 0
#define MPI_MIN                 1
#define MPI_SUM                 2
```

```
typedef struct {
    int MPI_SOURCE;
    int MPI_TAG;
    int MPI_ERROR;
    int cnt;
} MPI_Status;

typedef struct _mpi_request_s {
    void *buf;
    MPI_Status stat;
    MPI_Comm comm;
    int flag,tag,cnt,rank;
} *MPI_Request;

typedef int MPI_Errhandler;

#define MPI_ERRORS_ARE_FATAL        0
#define MPI_ERRORS_RETURN   1

#define MPI_MAX_ERROR_STRING        32

#define MPI_SUCCESS                 0
#define MPI_ERR_BUFFER              1
#define MPI_ERR_COUNT               2
#define MPI_ERR_TYPE                3
#define MPI_ERR_TAG                 4
#define MPI_ERR_COMM                5
#define MPI_ERR_RANK                6
#define MPI_ERR_REQUEST             7
#define MPI_ERR_ROOT                8
#define MPI_ERR_OP                  9
#define MPI_ERR_ARG                 10
#define MPI_ERR_UNKNOWN             11
#define MPI_ERR_TRUNCATE            12
#define MPI_ERR_OTHER               13
#define MPI_ERR_INTERN              14
#define MPI_ERR_LASTCODE            14

int MPI_Init(int *pargc,char ***pargv);
int MPI_Finalize(void);
int MPI_Abort(MPI_Comm comm,int code);

int MPI_Comm_size(MPI_Comm comm,int *psize);
int MPI_Comm_rank(MPI_Comm comm,int *prank);
int MPI_Comm_dup(MPI_Comm comm,MPI_Comm *pcomm);
int MPI_Comm_split(MPI_Comm comm,int col,int key,MPI_Comm *pcomm);

int MPI_Comm_get_errhandler(MPI_Comm comm,MPI_Errhandler *perrh);
int MPI_Comm_set_errhandler(MPI_Comm comm,MPI_Errhandler errh);
int MPI_Comm_call_errhandler(MPI_Comm comm,int code);
```

```
int MPI_Error_class(int code,int *pclass);
int MPI_Error_string(int code,char *str,int *plen);

int MPI_Type_size(MPI_Datatype dtype,int *psize);
int MPI_Get_count(MPI_Status *pstat,MPI_Datatype dtype,int *pcnt);

int MPI_Send(const void *buf,int cnt,MPI_Datatype dtype,
            int dest,int tag,MPI_Comm comm);
int MPI_Recv(void *buf,int cnt,MPI_Datatype dtype,
            int src,int tag,MPI_Comm comm,MPI_Status *pstat);

int MPI_Isend(const void *buf,int cnt,MPI_Datatype dtype,
            int dest,int tag,MPI_Comm comm,MPI_Request *preq);
int MPI_Irecv(void *buf,int cnt,MPI_Datatype dtype,
            int src,int tag,MPI_Comm comm,MPI_Request *preq);
int MPI_Test(MPI_Request *preq,int *pflag,MPI_Status *pstat);
int MPI_Wait(MPI_Request *preq,MPI_Status *pstat);

int MPI_Barrier(MPI_Comm comm);
int MPI_Bcast(void *buf,int cnt,MPI_Datatype dtype,int root,MPI_Comm comm);
int MPI_Reduce(const void *sbuf,void *rbuf,int cnt,MPI_Datatype dtype,
            MPI_Op op,int root,MPI_Comm comm);
int MPI_Allreduce(const void *sbuf,void *rbuf,int cnt,MPI_Datatype dtype,
            MPI_Op op,MPI_Comm comm);
int MPI_Gather(const void *sbuf,int scnt,MPI_Datatype sdtype,
            void *rbuf,int rcnt,MPI_Datatype rdtype,int root,MPI_Comm comm);
int MPI_Allgather(const void *sbuf,int scnt,MPI_Datatype sdtype,
            void *rbuf,int rcnt,MPI_Datatype rdtype,MPI_Comm comm);

double MPI_Wtime(void);
```

We will go through all these features one by one later on, and all questions you may have now will be answered in appropriate order. One can note already, however, that I do not really like these modern enum declarations. But the standard is a standard, and so I reluctantly added the const modifier to several pointer arguments related to the send buffers (see MPI_Send(), etc.).

4.3.1.2 Skeleton MPI Library
The skeleton library expands the contents of the aforementioned header file by empty function definitions of the kind we have come to know and love. We spare ourselves the sight of this listing here (see source file ext/v0/mpi.c at (Supalov A. , 2018).

4.3.1.3 Testing
The test program t0.c (see Listing 2.17) and the Makefile (see Listing 2.18) defined in Chapter 2 can be reused unchanged. The output is trivial and thus not shown here. Now, why do we go through all this over and over again? The explanation is simple:

if you want a predictable result, you need to follow a strict, repeatable development procedure. With this step behind us, we are reasonably certain of the constant definitions, function declarations, and the overall structure of the work to do.

4.3.2 Code Refactoring

Now we can continue our steady progression. The first thing to do is certainly the code refactoring. So far, we have been putting everything into one file for the library source code (mpi.c) and one file for the declarations (mpi.h). With the source code taking now about 16 pages in the print-out, we get into an area of complexity where different parts of the source code will look and feel better if they are kept separate. This will also help to pull away from each other the MPI header that should be visible to the MPI users, and the library source code that should normally not be visible to them. Likewise, the test programs that have been sharing the same folder with the library proper, can and should be moved away. The same fate awaits the resulting MPI library and the command files.

This yields the following canonical directory structure:
- bin/ for command files, if any
- include/mpi.h
- lib/libmpi.a
- src/ for library source files
- test/ for test programs

A top-level Makefile handles the routine tasks in the following manner (see Listing 4.2):

Listing 4.2: Refactored subset MPI: top-level Makefile (see ext/v1/Makefile)

```
all: lib test

lib:
        (cd src; make)

test:
        (cd test; make)

clean:
        (cd src; make clean)
        (cd lib; make clean)
        (cd test; make clean)

distclean:
```

```
            (cd src; make distclean)
            (cd lib; make distclean)
            (cd test; make distclean)

.PHONY: all lib test clean distclean
```

The matching Makefile from the source directory looks as follows (see Listing 4.3):

Listing 4.3: Refactored subset MPI: source Makefile (see ext/v1/src/Makefile)

```
include ../make.header

all: lib

lib: $(LIBDIR)/libmpi.a

$(LIBDIR)/libmpi.a: coll.o comm.o errh.o init.o time.o type.o xfer.o
        ar -cr $(LIBDIR)/libmpi.a *.o

%.o : %.c $(INCDIR)/mpi.h
        $(CC) -c $(CFLAGS) $(CPPFLAGS) $< -o $@

clean:
        -rm *.o

distclean: clean

.PHONY: all lib clean distclean
```

You can see that I split the library body into seven separate source files, each dedicated to a particular part of the functionality manifested by the file name. Matching this, the Makefile file uses so-called pattern rules to save on explicit instruction on how to make object files out of the respective sources. It also picks up common declarations in the make.header file living alongside the top-level Makefile (see Listing 4.4):

Listing 4.4: Refactored subset MPI: common Makefile declarations (see ext/v1/make.header)

```
INCDIR=../include
LIBDIR=../lib

CPPFLAGS=-I. -I$(INCDIR)
CFLAGS=-O2
LDFLAGS=-L$(LIBDIR) -lmpi
```

The test Makefile looks very similar to the source one shown above. Again, we spare ourselves the sight of it here (see ext/v1/test/Makefile).

4.3.3 Initialization and Termination

Now we have to add some meat to these healthy bones. Since we have split the code into bits and pieces, the initialization and termination part of it becomes rather simple (see Listing 4.5):

Listing 4.5: Initial subset MPI: initialization and termination (excerpt, see ext/v2/src/init.c)

```c
#include <stdio.h>        // NULL
#include <stdlib.h>       // exit(3)

#include "mpi.h"
#include "imp.h"

#include <assert.h>       // assert(3)

int _mpi_init = 0;

int MPI_Init(int *pargc,char ***pargv)
{
    int rc;

#ifdef DEBUG
    fprintf(stderr,"%d/%d: MPI_Init(%p,%p)\n",_mpi_rank,_mpi_size,pargc,pargv);
#endif

    _MPI_CHECK(!_mpi_init,MPI_COMM_WORLD,MPI_ERR_OTHER);
    _MPI_CHECK(pargc != NULL && *pargc > 0,MPI_COMM_WORLD,MPI_ERR_ARG);
    _MPI_CHECK(pargv != NULL,MPI_COMM_WORLD,MPI_ERR_ARG);

// TBD: initialize all configured subsystems
    if ((rc = _mpi_errh_init(pargc,pargv)) != MPI_SUCCESS)
        return _mpi_call_errhandler(MPI_COMM_WORLD,rc);
    if ((rc = _mpi_xfer_init(pargc,pargv)) != MPI_SUCCESS)
        return _mpi_call_errhandler(MPI_COMM_WORLD,rc);
    if ((rc = _mpi_comm_init(pargc,pargv)) != MPI_SUCCESS)
        return _mpi_call_errhandler(MPI_COMM_WORLD,rc);

    _mpi_init = 1;

#ifdef DEBUG
    fprintf(stderr,"%d/%d: -> success\n",_mpi_rank,_mpi_size);
#endif
    return MPI_SUCCESS;
}

int MPI_Finalize(void)
{
```

```
    int rc;

#ifdef DEBUG
    fprintf(stderr,"%d/%d: MPI_Finalize()\n",_mpi_rank,_mpi_size);
#endif

    _MPI_CHECK(_mpi_init,MPI_COMM_WORLD,MPI_ERR_OTHER);

// TBD: terminate subsystems as necessary
    if ((rc = _mpi_comm_fini()) != MPI_SUCCESS)
        return _mpi_call_errhandler(MPI_COMM_WORLD,rc);
    if ((rc = _mpi_xfer_fini()) != MPI_SUCCESS)
        return _mpi_call_errhandler(MPI_COMM_WORLD,rc);
    if ((rc = _mpi_errh_fini()) != MPI_SUCCESS)
        return _mpi_call_errhandler(MPI_COMM_WORLD,rc);

    _mpi_init = 0;

#ifdef DEBUG
    fprintf(stderr,"%d/%d: -> success\n",_mpi_rank,_mpi_size);
#endif
    return MPI_SUCCESS;
}

...
```

Skip all argument checking for now—we will deal with it shortly. Apart from this, what happens is just an orderly initialization of the minimum requisite subsystems we need for a functional first subset, namely, the error handling, the data transfer, and the communicator management. Indeed, before we start doing anything, we need to learn how to react to the eventual errors. After that, we have to establish communication between the processes. Finally, once the size of the job is known, the communicator subsystem needs to be initialized accordingly. On the termination side, we shut everything down in exactly the opposite order.

Now, where do all those declarations and global symbols come from? Since we have split the source code into separate smaller source files, you'd probably expect to see a matching set of small header files, too. Well, I thought about this approach, but then decided against it. As long as the header involved is needed by almost all (if not all) of the source files, and this header is relatively small in size, it is easier and more practical to keep all things in one internal header file rather than many smaller ones. Here's what it may look like in this case (see Listing 4.6):

Listing 4.6: Initial subset MPI: internal declarations, initialization and termination (excerpt, see ext/v2/src/imp.h)

```
extern int _mpi_init;
```

```
extern int _mpi_rank;
extern int _mpi_size;

int _mpi_xfer_init(int *pargc,char ***pargv);
int _mpi_xfer_fini(void);

int _mpi_comm_init(int *pargc,char ***pargv);
int _mpi_comm_fini(void);

...


int _mpi_errh_init(int *pargc,char ***pargv);
int _mpi_errh_fini(void);

...
```

Here we have our old friends _mpi_init, _mpi_rank, and _mpi_size that will be used throughout for library state checking, process identification, and debugging output. Again, I could have put all of them into a structure called, say, _mpi (surprise, surprise!), but this would simply make the typing a little more cumbersome than it needs to be.

The subsystem initialization calls get the set of program options provided by the MPI user to the main program. This way, those options that concern a particular subsystem can be processed and thrown away by this very subsystem, making option parsing a little more modular and manageable. Note that each internal call, with certain exceptions to be discussed later, returns the MPI_SUCCESS or an MPI error code immediately usable for the purposes of error handling.

4.3.4 Error Handling

So we predictably land on this sorry planet. Let's add the following features to the underlying MPI subset:
- Predefined error handlers MPI_ERRORS_ARE_FATAL and MPI_ERRORS_RETURN
- Functions MPI_Comm_get_errhandler() and MPI_Comm_set_errhandler()
- Functions MPI_Error_class() and MPI_Error_string()
- Predefined error classes requisite to the underlying MPI subset
- Function MPI_Abort()

If you look at it, here you have almost everything you need to gracefully terminate a program that ran astray, both from inside the MPI library and from inside an MPI application if you so desire. You can let the MPI library handle the *seppuku* by using the default error handler MPI_ERRORS_ARE_FATAL. If you however prefer to cut the thread yourself, you can select the error handler MPI_ERRORS_RETURN (that we have been using so far without as much as a thought), print out a well-formed last *haiku* using, among other things, the MPI_Error_string(), and then turn off the lights by calling the MPI_Abort(). Alternatively, you may try to deduce what happened using the MPI_Error_class(), take a corrective action if you dare, and try to run ahead, all fingers crossed. Moreover, having the facility of MPI_Comm_get_errhandler() and MPI_Comm_set_errhandler(), you can do this or that in different parts of the program. This is not all we will ever want to do, but this is a very good first approximation.

4.3.4.1 Design Considerations

One may think that error handling is a trivial matter. Actually, proper error detection and handling is an integral and distinctive part of a professional product. Apart from providing the user with the necessary information about the reason for the malfunction, well-behaved error handling also minimizes the damage done by the fault, as well as prevents wrong results getting accepted as correct ones.

Let's look at how this can be done in an MPI library (see Listing 4.7):

Listing 4.7: Initial subset MPI: internal declarations, error handling (excerpt, see ext/v2/src/imp.h)

```
...

int _mpi_abort(MPI_Comm comm,int code);

...

MPI_Errhandler _mpi_get_errhandler(MPI_Comm comm);

int _mpi_call_errhandler(MPI_Comm comm,int code);

#ifdef NCHECK
#define _MPI_CHECK(expr,comm,code)
#define _MPI_CHECK_COMM(comm)
#define _MPI_CHECK_ERRH(comm,errh)
#else
#define _MPI_CHECK(expr,comm,code)  (((expr)) ? MPI_SUCCESS :
_mpi_call_errhandler (comm,code))

void _mpi_check_comm(MPI_Comm comm);
void _mpi_check_errh(MPI_Comm comm,MPI_Errhandler errh);

#define _MPI_CHECK_COMM(comm) _mpi_check_comm(comm)
#define _MPI_CHECK_ERRH(comm,errh) _mpi_check_errh(comm,errh)
#endif
```

There are a couple of tricks here that have to be introduced due to the fact of splitting the source code into smaller files (yes, *everything*, any design decision you make has a downside; better get used to it and learn how to watch out for it).

Since we are going to define the internal structure of the MPI communicators and the way they are managed by the MPI library inside the file comm.c, and do the same for the error handlers inside the file errh.c, we need to teach the communicators how to check correctness of the error handlers and vice versa. In addition to this, checking correctness of a communicator, finding out that it is all wrong, and then calling an error handler in the name of this wrong communicator does not sound like a valid proposition, right? This is why we define functions _mpi_check_comm() and _mpi_check_errh() that handle the details of the checking and error handling hidden from the rest of the MPI library.

We also need to provide the communicators, in the name of which the error handlers are going to be invoked, with a clean way of doing so. This is what function _mpi_call_errhandler() is intended for. Of course, the error handling itself needs to extract the value of the error handler to call from the communicator that should remain an opaque MPI object even inside the MPI library—for the sake of modularity, extensibility, and all the good of the world. This is what function _mpi_get_errhanler() is there for. Note that it returns the error handler instead of having an extra argument, like a well-behaved external MPI call would do. Why? Well, this is the way things work better in C. Inside the MPI library we do not have the design constraints that influenced the whole of the MPI standard, namely, that MPI calls should look similar in the C and Fortran bindings.

Then, once in a while we may want to build an MPI library that does not check correctness of its arguments. Yes, this may make sense if we are certain that our application has been debugged to the point where there is only one error left in it (and there is always at least one error left there, as we know). In that case, checking for correctness of the arguments is simply a waste of time. This is why we introduce macros closely matching the respective internal argument checking calls both in names and in the arguments, and control their expansion into the actual calls by the macro NCHECK, the effect of which closely resembles that of the NDEBUG on the assert(3) macros.

And finally, as a result of error detection, we may want to terminate the program on the spot, which is what the default predefined error handler MPI_ERRORS_ARE_FATAL is supposed to do. Instead of calling the external function MPI_Abort() that may want to check its arguments as well, and thus possibly cause an endless recursion, we define an internal call _mpi_abort() that is less fussy about this stuff.

4.3.4.2 Error Codes, Classes, and Strings

The MPI standard makes a distinction between the *error codes* returned by the MPI calls and the *error classes* that are used to differentiate between the error conditions

encountered by the MPI implementation. The error codes are potentially implementation dependent. The error classes however are predefined by the MPI standard.

In our simple implementation, we make error classes equal to the error codes,
and define error strings so that they correspond directly to these error classes. Thus,
the respective calls become rather trivial (see Listing 4.8):

Listing 4.8: Initial subset MPI: error classes, codes, and strings (excerpt, see ext/v2/src/
errh.c)

```
#include <string.h>      // strncpy(3)
#include <stdio.h>       // NULL, fprintf(3)

#include "mpi.h"
#include "imp.h"

#include <assert.h>      // assert(3)

static char *_mpi_err_string[MPI_ERR_LASTCODE + 1] = {
    "success",                      // #define MPI_SUCCESS        0
    "wrong buffer",                 // #define MPI_ERR_BUFFER     1
    "wrong count",                  // #define MPI_ERR_COUNT      2
    "wrong type",                   // #define MPI_ERR_TYPE       3
    "wrong tag",                    // #define MPI_ERR_TAG        4
    "wrong communicator",           // #define MPI_ERR_COMM       5
    "wrong rank",                   // #define MPI_ERR_RANK       6
    "wrong request",                // #define MPI_ERR_REQUEST    7
    "wrong root",                   // #define MPI_ERR_ROOT       8
    "wrong reduction operation",    // #define MPI_ERR_OP         9
    "wrong argument",               // #define MPI_ERR_ARG        10
    "unknown error",                // #define MPI_ERR_UNKNOWN    11
    "buffer truncated",             // #define MPI_ERR_TRUNCATE   12
    "other error",                  // #define MPI_ERR_OTHER      13
    "internal error"                // #define MPI_ERR_INTERN     14
};

...

int MPI_Error_class(int code,int *pclass)
{
#ifdef DEBUG
    fprintf(stderr,"%d/%d: MPI_Error_class(%d,%p)\n",_mpi_rank,_mpi_size,code,pclass);
#endif

    _MPI_CHECK(_mpi_init,MPI_COMM_WORLD,MPI_ERR_OTHER);
    _MPI_CHECK(code >= MPI_SUCCESS && code <= MPI_ERR_LASTCODE,MPI_COMM_WORLD,
MPI_ERR_ARG);
    _MPI_CHECK(pclass != NULL,MPI_COMM_WORLD,MPI_ERR_ARG);
```

```
    *pclass = code;

#ifdef DEBUG
    fprintf(stderr,"%d/%d: -> success, *pclass = %d\n",_mpi_rank,_mpi_size,*pclass);
#endif
    return MPI_SUCCESS;
}

int MPI_Error_string(int code,char *str,int *plen)
{
#ifdef DEBUG
    fprintf(stderr,"%d/%d: MPI_Error_string(%d,%p,%p)\n",_mpi_rank,_mpi_size,
code,str,plen);
#endif

    _MPI_CHECK(_mpi_init,MPI_COMM_WORLD,MPI_ERR_OTHER);
    _MPI_CHECK(code >= MPI_SUCCESS && code <= MPI_ERR_LASTCODE,MPI_COMM_WORLD,
MPI_ERR_ARG);
    _MPI_CHECK(str != NULL,MPI_COMM_WORLD,MPI_ERR_ARG);
    _MPI_CHECK(plen != NULL,MPI_COMM_WORLD,MPI_ERR_ARG);

    *plen = strlen(strncpy(str,_mpi_err_string[code],MPI_MAX_ERROR_STRING));

#ifdef DEBUG
    fprintf(stderr,"%d/%d: -> success, str = %s, *plen = %d\n",_mpi_rank,_mpi_size,
str,*plen);
#endif
    return MPI_SUCCESS;
}
```

Note that we do check the arguments of the internal calls using the assert(3) macros. They will be deactivated in the streamlined library even before the MPI argument checking will. However, in the debugging build they provide an extra element of robustness to the whole structure. And believe me, this will help—if not here, then elsewhere! Never ever skimp on this. However, doing extra debugging output from the internal functions rarely makes sense, apart from the cases when an error is detected. Likewise, it is not recommended to invoke MPI error handlers and such in the lower level functions, for they may simply not have enough context to do this gracefully. It is better to pass an error code up and let it be dealt with there. This makes error detection a little more difficult, but helps a lot in keeping the code and its behavior under control.

4.3.4.3 Error Handlers
But how are the error handlers themselves defined and invoked? Here we go (see Listing 4.9):

Listing 4.9: Initial subset MPI: error handlers (excerpt, see ext/v2/src/errh.c)

```
...

typedef void MPI_Comm_errhandler_function(MPI_Comm *,int *,...);

void _mpi_errors_are_fatal(MPI_Comm *pcomm,int *pcode,...)
{
    fprintf(stderr,"%d/%d: %s\n",_mpi_rank,_mpi_size,_mpi_err_string[*pcode]);
    _mpi_abort(*pcomm,*pcode);
}

void _mpi_errors_return(MPI_Comm *pcomm,int *pcode,...)
{
}

#define _MPI_ERRH_MAX 2

static MPI_Comm_errhandler_function *_mpi_errh[_MPI_ERRH_MAX] = { _mpi_errors_are_fatal,
_mpi_errors_return };

#ifndef CHECK
void _mpi_check_errh(MPI_Comm comm,MPI_Errhandler errh)
{
    _MPI_CHECK(errh >= 0 && errh < _MPI_ERRH_MAX,comm,MPI_ERR_ARG);
}
#endif

int _mpi_call_errhandler(MPI_Comm comm,int code)
{
    MPI_Errhandler errh = _mpi_get_errhandler(comm);

    assert(errh >= 0 && errh < _MPI_ERRH_MAX);

    (*_mpi_errh[errh])(&comm,&code);

    return code;
}

int _mpi_errh_init(int *pargc,char ***pargv)
{
    assert(!_mpi_init);
    assert(pargc != NULL && *pargc > 0);
    assert(pargv != NULL);

    return MPI_SUCCESS;
}

int _mpi_errh_fini(void)
```

```
{
    assert(_mpi_init);

    return MPI_SUCCESS;
}
```

...

Instead of reinventing the wheel, we just take the error handler function definition from the MPI standard and align the rest of the design along it. Note that this calling convention with the pointers as arguments would also work in Fortran.

The error handler initialization and termination calls are effectively empty, because the initialization is handled by the static data definition. Note that here we start seeing that the decision to use zero as the default value was a good one: even if we do not initialize anything, we still get the right value out most of the time.

4.3.4.4 Abnormal Program Termination

What remains to be considered here is how we actually stop the program. Although this is normally done as part of the error handling (if *seppuku* can be considered as such), the MPI_Abort() code actually belongs elsewhere. The best place for a simple implementation that just turns off the lights is the initialization and termination part of the MPI library (see Listing 4.10, cf. Listing 4.5):

Listing 4.10: Initial subset MPI: error handlers (excerpt, see ext/v2/src/init.c)

...

```
int _mpi_abort(MPI_Comm comm,int code)
{
// TBD: terminate subsystems if possible

    _mpi_init = 0;

    exit(EXIT_FAILURE);
}

int MPI_Abort(MPI_Comm comm,int code)
{
#ifdef DEBUG
    fprintf(stderr,"%d/%d: MPI_Abort(%d,%d)\n",_mpi rank,_mpi_size,comm,code);
#endif

    _MPI_CHECK(_mpi_init,MPI_COMM_WORLD,MPI_ERR_OTHER);
    _MPI_CHECK_COMM(comm);

    _mpi_abort(comm,code);
```

```
#ifdef DEBUG
    fprintf(stderr,"%d/%d: -> success\n",_mpi_rank,_mpi_size);
#endif
    return MPI_SUCCESS;
}
```

As you can see, we simply terminate the current process and pray that the rest of the
MPI processes will take notice, which they normally do as long as communication is
underway. If this does not happen, you will have to kill the remaining processes yourself. Note that we basically ignore the communicator argument and do not even try to
pretend that we may want to kill only those processes involved. Well, this is what most
MPI implementations do anyway, and we are certainly not going to waste any time in
a vain attempt to look holier than the Pope.

4.3.5 Process Start-up and Data Transfer

What is an MPI library without MPI processes? A toy not worthy of its name. Fortunately, and thanks to the work done well in Chapter 3, this step will be surprisingly easy.

4.3.5.1 Design Considerations

If you know some of the MPI internals from your past experience, you will probably
expect me to introduce an MPI device layer right in this section. Indeed, this would
abstract out the specific medium used to transfer data back and forth and move us
still closer to the Real Thing, right? Right?!

No, we will not succumb to this natural temptation here and now. Why? We want
to extend the MPI implementation rather than make it run faster or look properly
structured in one go. Extension is what this chapter is about, and any additional load
would probably break it by drawing our attention away from the task at hand.

However, we will introduce the device layer in the next chapter and do that for
very good reasons: adding support for multifabric functionality (say, shared memory
and sockets combined) and optionally enabling fast networks (say, InfiniBand), not
to mention many other optimizations *internal* to the MPI library.

Here, to the contrary, we will keep it simple by reusing the sockets-based data
transfer layer developed in Chapter 3 and making it work in a slightly different setting.

4.3.5.2 Start-up and Termination

Based on this, start-up and termination are taken verbatim out of Chapter 3. We just
move the mpiexec command file into the bin directory, and reformulate the library

internals slightly to establish the connection out of the new data transfer layer (see Listing 4.11):

Listing 4.11: Initial subset MPI: process start-up and connection establishment (excerpt, see ext/ v2/src/xfer.c)

```
...

int _mpi_rank = 0;
int _mpi_size = 0;

static char **_mpi_hosts,**_mpi_ports;
static int *_mpi_fds;

...

int _mpi_xfer_init(int *pargc,char ***pargv)
{
    int rc;

    assert(_mpi_init == 0);
    assert(pargc != NULL && *pargc > 0);
    assert(pargv != NULL);

    _mpi_size = _mpi_getint("MPI_SIZE",_mpi_size);
    assert(_mpi_size > 0);
    _mpi_rank = _mpi_getint("MPI_RANK",_mpi_rank);
    assert(_mpi_rank >= 0 && _mpi_rank < _mpi_size);

    _mpi_hosts = _mpi_getarr("MPI_HOSTS",_mpi_size);
    assert(_mpi_hosts != (char **)NULL);
    _mpi_ports = _mpi_getarr("MPI_PORTS",_mpi_size);
    assert(_mpi_ports != (char **)NULL);

    if ((rc = _mpi_connect_all()) != MPI_SUCCESS) {
#ifdef DEBUG
        fprintf(stderr,"%d/%d: -> failure\n",_mpi_rank,_mpi_size);
#endif
        return rc;
    }

    if ((rc = _mpi_nonblock_all()) != MPI_SUCCESS) {
#ifdef DEBUG
        fprintf(stderr,"%d/%d: -> failure\n",_mpi_rank,_mpi_size);
#endif
        return rc;
    }

    return MPI_SUCCESS;
```

```
}

int _mpi_xfer_fini(void)
{
    assert(_mpi_init);

    return MPI_SUCCESS;
}

...
```

Since we learn the size of the job and identify MPI processes during the connection establishment phase, the respective global variables _mpi_size and _mpi_rank are defined here. The code from the earlier MPI_Init() and MPI_Finalize() functions migrate into the internal functions _mpi_xfer_init() and _mpi_xfer_fini(), respectively. The rest of the start-up and termination code remains unchanged, except for returning fitting MPI error codes instead of the !MPI_SUCCESS placeholder used earlier. The respective error codes are consistently passed up to the caller.

The termination sequence is empty: we still count on the operating system to clean up the mess we may accidentally leave behind.

4.3.5.3 Point-to-Point Communication
The point-to-point communication is dealt with as before. We even keep the limitation of at most one pending send, one pending posted receive, and one unexpected receive request, which will be sufficient for the purposes of this initial subset. The rest of the code remains essentially unchanged, but for replacement of the MPI argument checking by the respective macros defined in Section 4.3.1. Here we consistently avoid double argument checking, passing this task to the lowest externally visible MPI function layer and letting it invoke the MPI error handler on the site of failure. For example, in case of the MPI_Send() this will be the MPI_Isend() and the MPI_Wait() functions that this former call is decomposed into (see Listing 4.12):

Listing 4.12: Initial subset MPI: blocking send (excerpt, see ext/v2/src/xfer.c)

```
int MPI_Send(const void *buf,int cnt,MPI_Datatype dtype,int dest,int tag,MPI_Comm
comm)
{
    MPI_Request req;
    int rc;

#ifdef DEBUG
    fprintf(stderr,"%d/%d: MPI_Send(%p,%d,%d,%d,%d,%d)\n",_mpi_rank,_mpi_
size,buf,cnt,dtype,dest,tag,comm);
#endif
```

```
    if ((rc = MPI_Isend(buf,cnt,dtype,dest,tag,comm,&req)) != MPI_SUCCESS)
        return rc;
    if ((rc = MPI_Wait(&req,NULL)) != MPI_SUCCESS)
        return rc;

    return MPI_SUCCESS;
}
```

You should look up the rest of the code in the respective source files. Just to make sure you get the drift, here is how MPI_Wait() and MPI_Test() are layered on top of each other (see Listing 4.13):

Listing 4.13: Initial subset MPI: message completion calls (excerpt, see ext/v2/src/xfer.c)

```
int MPI_Test(MPI_Request *preq,int *pflag,MPI_Status *pstat)
{
    int rc;

#ifdef DEBUG
    fprintf(stderr,"%d/%d: MPI_Test(%p,%p,%p)\n",_mpi_rank,_mpi_size,preq,pflag,
pstat);
#endif

    _MPI_CHECK(_mpi_init,MPI_COMM_WORLD,MPI_ERR_OTHER);
    _MPI_CHECK(preq != NULL,MPI_COMM_WORLD,MPI_ERR_REQUEST);
    _MPI_CHECK(pflag != NULL,MPI_COMM_WORLD,MPI_ERR_ARG);

    if ((rc = _mpi_progress()) != MPI_SUCCESS)
        return _mpi_call_errhandler(MPI_COMM_WORLD,rc);

    *pflag = 0;
    if (*preq != NULL)
    {
        MPI_Request req = *preq;

        if (req->flag) {
            *pflag = 1;
            if (pstat != NULL)
                _mpi_statcpy(pstat,&(req->stat));
            req->flag = 0;
            _mpi_freereq(_mpi_findq(req),preq);
        }
    }

#ifdef DEBUG
    if (pstat != NULL)
        fprintf(stderr,"%d/%d: -> success, *pflag = %d, *pstat = { %d, %d }\n",
_mpi_rank,_mpi_size,*pflag,pstat->MPI_SOURCE,pstat->MPI_TAG);
    else
```

```
        fprintf(stderr,"%d/%d: -> success, *pflag = %d\n",_mpi_rank,_mpi_size,*pflag);
#endif
    return MPI_SUCCESS;
}

int MPI_Wait(MPI_Request *preq,MPI_Status *pstat)
{
    int rc,flag = 0;

#ifdef DEBUG
    fprintf(stderr,"%d/%d: MPI_Wait(%p,%p)\n",_mpi_rank,_mpi_size,preq,pstat);
#endif

    while (!flag) {
        if ((rc = MPI_Test(preq,&flag,pstat)) != MPI_SUCCESS)
            return rc;
    }

    return MPI_SUCCESS;
}
```

Here you see that the MPI_Wait() uses the MPI_Test() as it did before. The MPI_Test() in turn checks the arguments, calls the respective internal functions, and invokes the active MPI error handler in case of failure. This strikes a reasonable balance between the volume of the necessary code changes on one hand, and the usefulness of the resulting error diagnostics, should it come to that, on the other.

Note that we still keep the code limited to the MPI_COMM_WORLD only. We will remove this limitation in the next section, once we make sure that what we have done so far works well and runs fast.

4.3.6 Testing

Well, so far so good. It took me about 10 smaller steps, not detailed in this book, and as many days, not all spent programming, to get the current subset up and running. Of course, testing was being done all the way, first by building the intermediate sub-versions, each bringing the whole thing a step forward, and then by adding the test programs we developed earlier.

To make testing a little better structured, I changed the respective Makefile in the test directory as follows (see Listing 4.14):

Listing 4.14: Initial subset MPI: test Makefile (excerpt, see ext/v2/test/Makefile)

```
include ../make.header

all: build test time
```

```
build: t0 t1 t1s t1x t2 t2i t2o pingping pongpong pingpong ping pong dummy

test:
        export MPI_PORTBASE=10111; $(BINDIR)/mpiexec -n 2 $(PWD)/t0

...

time:
        export MPI_PORTBASE=10321; $(BINDIR)/mpiexec -n 2 $(PWD)/pingpong

...

t0: t0.c $(INCDIR)/mpi.h $(LIBDIR)/libmpi.a
        cc $(CFLAGS) $(CPPFLAGS) -o t0 t0.c $(LDFLAGS)

pingpong: b.c $(INCDIR)/mpi.h $(LIBDIR)/libmpi.a
        cc -DPING -DPONG $(CFLAGS) $(CPPFLAGS) -o pingpong b.c $(LDFLAGS)

...

clean:
        -rm *.o

distclean: clean
        -rm t0 t1 t1s t1x t2 t2i t2o pingping pongpong pingpong ping pong dummy

.PHONY: all build test time clean distclean
```

Fundamentally, now it is possible to build the test programs, and then to test and benchmark the library separately.

4.3.7 Benchmarking

As expected, benchmarking did not reveal any noticeable and systematic loss of performance compared to the original code from Chapter 3.

4.3.8 Exercises

Exercise 4.4 Do you fill lost in all these short program names? If so, introduce more descriptive names and see what happens.

4.4 Subset Extension 1: Communicators, Datatypes, and Collectives

The next logical step mapped out earlier is to add the following features:
- Communicator management
- Elementary datatypes
- Collective operations

Here again we will not go all the way through but rather add only those parts of the implementation that are essential for the continued advance to the final goal.

4.4.1 Extension Directions

Unlike the previous section, where we had to perform a big leap, here we have only one slight interdependency between the communicators and collective operations. So, with due foresight related to the communication contexts (see Section 4.4.2.2), we can proceed relatively independently along all three extension paths mentioned above.

4.4.2 Communicator Management

Group and communicator management is a huge part of the MPI standard. We will not target the whole of it and instead only show on a set of examples how to approach this largely trivial but still important job. Let's add the following features to the underlying MPI subset:
- Predefined communicator `MPI_COMM_SELF`

This simple step will have huge repercussions across the code, because instead of the single communicator handle `MPI_COMM_WORLD`, we will have to expect another communicator, like `MPI_COMM_SELF`.

4.4.2.1 Design Considerations

We do not target extreme scalability in this book. Due to this, the easiest way of representing a process group is a simple list of global process ranks ordered by the ranks in the group (as usual, from 0 to the group size minus 1). All operations upon the groups and communicators are then reduced to the set operations upon the respective process rank lists on one hand, and to figuring out a proper context value that is unique in all communicators a process happens to be participating in on the other hand.

4.4.2.2 Initialization and Termination

The contents of the predefined communicators MPI_COMM_SELF and MPI_COMM_WORLD consist of two parts:

- Job-independent values like own rank and communicator size (for MPI_COMM_SELF) or context(s)
- Job-dependent values like own rank and communicator size (for MPI_COMM_WORLD) or rank mapping

Because of this, we will need non-empty initialization and termination sequences (see Listing 4.15):

Listing 4.15: Extended subset MPI: communicator initialization and termination (excerpt, see ext/ v3/src/comm.c)

```
#include <stdio.h>        // NULL, fprintf(3)
#include <stdlib.h>       // malloc(3), free(3)

#include "mpi.h"
#include "imp.h"

#include <assert.h>       // assert(3)

struct _mpi_comm_s {
    int *prank;
    MPI_Errhandler errh;
    int size,rank,ctxt;
};

#define _MPI_COMM_MAX 3
#define _MPI_CTXT_INC 2

static struct _mpi_comm_s _mpi_comm[_MPI_COMM_MAX];

static int _mpi_comm_max = 0;
static int _mpi_ctxt_max = 0;

...

static int _mpi_comm_create_self()
{
    int i;

    assert(!_mpi_init);

    _mpi_comm[MPI_COMM_SELF].size = 1;
    _mpi_comm[MPI_COMM_SELF].rank = 0;
    _mpi_comm[MPI_COMM_SELF].ctxt = _mpi_ctxt_max;
```

```
    _mpi_comm[MPI_COMM_SELF].prank = &_mpi_rank;

    _mpi_comm_max++;
    _mpi_ctxt_max += _MPI_CTXT_INC;

    return MPI_SUCCESS;
}

static int _mpi_comm_create_world()
{
    int i;

    assert(!_mpi_init);

    _mpi_comm[MPI_COMM_WORLD].size = _mpi_size;
    _mpi_comm[MPI_COMM_WORLD].rank = _mpi_rank;
    _mpi_comm[MPI_COMM_WORLD].ctxt = _mpi_ctxt_max;

    if ((_mpi_comm[MPI_COMM_WORLD].prank = (int *)malloc(sizeof(int)*_mpi_size)) ==
(int *)NULL)
        return MPI_ERR_OTHER;

    for (i = 0; i < _mpi_size; i++)
        _mpi_comm[MPI_COMM_WORLD].prank[i] = i;

    _mpi_comm_max++;
    _mpi_ctxt_max += _MPI_CTXT_INC;

    return MPI_SUCCESS;
}

int _mpi_comm_init(int *pargc,char ***pargv)
{
    assert(!_mpi_init);
    assert(pargc != NULL && *pargc > 0);
    assert(pargv != NULL);

    if (_mpi_comm_create_self() == MPI_SUCCESS &&
        _mpi_comm_create_world() == MPI_SUCCESS)
        return MPI_SUCCESS;

    return !MPI_SUCCESS;
}

int _mpi_comm_fini(void)
{

    assert(_mpi_init);
```

```
    return MPI_SUCCESS;
}
```

...

You can see that we select the simplest possible way of storing the communicator data. Indeed, since we plan to provide only MPI communicator creation calls and no deletion calls, adding communicators to an array of redefined maximum length will serve us well. Adding a library for handling the array of arbitrary sized structures could help the overall design, but do we really need this fuss for holding a couple of trivial bits of data?

The data is initialized by default to acceptable initial values. The only nontrivial bit here is the reservation of two contexts rather than one per communicator. The wisdom of this move will reveal itself once we get to the collective operations (see Section 4.4.7). Note that without collective operations and especially the communicator splitting, we could have relied on the communicator value itself to provide a unique communication context.

4.4.2.3 Query Functions

Standard query functions MPI_Comm_size() and MPI_Comm_rank() become just a little more involved in this code incarnation (see Listing 4.16):

Listing 4.16: Extended subset MPI: communicator query functions (excerpt, see ext/v3/src/ comm.c)

```
int MPI_Comm_size(MPI_Comm comm,int *psize)
{
#ifdef DEBUG
    fprintf(stderr,"%d/%d: MPI_Comm_size(%d,%p)\n",_mpi_rank,_mpi_size,comm,psize);
#endif

    _MPI_CHECK(_mpi_init,MPI_COMM_WORLD,MPI_ERR_OTHER);
    _MPI_CHECK_COMM(comm);
    _MPI_CHECK(psize != NULL,comm,MPI_ERR_ARG);

    *psize = _mpi_comm[comm].size;

#ifdef DEBUG
    fprintf(stderr,"%d/%d: -> success, *psize = %d\n",_mpi_rank,_mpi_size,*psize);
#endif
    return MPI_SUCCESS;
}

int MPI_Comm_rank(MPI_Comm comm,int *prank)
{
#ifdef DEBUG
```

```
    fprintf(stderr,"%d/%d: MPI_Comm_rank(%d,%p)\n",_mpi_rank,_mpi_size,comm,prank);
#endif

    _MPI_CHECK(_mpi_init,MPI_COMM_WORLD,MPI_ERR_OTHER);
    _MPI_CHECK_COMM(comm);
    _MPI_CHECK(prank != NULL,comm,MPI_ERR_ARG);

    *prank = _mpi_comm[comm].rank;

#ifdef DEBUG
    fprintf(stderr,"%d/%d: -> success, *prank = %d\n",_mpi_rank,_mpi_size,*prank);
#endif
    return MPI_SUCCESS;
}
```

Note that MPI_Comm_rank() returns a rank local to the communicator, as it should.

4.4.2.4 Error Handling

With the advent of non-trivial communicators, we have to slightly extend the error handling. The corresponding internals look as follows (see Listing 4.17):

Listing 4.17: Extended subset MPI: communicator error handling (excerpt, see ext/v3/src/comm.c)

...

```
#ifndef NCHECK
int _mpi_check_comm(MPI_Comm comm)
{
    return _MPI_CHECK(comm >= 0 && comm < _mpi_comm_max,MPI_COMM_WORLD,MPI_ERR_COMM);
}

int _mpi_check_rank(MPI_Comm comm,int rank)
{
    assert(comm >= 0 && comm < _mpi_comm_max);

    return _MPI_CHECK(rank >= 0 && rank < _mpi_comm[comm].size,comm,MPI_ERR_RANK);
}
#endif


...


MPI_Errhandler _mpi_get_errhandler(MPI_Comm comm)
{
    assert(comm >= 0 && comm < _MPI_COMM_MAX);

    return _mpi_comm[comm].errh;
}
```

```
int MPI_Comm_get_errhandler(MPI_Comm comm,MPI_Errhandler *perrh)
{
#ifdef DEBUG
    fprintf(stderr,"%d/%d: MPI_Comm_get_errhandler(%d,%p)\n",_mpi_rank,_mpi_size,
comm,perrh);
#endif

    _MPI_CHECK(_mpi_init,MPI_COMM_WORLD,MPI_ERR_OTHER);
    _MPI_CHECK_COMM(comm);
    _MPI_CHECK(perrh != NULL,comm,MPI_ERR_ARG);

    *perrh = _mpi_get_errhandler(comm);

#ifdef DEBUG
    fprintf(stderr,"%d/%d: -> success, *perrh = %d\n",_mpi_rank,_mpi_size,*perrh);
#endif
    return MPI_SUCCESS;
}

int MPI_Comm_set_errhandler(MPI_Comm comm,MPI_Errhandler errh)
{
#ifdef DEBUG
    fprintf(stderr,"%d/%d: MPI_Comm_set_errhandler(%d,%d)\n",_mpi_rank,_mpi_size,
comm,errh);
#endif

    _MPI_CHECK(_mpi_init,MPI_COMM_WORLD,MPI_ERR_OTHER);
    _MPI_CHECK_COMM(comm);
    _MPI_CHECK_ERRH(comm,errh);

    _mpi_comm[comm].errh = errh;

#ifdef DEBUG
    fprintf(stderr,"%d/%d: -> success\n",_mpi_rank,_mpi_size);
#endif
    return MPI_SUCCESS;
}
```

Here we get a taste of the tedious, almost placeholder source code that fills most of big MPI implementations. Oh well, this will be about the only such sacrifice we have to make to keep things in order.

4.4.3 Datatypes

Let's add the following features to the underlying MPI subset:
- Elementary datatypes MPI_INT and MPI_DOUBLE

- Function `MPI_Get_count()`

Note that the `MPI_Get_count()` in particular will interact with the `MPI_Status` objects. However, the fact of having more than `MPI_BYTE` will make itself felt in a big way in the point-to-point communication.

4.4.3.1 Design Considerations
Well, we could have made life complicated for ourselves by going for a full-scale data-type implementation—but we do not need to do this in order to get off the ground.

4.4.3.2 Predefined Datatypes
Since we are only going to support a handful of elementary datatypes in a homogeneous environment, a very simple implementation will suffice (see Listing 4.18):

Listing 4.18: Extended subset MPI: datatypes (excerpt, see ext/v3/src/type.c)

```
#include <stdio.h>        // NULL, fprintf(3)

#include "mpi.h"
#include "imp.h"

#include <assert.h>       // assert(3)

#ifndef CHECK
int _mpi_check_type(MPI_Comm comm,MPI_Datatype type)
{
    return _MPI_CHECK(type == MPI_BYTE || type == MPI_INT || type == MPI_DOUBLE,
comm,MPI_ERR_TYPE);
}
#endif

int _mpi_type_init(int *pargc,char ***pargv)
{
    assert(!_mpi_init);
    assert(pargc != NULL && *pargc > 0);
    assert(pargv != NULL);

    return MPI_SUCCESS;
}

int _mpi_type_fini(void)
{
    assert(_mpi_init);

    return MPI_SUCCESS;
}
```

```
int _mpi_type_size(MPI_Datatype type)
{
    assert(_mpi_init);
    assert(type == MPI_BYTE || type == MPI_INT || type == MPI_DOUBLE);

    return type;
}

int MPI_Type_size(MPI_Datatype dtype,int *psize)
{
#ifdef DEBUG
    fprintf(stderr,"%d/%d: MPI_Type_size(%d,%p)\n",_mpi_rank,_mpi_size,dtype,psize);
#endif

    *psize = _mpi_type_size(dtype);

#ifdef DEBUG
    fprintf(stderr,"%d/%d: -> success, *psize = %d\n",_mpi_rank,_mpi_size,*psize);
#endif
    return MPI_SUCCESS;
}

...
```

Yes, the datatype handle is also the elementary datatype size in bytes! This is a perfectly legal approach under the circumstances, since we are not going to do any fancy stuff with the datatypes in this chapter. So, why bother?

4.4.4 Point-to-Point Communication

The changes accumulated so far in this section make it necessary to adapt some bits of the data transfer mechanism. Indeed, instead of working on all processes inside the MPI_COMM_WORLD, and handling only contiguous byte arrays, now we have to handle more communicators with arbitrary rank mapping and pass back and forth arrays of integer or even double precision values!

4.4.4.1 Design Considerations
There is always more than one way to cut the cake. However, there is always only one way to do this right. If this self-created maxim is applied to the current situation, it means that we should try to avoid changing too much of the transfer "just" because we have introduced communicator ranks and contexts, as well as elementary datatypes.

How can this be done? The most natural way is to let the high-level MPI calls like `MPI_Isend()` and `MPI_Irecv()` work in terms of communicator-specific ranks and contexts, as well as datatypes, and let the lower level data transfer machinery remain in blissful ignorance of those notions. In other words, the underlying layers should keep thinking that they still work inside the `MPI_COMM_WORLD` and have only contiguous byte sequences to handle.

So, the high-level calls should map communicator-specific ranks to global (i.e., `MPI_COMM_WORLD` specific) ones on entry and transform these global ranks back to local on exit. Likewise, contiguous buffers expressed in terms of counts and datatypes should be transformed into appropriately sized contiguous byte arrays on the way in and backward on the way out. This will be easy, since all of them are contiguous, after all.

Now, what should we do about contexts? Messages being passed inside one communicator should never collide with messages passed inside another. Well, let me be frank: this is a well-known problem with a well-known solution. Just recall that we keep the message tag in 4-byte `int` variables but use only 15 bits of them (from 0 to 32767, the lowest maximum value prescribed by the standard). So we can put the context value into the upper 16 bits of the tag, and let the existing machinery handle the tag matching as before. Again, we will have to transform the tags on the way in and out, but since we are doing this to the ranks and buffer descriptions anyway, this will be just a minor addition.

4.4.4.2 Way In
With this in mind, this is what we get on the way in, as exemplified by the `MPI_Isend()` function (see Listing 4.19):

Listing 4.19: Extended subset MPI: argument conversion on entry (excerpt, see `ext/v3/src/xfer.c`)

```
...

    _MPI_CHECK(_mpi_init,MPI_COMM_WORLD,MPI_ERR_OTHER);
    _MPI_CHECK(comm == MPI_COMM_WORLD,comm,MPI_ERR_COMM);
    _MPI_CHECK(buf != NULL,comm,MPI_ERR_BUFFER);
    _MPI_CHECK(preq != NULL,comm,MPI_ERR_REQUEST);

    _MPI_CHECK(cnt >= 0,comm,MPI_ERR_COUNT);
    _MPI_CHECK_TYPE(comm,dtype);
    cnt *= _mpi_type_size(dtype);

    _MPI_CHECK_RANK(comm,dest);
    dest = _mpi_get_grank(comm,dest);
    _MPI_CHECK(dest != _mpi_rank,comm,MPI_ERR_INTERN);
    _MPI_CHECK(tag >= 0 && tag <= 32767,comm,MPI_ERR_TAG);
```

```
tag = _mpi_get_gtag(comm,tag);
```

...

Let's assume for now the internal functions shown (_mpi_type_size(), _mpi_get_grank(), and _mpi_get_gtag()) know their job.

4.4.4.3 Way Out

The argument conversion on the way out may then look as follows, and this can be done in one place—the MPI_Test() function (see Listing 4.20):

Listing 4.20: Extended subset MPI: argument conversion on entry (excerpt, see ext/v3/src/ xfer.c)

...

```
        if (pstat != NULL) {
            pstat->MPI_SOURCE = _mpi_get_lrank(req->comm,req->stat.MPI_SOURCE);
            pstat->MPI_TAG    = _mpi_get_ltag(req->comm,req->stat.MPI_TAG);
            pstat->cnt        = req->stat.cnt;
        }
```

...

Note also that I dropped the _mpi_statcpy() call (and its definition) in order to prop-erly handle (i.e., do not touch) the MPI_ERROR field. We can keep the number of bytes transferred in the MPI_Status field cnt (see Listing 4.1), and compute the number of items by dividing the number of bytes by the datatype size in an easy way, still iso-lated from the rest of the code by the corresponding internal query call (see Listings 4.19 and 4.21):

Listing 4.21: Extended subset MPI: MPI_Status count query (excerpt, see ext/v3/src/type.c)

```
int MPI_Get_count(MPI_Status *pstat,MPI_Datatype type,int *pcnt)
{
#ifdef DEBUG
    fprintf(stderr,"%d/%d: MPI_Get_count(%p,%d,%p)\n",_mpi_rank,_mpi_size,pstat,
type,pcnt);
#endif

    _MPI_CHECK(_mpi_init,MPI_COMM_WORLD,MPI_ERR_OTHER);
    _MPI_CHECK(pstat != NULL,MPI_COMM_WORLD,MPI_ERR_ARG);
    _MPI_CHECK_TYPE(MPI_COMM_WORLD,type);
    _MPI_CHECK(pcnt != NULL,MPI_COMM_WORLD,MPI_ERR_ARG);

    *pcnt = pstat->cnt/_mpi_type_size(type);
```

```
#ifdef DEBUG
    fprintf(stderr,"%d/%d: -> success, *pcnt = %d\n",_mpi_rank,_mpi_size,*pcnt);
#endif
    return MPI_SUCCESS;
}
```

Well, as we will see elsewhere, this is not the best way to do this (strictly speaking, we completely ignore here any possible holes in the datatype), but this will do for now.

4.4.4.4 Mapping
The respective internal mapping calls look as follows (see Listing 4.22):

Listing 4.22: Extended subset MPI: internal mapping calls (excerpt, see ext/v3/src/comm.c)

```
int _mpi_get_grank(MPI_Comm comm,int lrank)
{
    assert(_mpi_init);
    assert(comm >= 0 && comm < _mpi_comm_max);
    assert(lrank >= 0 && lrank < _mpi_comm[comm].size);

    return _mpi_comm[comm].prank[lrank];
}

int _mpi_get_lrank(MPI_Comm comm,int grank)
{
    int i;

    assert(_mpi_init);
    assert(comm >= 0 && comm < _mpi_comm_max);
    assert(grank >= 0 && grank < _mpi_size);

    for (i = 0; i < _mpi_comm[comm].size; i++)
        if (_mpi_comm[comm].prank[i] == grank)
            return i;

    assert(1);
    return -1;
}

int _mpi_get_gtag(MPI_Comm comm,int ltag)
{
    assert(_mpi_init);
    assert(comm >= 0 && comm < _mpi_comm_max);
    assert(ltag >= 0 && ltag <= 32767);

    return (_mpi_comm[comm].ctxt << 16) | (ltag & 0x7FFF);
}
```

```
int _mpi_get_ltag(MPI_Comm comm,int gtag)
{
    assert(_mpi_init);
    assert(comm >= 0 && comm < _mpi_comm_max);

    return gtag & 0x7FFF;
}
```

We do not make a big issue of the backward conversion from the global rank to the local rank, because our communicators are not going to be overly big, and so a simple linear search should work well. If we ever find out that this becomes a performance bottleneck, we can always introduce something fancy, like binary search or even a hash function.

4.4.5 Collective Operations

Let's add the following features to the underlying MPI subset:
–   Predefined reduction operations MPI_MAX, MPI_MIN, and MPI_SUM
–   Functions   MPI_Barrier(),   MPI_Bcast(),   MPI_Reduce(),   MPI_Allreduce(),
    MPI_Gather(), and MPI_Allgather()

For starters, let's make our task a little easier by limiting the number of processes involved to two at most. This will also allow for a quite visual comparison of the point-to-point and collective call internals.

4.4.5.1 Design Considerations
An operation done over MPI_COMM_SELF is basically a NOP, apart from the MPI_Reduce() that should copy its send buffer into the receive buffer.
        The communication pattern of an operation on a two-process communicator depends on the operation class:
–   One-to-many (like MPI_Bcast()) fans the data out from the root process, in this
    case sending it from the root process to the leaf process
–   Many-to-one (like MPI_Reduce() and MPI_Gather()) fans the data in from the leaf
    process(es) to the root
–   Many-to-many (like MPI_Allreduce() and MPI_Allgather()) fans the data in to the
    root, and then out to the leaves, at least conceptually

A synchronization call like MPI_Barrier() normally includes a fan-in stage to an implicit root, followed by a fan-out stage. In the case of two processes, this looks like the well-known ping-pong pattern (or pong-ping, if you wish).

How do we go about actually doing the communication? The easiest way is to map the collective calls upon the existing point-to-point ones. This will automatically handle the rank, context, and tag conversion, take care of the datatypes, and do most of the argument checking as well. The only catch here is that the messages representing the collective communication should be kept separate from those passed as plain point-to-point messages by the user. Indeed, it is imaginable, although highly unusual, to have a couple nonblocking point-to-point messages in transit while the processes, say, synchronize themselves after a particular computational stage, like an iteration or the like. Note that this kind of synchronization should not be necessary, but MPI users like to do this anyway, because this gives them peace of mind. Of course, this becomes a pain in the neck for the MPI implementors.

Fortunately, we have been thinking ahead and reserved two context values instead of one for each communicator. So, if we somehow pass another context value to the underlying point-to-point calls, we will effectively shift the entire communication into another, parallel universe. And in a single threaded MPI implementation like ours, we can do this by simply incrementing the communicator context value before the communication begins and decrementing it afterwards. In the more involved implementations, people either define so-called shadow communicators to handle the collective transfer (like MPICH does) or support several contexts per communicator. In any case, the appropriate context is passed somehow to the lower level communication calls that normally have an interface different from the high-level MPI functions.

4.4.5.2 Initialization and Termination
The initialization and termination of the collective subsystem is trivial, as is the respective extension of the internal header file imp.h (not shown).

4.4.5.3 Barrier
This said, a two-process barrier looks like this (see Listing 4.23):

Listing 4.23: Extended subset MPI: two-process barrier (excerpt, see ext/v3/src/coll.c)

```
#include <stdio.h>          // NULL, fprintf(3)

#include "mpi.h"
#include "imp.h"

#include <assert.h>         // assert(3)

#define _MPI_TAG_BARRIER   0
#define _MPI_TAG_BCAST             1
#define _MPI_TAG_REDUCE            2

...
```

```
int MPI_Barrier(MPI_Comm comm)
{
    int rank,size,rc;
    char buf[1];
    MPI_Status stat;

#ifdef DEBUG
    fprintf(stderr,"%d/%d: MPI_Barrier(%d)\n",_mpi_rank,_mpi_size,comm);
#endif

    if ((rc = MPI_Comm_size(comm,&size)) != MPI_SUCCESS ||
        (rc = MPI_Comm_rank(comm,&rank)) != MPI_SUCCESS)
        return rc;

    if (size == 1) {
#ifdef DEBUG
        fprintf(stderr,"%d/%d: -> success\n",_mpi_rank,_mpi_size);
#endif
        return MPI_SUCCESS;
    }
    else if (size > 2) {
        MPI_Comm_call_errhandler(comm,MPI_ERR_INTERN);
        return MPI_ERR_INTERN;
    }

    _mpi_inc_ctxt(comm);

    if (rank == 0) {
        if ((rc = MPI_Send(buf,0,MPI_BYTE,1,_MPI_TAG_BARRIER,comm)) == MPI_SUCCESS)
            rc = MPI_Recv(buf,0,MPI_BYTE,1,_MPI_TAG_BARRIER,comm,&stat);
    }
    else {
        if ((rc = MPI_Recv(buf,0,MPI_BYTE,0,_MPI_TAG_BARRIER,comm,&stat)) ==
MPI_SUCCESS)
            rc = MPI_Send(buf,0,MPI_BYTE,0,_MPI_TAG_BARRIER,comm);
    }

    _mpi_dec_ctxt(comm);

    return rc;
}
```

Messages are tagged using the _MPI_TAG_BARRIER constant to separate them from other collective operations on the same communicator. Even though the MPI standard does not allow more than one blocking collective operation per process at a time (across all communicators it is participating in), MPI users are prone to making this mistake once in a while, and it makes a lot of sense to guard the MPI library from the consequences of this in advance. Note that we consistently let the underlying communication calls handle the eventual error conditions and report them. We also avoid bracketing the debugging output in the case of success that involved some communication, just as we did in the case of layered point-to-point calls like MPI_Send().

You see that we use calls _mpi_ctxt_int() and _mpi_ctxt_dec() to increment and decrement the communicator context. They look as follows (see Listing 4.24):

Listing 4.24: Extended subset MPI: context manipulation (excerpt, see ext/v3/src/comm.c)

```
int _mpi_inc_ctxt(MPI_Comm comm)
{
    assert(_mpi_init);
    assert(comm >= 0 && comm < _mpi_comm_max);

    return _mpi_comm[comm].ctxt++;
}

int _mpi_dec_ctxt(MPI_Comm comm)
{
    assert(_mpi_init);
    assert(comm >= 0 && comm < _mpi_comm_max);

    return _mpi_comm[comm].ctxt--;
}
```

4.4.5.4 Broadcast

After MPI_Barrier(), the MPI_Bcast() is a snap, with the added bonus of passing something tangible rather than empty messages (see Listing 4.25):

Listing 4.25: Extended subset MPI: 2-process broadcast (excerpt, see ext/v3/src/coll.c)

```
int MPI_Bcast(void *buf,int cnt,MPI_Datatype dtype,int root,MPI_Comm comm)
{
    int rank,size,rc;
    MPI_Status stat;

#ifdef DEBUG
    fprintf(stderr,"%d/%d: MPI_Bcast(%p,%d,%d,%d,%d)\n",_mpi_rank,_mpi_size,buf,cnt,
dtype,root,comm);
#endif

    if ((rc = MPI_Comm_size(comm,&size)) != MPI_SUCCESS ||
        (rc = MPI_Comm_rank(comm,&rank)) != MPI_SUCCESS)
        return rc;

    _MPI_CHECK(root >= 0 && root < size,comm,MPI_ERR_ROOT);

    if (size == 1) {
```

```
#ifdef DEBUG
        fprintf(stderr,"%d/%d: -> success\n",_mpi_rank,_mpi_size);
#endif
        return MPI_SUCCESS;
    }
    else if (size > 2) {
        MPI_Comm_call_errhandler(comm,MPI_ERR_INTERN);
        return MPI_ERR_INTERN;
    }

    _mpi_inc_ctxt(comm);

    if (rank == root)
        rc = MPI_Send(buf,cnt,dtype,!root,_MPI_TAG_BCAST,comm);
    else
        rc = MPI_Recv(buf,cnt,dtype,root,_MPI_TAG_BCAST,comm,&stat);

    _mpi_dec_ctxt(comm);

    return rc;
}
```

Unlike MPI_Barrier(), the MPI_Bcast() has an explicit root process that we have to check for correctness once the query calls verify the communicator. Of course, MPI users are prone to using different root values in different MPI processes, but this is an error that is normally ignored by the MPI implementations until they get stuck in the process of communication. There are special, heavy-checking versions of some MPI libraries, however, that do catch this and many other less trivial error conditions if asked to.

4.4.5.5 Reduce

Reductions are a little trickier because they not only send the data around, but also operate on it in the prescribed way. Let's assume that we know how the operation works, and look at the code (see Listing 4.26):

Listing 4.26: Extended subset MPI: two-process reduction (excerpt, see ext/v3/src/coll.c)

```
...

#ifndef CHECK
int _mpi_check_op(MPI_Op op,MPI_Comm comm)
{
    return _MPI_CHECK(op == MPI_MAX || op == MPI_MIN || op == MPI_SUM,comm,
MPI_ERR_OP);
}
#endif
```

...

```
int MPI_Reduce(const void *sbuf,void *rbuf,int cnt,MPI_Datatype dtype,MPI_Op op,int
root,MPI_Comm comm)
{
    int rank,size,rc;
    MPI_Status stat;

#ifdef DEBUG
    fprintf(stderr,"%d/%d: MPI_Reduce(%p,%p,%d,%d,%d,%d,%d)\n",_mpi_rank,_mpi_size,
sbuf,rbuf,cnt,dtype,op,root,comm);
#endif

    if ((rc = MPI_Comm_size(comm,&size)) != MPI_SUCCESS ||
        (rc = MPI_Comm_rank(comm,&rank)) != MPI_SUCCESS)
        return rc;

    _MPI_CHECK_OP(op,comm);
    _MPI_CHECK(root >= 0 && root < size,comm,MPI_ERR_ROOT);

    if (size == 1) {
        _MPI_CHECK(rbuf != NULL,comm,MPI_ERR_BUFFER);
        _MPI_CHECK(sbuf != NULL,comm,MPI_ERR_BUFFER);
        _MPI_CHECK(cnt >= 0,comm,MPI_ERR_COUNT);
        _MPI_CHECK_TYPE(dtype,comm);

        _mpi_type_copy(rbuf,(void *)sbuf,cnt,dtype);

#ifdef DEBUG
        fprintf(stderr,"%d/%d: -> success\n",_mpi_rank,_mpi_size);
#endif
        return MPI_SUCCESS;
    }
    else if (size > 2) {
        MPI_Comm_call_errhandler(comm,MPI_ERR_INTERN);
        return MPI_ERR_INTERN;
    }

    _mpi_inc_ctxt(comm);

    if (rank == root) {
        if ((rc = MPI_Recv(rbuf,cnt,dtype,!root,_MPI_TAG_REDUCE,comm,&stat)) ==
MPI_SUCCESS)
            (*_mpi_op[op])((void *)sbuf,rbuf,&cnt,&dtype);
    }
    else
        rc = MPI_Send(sbuf,cnt,dtype,root,_MPI_TAG_REDUCE,comm);

    _mpi_dec_ctxt(comm);

    return rc;
```

```
}
```

We have to check the operation and the root rank. The rest will be handled by the
underlying communication routines. Note that the rbuf argument is significant only
on the root process.

In the one-process case, we just copy the send buffer into the receive buffer on the
root process, paying attention to the datatype (see Listing 4.27):

Listing 4.27: Extended subset MPI: datatype driven memory copy function (excerpt, see ext/v3/
src/type.c)

```
void *_mpi_type_copy(void *po,void *pi,int cnt,MPI_Datatype dtype)
{
    assert(_mpi_init);
    assert(po != NULL);
    assert(pi != NULL);
    assert(cnt >= 0);
    assert(dtype == MPI_BYTE || dtype == MPI_INT || dtype == MPI_DOUBLE);

    return memcpy(po,pi,cnt*_mpi_type_size(dtype));
}
```

Otherwise, in the two-process case, we call the operation defined as if it were yet
another reduction function in the way foreseen by the MPI standard. In this manner,
if we ever decide to extend the implementation, we will be able to reuse everything
other people have ever done (see Listing 4.28):

Listing 4.28: Extended subset MPI: reduction function (excerpt, see ext/v3/src/coll.c)

```
typedef void MPI_User_function(void *,void *,int *,MPI_Datatype *);

...

//#define FAST_SUM

static void _mpi_sum_byte(char *pi,char *pio,int cnt)
{
#ifdef FAST_SUM
    while (cnt--)
        *pio++ += *pi++;
#else
    int i;

    for (i = 0; i < cnt; i++)
        pio[i] += pi[i];
#endif
}

static void _mpi_sum_int(int *pi,int *pio,int cnt)
```

```
{
#ifdef FAST_SUM
    while (cnt--)
        *pio++ += *pi++;
#else
    int i;

    for (i = 0; i < cnt; i++)
        pio[i] += pi[i];
#endif
}

static void _mpi_sum_double(double *pi,double *pio,int cnt)
{
#ifdef FAST_SUM
    while (cnt--)
        *pio++ += *pi++;
#else
    int i;

    for (i = 0; i < cnt; i++)
        pio[i] += pi[i];
#endif
}

static void _mpi_op_sum(void *pi,void *pio,int *pcnt,MPI_Datatype *pdtype)
{
    switch(*pdtype) {
    case MPI_BYTE:
        _mpi_sum_byte(pi,pio,*pcnt);
        break;
    case MPI_INT:
        _mpi_sum_int(pi,pio,*pcnt);
        break;
    case MPI_DOUBLE:
        _mpi_sum_double(pi,pio,*pcnt);
        break;
    default:
        break;
    }
}

static MPI_User_function *_mpi_op[] = { _mpi_op_max,_mpi_op_min,_mpi_op_sum };

...
```

We describe only the sum operation here, but the rest is defined similarly. We can use pointer arithmetic or conventional indexed arrays. Which option is better is determined by the compiler that may be able to recognize one pattern better than the other and generate a better binary as a result. In principle, one can also use some acceler-ated functions here, e.g., BLAS routines (Netlib.org, 2017) of equal intent.

The interface of the top-level reduction function is defined in a way that allows it to be implemented in Fortran, if the respective compiler does a better job than the C compiler on arithmetic vector operations. Note that this provides a hint as to what the Fortran binding may look like if expressed in terms of the C language. We will rely on this sideway insight when we get to define the Fortran binding for our library (see Section 4.6).

In any case, it may make sense to experiment with the compiler options and look into the resulting assembler output. Note that sometimes you will hardly be able to recognize the original vector sum, so strange is its mapping upon the new com-plicated vector-oriented operations provided by the typical modern processors. In extreme cases, the reduction operations can even be moved out into a separate file, to be compiled at a much higher optimization level than the rest of the code, with addi-tional fancy options that make the compiler forget all decency and dance lambada with the underlying hardware.

4.4.5.6 Allreduce
In its simplest incarnation, the MPI_Allreduce() is mapped upon the MPI_Reduce() and the immediately following MPI_Bcast() for the resulting array (see Listing 4.29):

Listing 4.29: Extended subset MPI: two-process global reduction (excerpt, see ext/v3/src/ coll.c)

```
int MPI_Allreduce(const void *sbuf,void *rbuf,int cnt,MPI_Datatype dtype,MPI_Op
op,MPI_Comm comm)
{
    int rc;

#ifdef DEBUG
    fprintf(stderr,"%d/%d: MPI_Allreduce(%p,%p,%d,%d,%d,%d)\n",_mpi_rank,_mpi_size,
sbuf,rbuf,cnt,dtype,op,comm);
#endif

    if ((rc = MPI_Reduce(sbuf,rbuf,cnt,dtype,op,0,comm)) == MPI_SUCCESS)
        rc = MPI_Bcast(rbuf,cnt,dtype,0,comm);

    return rc;
}
```

Note that here all argument checking is again done by the underlying MPI calls. The operation itself is ready for more than two processes as soon as the component func-tions are capable of handling that situation.

Of course, this is not the fastest way to do a global reduction. There are fancy algorithms that do such an operation directly, cleverly reusing the data available on the spot anyway, but this is an advanced matter we will address in Chapter 5.

4.4.5.7 Gather

The four collective operations we have introduced so far (MPI_Barrier(), MPI_Bcast(), MPI_Reduce(), and MPI_Allreduce()) are usually considered to be the most often used ones (Rabenseifner, 2000). Beyond them, there are only the MPI_Allgather() and the inaptly named MPI_Reduce_scatter()—we will implement only the former. To this end, we will use the same approach as with the MPI_Allreduce(), namely, we will first implement the MPI_Gather() operation, and then express the MPI_Allgather() in terms of the MPI_Gather() and MPI_Bcast(). By the way, we will indeed use MPI_Allgather() somewhat later (see Section 4.5.2.3.2), this is why we go to all this trouble at all.

The MPI_Gather() is a little tricky because it needs to work correctly in several corner cases (see Listing 4.30):

Listing 4.30: Extended subset MPI: two-process gather (excerpt, see ext/v3/src/coll.c)

```
int MPI_Gather(const void *sbuf,int scnt,MPI_Datatype sdtype,void *rbuf,int rcnt,
MPI_Datatype rdtype,int root,MPI_Comm comm)
{
    int rank,size,rc;
    MPI_Status stat;

#ifdef DEBUG
    fprintf(stderr,"%d/%d: MPI_Gather(%p,%d,%d,%p,%d,%d,%d,%d)\n",_mpi_rank,_mpi_size,
sbuf,scnt,sdtype,rbuf,rcnt,rdtype,root,comm);
#endif

    if ((rc = MPI_Comm_size(comm,&size)) != MPI_SUCCESS ||
        (rc = MPI_Comm_rank(comm,&rank)) != MPI_SUCCESS)
        return rc;

    _MPI_CHECK(root >= 0 && root < size,comm,MPI_ERR_ROOT);

    if (size == 1) {
        _MPI_CHECK(sbuf != NULL,comm,MPI_ERR_BUFFER);
        _MPI_CHECK(scnt >= 0,comm,MPI_ERR_COUNT);
        _MPI_CHECK_TYPE(sdtype,comm);
        _MPI_CHECK(rbuf != NULL,comm,MPI_ERR_BUFFER);
        _MPI_CHECK(rcnt >= 0,comm,MPI_ERR_COUNT);
        _MPI_CHECK_TYPE(rdtype,comm);

        if (_mpi_type_copy2(rbuf,rcnt,rdtype,(void *)sbuf,scnt,sdtype) == NULL) {
            MPI_Comm_call_errhandler(comm,MPI_ERR_INTERN);
```

```
            return MPI_ERR_INTERN;
        }

#ifdef DEBUG
        fprintf(stderr,"%d/%d: -> success\n",_mpi_rank,_mpi_size);
#endif
        return MPI_SUCCESS;
    }
    else if (size > 2) {
        MPI_Comm_call_errhandler(comm,MPI_ERR_INTERN);
        return MPI_ERR_INTERN;
    }

    _mpi_inc_ctxt(comm);

    if (rank == root)
    {
        int rext;

        MPI_Type_size(rdtype,&rext);
        rext *= rcnt;

        if ((rc = MPI_Recv((char *)rbuf + rext*(!root),rcnt,rdtype,!root,
_MPI_TAG_GATHER,comm,&stat)) == MPI_SUCCESS) {
            if (_mpi_type_copy2((char *)rbuf + rext*root,rcnt,rdtype,(void *)
sbuf,scnt,sdtype) == NULL) {
                _mpi_dec_ctxt(comm);
                MPI_Comm_call_errhandler(comm,MPI_ERR_INTERN);
                return MPI_ERR_INTERN;
            }
        }
    }
    else
        rc = MPI_Send(sbuf,scnt,sdtype,root,_MPI_TAG_GATHER,comm);

    _mpi_dec_ctxt(comm);

    return rc;
}
```

- First, it needs to copy the contents of the send buffer to the receive buffer if called on a communicator with only one process. In this, the MPI_Gather() is similar to the MPI_Reduce(), with the added complexity of using two buffer descriptions rather than one. Hence, we need a new function, called _mpi_type_copy2(), to cover this scenario. Note that here we need to check all arguments, because we cannot rely on any underlying MPI function to do this for us.
- Second, in case of more than one process, the MPI_Gather() on the root process again needs to copy its send buffer into the portion of the receive buffer corre-

sponding to the root process. The aforementioned function _mpi_type_copy2()
helps us out here, too.

We implement this internal function in a limited manner, enforcing full similarity of
both buffer descriptions for now. We can revisit this matter if it ever becomes an issue
(see Listing 4.31):

Listing 4.31: Extended subset MPI: two-datatype driven memory copy function (excerpt, see ext/
v3/src/type.c)

```
void *_mpi_type_copy2(void *po,int ocnt,MPI_Datatype odtype,void *pi,int icnt,
MPI_Datatype idtype)
{
    assert(_mpi_init);
    assert(po != NULL);
    assert(pi != NULL);
    assert(icnt >= 0);
    assert(idtype == MPI_BYTE || idtype == MPI_INT || idtype == MPI_DOUBLE);
    assert(ocnt == icnt);
    assert(odtype == idtype);

    return memcpy(po,pi,icnt*_mpi_type_size(idtype));
}
```

Note that the rbuf argument and the associated receive buffer description are significant only on the root process.

4.4.5.8 Allgather
The MPI_Allgather() becomes a snap once the MPI_Gather() is done (see Listing 4.32):

Listing 4.32: Extended subset MPI: two-process global gather (excerpt, see ext/v3/src/coll.c)

```
int MPI_Allgather(const void *sbuf,int scnt,MPI_Datatype sdtype,void *rbuf,int
rcnt,MPI_Datatype rdtype,MPI_Comm comm)
{
    int size,rc;

#ifdef DEBUG
    fprintf(stderr,"%d/%d: MPI_Allgather(%p,%d,%d,%p,%d,%d,%d)\n",_mpi_rank,
_mpi_size,sbuf,scnt,sdtype,rbuf,rcnt,rdtype,comm);
#endif

    if ((rc = MPI_Comm_size(comm,&size)) != MPI_SUCCESS)
        return rc;

    if ((rc = MPI_Gather(sbuf,scnt,sdtype,rbuf,rcnt,rdtype,0,comm)) == MPI_SUCCESS)
```

```
        rc = MPI_Bcast(rbuf,rcnt*size,rdtype,0,comm);

    return rc;
}
```

That's all. One more bit of code had to be changed compared to Section 4.3, though: the communicator and target handle arguments of the _mpi_check_*() functions and the respective macros were swapped, so that the calling sequences of all assertion and checking calls had a similar form: expression first, communicator next, if any.

This kind of minor code reshuffling happens now and then despite all forethought. In fact, I love programming in part due to this feeling of omnipotence: whatever has been done, can be improved later on. There is no qualitative jump here characteristic of real life which does have irreversible processes and irretrievable losses. All you need to do things right in programming is courage—and enough time for yet another release cycle.

4.4.6 Testing

Testing datatypes and then also collective operations required modification of the existing test programs, an addition of an extra test program, as well as further restructuring of the respective Makefile.

4.4.6.1 Unified Collective Benchmark

Instead of adding collectives to the unified point-to-point benchmark we have been using so far, I decided to start a new unified collective benchmark (see Listing 4.33):

Listing 4.33: Extended subset MPI: unified collective benchmark (see ext/v3/test/c.c)

```
#include <stdlib.h>        // malloc(3)
#include <stdio.h>         // NULL, printf(3)

#include "mpi.h"
#include "b.h"

#ifndef ROOT
#define ROOT    0
#endif

int main(int argc,char **argv)
{
    double t;
#ifndef BARRIER
    TYPE *b1,*b2 = NULL;
#endif
```

```
    int i,imax = IMAX,r,s;
#ifndef BARRIER
    int l,lmax = LMAX,kmax;
#ifdef CHECK
    int k,m;
#endif
#endif

    MPI_Init(&argc,&argv);

    MPI_Comm_size(MPI_COMM_WORLD,&s);
    MPI_Comm_rank(MPI_COMM_WORLD,&r);

#ifndef BARRIER
    b1 = malloc(lmax*sizeof(TYPE));
#if defined(GATHER) || defined(ALLGATHER)
#ifdef GATHER
    if (r == ROOT)
#endif
    b2 = malloc((kmax = s*lmax)*sizeof(TYPE));
#else
#ifdef REDUCE
    if (r == ROOT)
#endif
    b2 = malloc((kmax = lmax)*sizeof(TYPE));
#endif

#ifdef CHECK
    for (k = 0; k < lmax; k++)
        b1[k] = (TYPE)k;

#if defined(GATHER) || defined(REDUCE)
    if (r == ROOT)
#endif
    for (k = 0; k < kmax; k++)
        b2[k] = (TYPE)(kmax - k);
#endif

    for (l = 0; l <= lmax; l = (l) ? l << 1 : 1) {
#endif

    t = MPI_Wtime();
    for (i = 0; i < imax; i++) {
#ifdef BARRIER
        MPI_Barrier(MPI_COMM_WORLD);
#endif
#ifdef BCAST
        if (r == ROOT)
            MPI_Bcast(b1,l,DTYPE,0,MPI_COMM_WORLD);
        else {
```

```
            MPI_Bcast(b2,l,DTYPE,0,MPI_COMM_WORLD);
#ifdef CHECK
        for (k = 0; k < l; k++)
            if (b2[k] != b1[k])
                fprintf(stderr,"ERROR: r = %d, l = %d, i = %d, b2[%d] = "FORMAT",
b1[%d] = "FORMAT"\n",r,l,i,k,b2[k],k,b1[k]);
#endif
        }
#endif
#ifdef REDUCE
        MPI_Reduce(b1,b2,l,DTYPE,MPI_SUM,ROOT,MPI_COMM_WORLD);
#ifdef CHECK
        if (r == ROOT)
            for (k = 0; k < l; k++)
                if (b2[k] != (TYPE)(s*b1[k]))
                    fprintf(stderr,"ERROR: r = %d, l = %d, i = %d, b2[%d] = "FORMAT",
b1[%d] = "FORMAT"\n",r,l,i,k,b2[k],k,b1[k]);
#endif
#endif
#ifdef ALLREDUCE
        MPI_Allreduce(b1,b2,l,DTYPE,MPI_SUM,MPI_COMM_WORLD);
#ifdef CHECK
        for (k = 0; k < l; k++)
            if (b2[k] != (TYPE)(s*b1[k]))
                fprintf(stderr,"ERROR: r = %d, l = %d, i = %d, b2[%d] = "FORMAT",
b1[%d] = "FORMAT"\n",r,l,i,k,b2[k],k,b1[k]);
#endif
#endif
#ifdef GATHER
        MPI_Gather(b1,l,DTYPE,b2,l,DTYPE,ROOT,MPI_COMM_WORLD);
#ifdef CHECK
        if (r == ROOT)
            for (m = 0, kmax = s*l; m < kmax; m += l)
                for (k = 0; k < l; k++)
                    if (b2[k + m] != b1[k])
                        fprintf(stderr,"ERROR: r = %d, l = %d, i = %d, b2[%d] =
"FORMAT", b1[%d] = "FORMAT"\n",r,l,i,k + m,b2[k + m],k,b1[k]);
#endif
#endif
#ifdef ALLGATHER
        MPI_Allgather(b1,l,DTYPE,b2,l,DTYPE,MPI_COMM_WORLD);
#ifdef CHECK
        for (m = 0, kmax = s*l; m < kmax; m += l)
            for (k = 0; k < l; k++)
                if (b2[k + m] != b1[k])
                    fprintf(stderr,"ERROR: r = %d, l = %d, i = %d, b2[%d] = "FORMAT",
b1[%d] = "FORMAT"\n",r,l,i,k + m,b2[k + m],k,b1[k]);
#endif
#endif
        }
```

```
    t = MPI_Wtime() - t;

#ifndef CHECK
#ifdef BARRIER
    printf("r = %d\titers = %-8d\ttime = %-12.6g\t"
            "lat = %-12.6g\n",r,i,t,t/i);
#else
    printf("r = %d\tbytes = %-8d\titers = %-8d\ttime = %-12.6g\t"
            "lat = %-12.6g\n",r,(int)(l*sizeof(TYPE)),i,t,t/i);
#endif
#endif

#ifndef BARRIER
    }
#endif

    MPI_Finalize();

    return 0;
}
```

The source code is self-explanatory. It grew step by step, providing testbed for the operations that were being added to our MPI subset. If you do not feel comfortable with conditional compilation, which is ubiquitous in system programming in general and the MPI internals in particular, copy the source code into your favorite text editor and cut those bits that get in the way. With some practice, you will learn to do this mentally while reviewing the code.

One way or another, you will see that depending on the target MPI collective function, data buffers are allocated or not, filled out by data items or not, transferred or not, and then checked if necessary for correctness. Note that we have to go to the extra trouble of allocating and checking longer receive buffers in the case of MPI_Gather() and MPI_Allgather() calls. We do not output bandwidth anywhere, because its definition depends on the communication pattern used inside the operation. We rely on latency instead, which is directly comparable to that of the point-to-point operations, minding the traditional halving of the ping-pong latency. Likewise, we do not output any message length for the barrier timing, because this would make no sense apart from unifying the benchmark output format across all collectives.

Common definitions, used also in the appropriately updated unified point-to-point benchmark b.c (not shown), adapt the runtime options to the size of the elementary datatype items involved (see Listing 4.34):

Listing 4.34: Extended subset MPI: unified benchmark header (see ext/v3/test/b.h)

```
#ifndef UNIT
#define UNIT 1
#endif

#if UNIT == 1
#define TYPE char
#define DTYPE MPI_BYTE
#define FORMAT "%02x"
#elif UNIT == 4
#define TYPE int
#define DTYPE MPI_INT
#define FORMAT "%04x"
#elif UNIT == 8
#define TYPE double
#define DTYPE MPI_DOUBLE
#define FORMAT "%-12.6g"
#endif

#ifndef IMAX
#define IMAX        10000
#endif

#ifndef LMAX
#define LMAX        8*1024*1024/sizeof(TYPE)
//#define LMAX       32*1024*1024/sizeof(TYPE)
#endif
```

The Makefile was modified accordingly (not shown). In order to make the library and the test build procedures less interdependent, the CFLAGS and other related options were moved into the respective Makefiles out of the make.header file that contains only the path information now (not shown).

4.4.6.2 Issue Fixing and Backporting

During testing I was able to detect and fix an issue that led to intermittent failures, especially on the newly introduced collective operations. I would not have mentioned this if it were not important not only for this book, but also for the development methodology.

Benchmarks would fail once in a while, either hanging or breaking, reporting strange sightings of data having been corrupted inside the channel. Looking into another place first on pure foreboding, since I have always fretted a little about the short supply of the requests in the queues (still only one send, one posted receive, and one unexpected receive at most at any given moment), I finally homed in on the progress engine. This required some small changes to the error handling and such (not shown).

In the end, it was an old wrong decision biting back in a way that I knew was possible, but which I had discounted as unlikely. If you may remember, when dealing with sockets (see Section 3.3.2.3), I bet that the first 8 bytes—the message header— would always be transferred in one piece. However, it turned out that once in a while this does not happen! In fact, I have seen all kinds of splits, from 1 to 7 bytes being transferred in one piece, with the rest hanging about somewhere in the channel for reasons of their own.

In the end, the required change to the sending part of the progress engine was rather nontrivial (see Listing 4.35, cf. Listing 3.32):

Listing 4.35: Extended subset MPI: send progress fix (excerpt, see ext/v3/src/xfer.c)

```
static int _mpi_progress_send(MPI_Request req)
{
    char *buf;
    struct iovec iov[2];
    int m,l = 0,n = 2*sizeof(int);

    if (req->flag)
        return MPI_SUCCESS;

    iov[0].iov_base = &(req->tag);
    iov[0].iov_len  = n;
    iov[1].iov_base = req->buf;
    iov[1].iov_len  = req->cnt;

again:
    switch ((m = writev(_mpi_fds[req->rank],iov,2))) {
    case -1:
#ifdef DEBUG
        fprintf(stderr,"%d/%d: write tag: %s\n",_mpi_rank,_mpi_size,strerror(errno));
#endif
        if (errno == EWOULDBLOCK || errno == EAGAIN) {
            if (1)
                goto again;

            return MPI_SUCCESS;
        }
        return MPI_ERR_OTHER;
    case 0:
#ifdef DEBUG
        fprintf(stderr,"%d/%d: write tag: unexpected end of file\n",_mpi_rank,
_mpi_size);
#endif
        return MPI_ERR_OTHER;
    default:
        if (m >= n) {
            m -= n;
```

```
            break;
        }
#ifdef DEBUG
        fprintf(stderr,"%d/%d: write tag: partial transter, done = %d\n",
_mpi_rank,_mpi_size,m);
#endif
        l += m;
        n -= m;
        iov[0].iov_base = ((char *)&(req->tag)) + l;
        iov[0].iov_len  = n;
        goto again;
    }
```
...

In essence, if we get caught with our pants down midway through the header, we need to repeat the reading attempt, hopefully getting the remaining bytes and possibly more this time round. This is relatively clear and comparable to the handling of the rest of the message body that did not require any changes this time.

What is remarkable, however, is that we need to deal with the progress engine getting stuck and returning control back to the caller if there is nothing to be read on the second attempt! This is what the test on the nonzero value of variable l does above. Indeed, if we fail to read a complete message header, the probability is rather high that an immediate attempt to read the rest of it will lead to the EWOULDBLOCK (or equivalent EAGAIN) error condition. However, if we are midway through the header, we have high hopes of the data arriving if not immediately, then relatively soon. This is why we can afford a very tight goto based loop basically polling for more data to come, at the risk of hanging the program here in case something goes really terribly wrong. So this is not a 100% clean solution still, but we bet it will work. You've been warned.

The corresponding part of the receive progress was comparable but less tricky because there we first get the header, and only then, having checked it and decided on the way of handling the data, proceed to the message body (see Listing 4.36, cf. Listing 3.33):

Listing 4.36: Extended subset MPI: send progress fix (excerpt, see ext/v3/src/xfer.c)

```
static int _mpi_progress_recv(MPI_Request req)
{
    char *buf;
    int hdr[2],m,l = 0,n = 2*sizeof(int),unexp = 0;

    if (req->flag)
        return MPI_SUCCESS;

again:
```

```
    switch ((m = read(_mpi_fds[req->rank],((char *)hdr) + 1,n))) {
    case -1:
#ifdef DEBUG
        fprintf(stderr,"%d/%d: read itag: %s\n",_mpi_rank,_mpi_size,strerror(errno));
#endif
        if (errno == EWOULDBLOCK || errno == EAGAIN) {
            if (1)
                goto again;

            return MPI_SUCCESS;
        }
        return MPI_ERR_OTHER;
    case 0:
#ifdef DEBUG
        fprintf(stderr,"%d/%d: read itag: unexpected end of file\n",_mpi_rank,
_mpi_size);
#endif
        return MPI_ERR_OTHER;
    default:
        if (m == n)
            break;
#ifdef DEBUG
        fprintf(stderr,"%d/%d: read itag: partial transfer, done = %d\n",_mpi_rank,
_mpi_size,m);
#endif
        l += m;
        n -= m;
        goto again;
    }

...
```

Now, should we backport this important fix into the sockets code developed in Chapter 3? We have never seen this issue there, perhaps we never loaded the system well enough then, or, possibly, because some system update in the meantime changed certain timings, and the hidden issue only now became apparent. In any case, a backport appears warranted, as does yet another reiteration of the rule that work done sloppily will always come back to bite you.

4.4.7 Benchmarking

Benchmarking done using the unified point-to-point and unified collective benchmarks was unremarkable, confirming the reasonable estimates of the expected performance based on the understanding of the point-to-point patterns we have investigated sufficiently well in the past, and the relative slowness of the sockets.

One observation worth mentioning is that, just as before, process pinning does make a lot of difference, especially for the reduction operations. Here is a short summary of the results observed for the operations involved (see Table 4.1):

Table 4.1: Two-process benchmarks: minimum latency at 16 bytes, 16 KiB, 128 KiB, and 4 MiB (seconds)

Pattern	16 bytes	16 KiB	128 KiB	4 MiB
Ping-pong	2.30e-06	4.23e-06	1.36e-05	0.00063
Ping-ping	3.00e-06	4.95e-06	1.83e-05	0.00145
Pong-pong	3.28e-06	5.19e-06	1.89e-05	0.00118
Barrier*	4.68e-06			
Bcast	2.18e-06	3.64e-06	1.75e-05	0.00055
Reduce	**0.65e-06**	7.12e-06	4.05e-05	0.00169
Allreduce	4.64e-06	0.11e-06	5.09e-05	0.00221

* Barrier is special: it's shown as a 16-byte result even though it uses zero-sized messages

Note the Reduce result for 16 bytes (in **bold**): here the process pinning did its usual trick. Other than this anomaly, all numbers look reasonable and basically represent the respective communication pattern plus the overhead for the arithmetic operation, if involved.

4.5 Subset Extension 2: Communicators and Collectives Revisited

Well, it's time to add:
– Communicator management functions
– Collective operations for any number of processes

4.5.1 Extension Directions

These two extension paths are again largely independent of each other.

4.5.2 Communicator Management Revisited

Group and communicator management is a huge part of the MPI standard. We will not target the whole of it and instead only show with a set of examples how to approach this largely trivial but still important job. Let's add the following features to the underlying MPI subset:

– Functions `MPI_Comm_dup()` and `MPI_Comm_split()`
– Constants `MPI_UNDEFINED` and `MPI_COMM_NULL`

The aforementioned constants are necessary to cover all possible argument values of the communicator creation calls.

4.5.2.1 Design Considerations
We use a pointer to the rank array inside every communicator, and every externally visible communicator has a corresponding internal data structure. Instead of introducing reference counting that helps reduce memory consumption by keeping track of the multiply used data structures, whether rank arrays or something else, we will cut a couple corners and simply copy data over. This is why, in part, we do not include any communicator deletion calls into the subset.

4.5.2.2 Communicator Duplication
Communicator duplication is a global operation that normally requires communication in order to figure out a safe context to use. Indeed, some, and not all, of the processes involved in the old communicator may also participate in other communicators. If we were to inadvertently reuse one of their contexts, the MPI library would no longer be able to differentiate between the messages sent within different communicators, and a complete chaos would ensue.

With this in mind, communicator duplication is rather trivial (see Listing 4.37):

Listing 4.37: Extended subset MPI: communicator duplication (see `ext/v3/src/comm.c`)

```
static int _mpi_comm_dup(MPI_Comm comm,MPI_Comm *pcomm)
{
    int rc,ctxt_max;

    assert(_mpi_init);
    assert(comm >= 0 && comm < _mpi_comm_max);
    assert(pcomm != NULL);

    if (_mpi_comm_max == _MPI_COMM_MAX)
        return MPI_ERR_INTERN;

    if ((rc = MPI_Allreduce(&_mpi_ctxt_max,&ctxt_max,1,MPI_INT,MPI_MAX,comm)) !=
MPI_SUCCESS)
        return rc;

    _mpi_comm[_mpi_comm_max] = _mpi_comm[comm];
    _mpi_comm[_mpi_comm_max].ctxt = ctxt_max;

    *pcomm = _mpi_comm_max++;
```

```
    _mpi_ctxt_max = ctxt_max + _MPI_CTXT_INC;

    return MPI_SUCCESS;
}

int MPI_Comm_dup(MPI_Comm comm,MPI_Comm *pcomm)
{
    int rc;

#ifdef DEBUG
    fprintf(stderr,"%d/%d: MPI_Comm_dup(%d,%p)\n",_mpi_rank,_mpi_size,comm,pcomm);
#endif

    _MPI_CHECK(_mpi_init,MPI_COMM_WORLD,MPI_ERR_OTHER);
    _MPI_CHECK_COMM(comm);
    _MPI_CHECK(pcomm != NULL,comm,MPI_ERR_ARG);

    if ((rc = _mpi_comm_dup(comm,pcomm)) != MPI_SUCCESS)
        _mpi_call_errhandler(comm,rc,__FILE__,__LINE__);

#ifdef DEBUG
    fprintf(stderr,"%d/%d: -> success, *pcomm = %d\n",_mpi_rank,_mpi_size,*pcomm);
#endif
    return MPI_SUCCESS;
}
```

Taking the maximum of all maximum contexts across the processes participating in
the old communicator is a standard way out of the situation described above. A call
to MPI_Allreduce() does this for us. Note that if the old communicator has only one
process in it, calling the MPI_Allreduce() becomes quite a bit of overkill, and we may
want to handle this corner case explicitly later on.

4.5.2.3 Communicator Splitting
This will be the heaviest lifting so far, and one of the most complicated calls in the
MPI library as far as the algorithmic aspect is concerned. Let's deal with it in reverse
fashion: first the external function, then the internal one, and finally the corresponding paraphernalia.

4.5.2.3.1 External Function
The external function MPI_Comm_split() follows the well-known pattern (see Listing
4.38):

Listing 4.38: Extended subset MPI: communicator splitting (excerpt, see ext/v3/src/comm.c)

```
int MPI_Comm_split(MPI_Comm comm,int col,int key,MPI_Comm *pcomm)
```

```
{
    int rc;

#ifdef DEBUG
    fprintf(stderr,"%d/%d: MPI_Comm_split(%d,%d,%d,%p)\n",_mpi_rank,_mpi_size,
comm,col,key,pcomm);
#endif

    _MPI_CHECK(_mpi_init,MPI_COMM_WORLD,MPI_ERR_OTHER);
    _MPI_CHECK_COMM(comm);
    _MPI_CHECK(col >= 0 || col == MPI_UNDEFINED,comm,MPI_ERR_ARG);
    _MPI_CHECK(pcomm != NULL,comm,MPI_ERR_ARG);

    if ((rc = _mpi_comm_split(comm,col,key,pcomm)) != MPI_SUCCESS)
        _mpi_call_errhandler(comm,rc,__FILE__,__LINE__);

#ifdef DEBUG
    fprintf(stderr,"%d/%d: -> success, *pcomm = %d\n",_mpi_rank,_mpi_size,*pcomm);
#endif
    return MPI_SUCCESS;
}
```

Note that this time, since we are dealing with externally prescribed values, we check the color value for validity, explicitly including the MPI_UNDEFINED value, which is negative by the MPI standard. The key value, to the contrary, is arbitrary.

4.5.2.3.2 Internals
This is the first and hopefully the last internal function that does need some comments inside (see Listing 4.39):

Listing 4.39: Extended subset MPI: communicator splitting, internal function (excerpt, see ext/v3/src/comm.c)

```
struct _mpi_comm_split_s {
    int col,key,rank,grank,ctxt_max;
};

int _mpi_comm_split_cmp(const struct _mpi_comm_split_s **h1,const struct
_mpi_comm_split_s **h2)
{
    int keydiff = (*h1)->key - (*h2)->key;

    if (keydiff)
        return keydiff;

    return (*h1)->rank - (*h2)->rank;
}
```

```
int _mpi_comm_split(MPI_Comm comm,int col,int key,MPI_Comm *pcomm)
{
    int rc,i,n,ncol = 0,ctxt_max = 0;
    struct _mpi_comm_split_s ent,*pent,**hent;

    assert(_mpi_init);
    assert(comm >= 0 && comm < _mpi_comm_max);
    assert(col >= 0 || col == MPI_UNDEFINED);
    assert(pcomm != NULL);

// spread information to all processes involved
    if ((pent = malloc(_mpi_comm[comm].size*sizeof(struct _mpi_comm_split_s))) ==
NULL)
        return MPI_ERR_OTHER;

    ent.col      = col;
    ent.key      = key;
    ent.rank     = _mpi_comm[comm].rank;
    ent.grank    = _mpi_rank;
    ent.ctxt_max = _mpi_ctxt_max;
#ifdef DEBUG
    fprintf(stderr,"%d/%d: ent = { %d,%d,%d,%d,%d }\n",_mpi_rank,_mpi_size,ent.col,
ent.key,ent.rank,ent.grank,ent.ctxt_max);
#endif
    if ((rc = MPI_Allgather(&ent,5,MPI_INT,pent,5,MPI_INT,comm)) != MPI_SUCCESS) {
        free(pent);

        return rc;
    }
#ifdef DEBUG
    for (i = 0; i < _mpi_comm[comm].size; i++)
        fprintf(stderr,"%d/%d: pent[%d] = { %d,%d,%d,%d,%d }\n",_mpi_rank,_mpi_size,
i,pent[i].col,pent[i].key,pent[i].rank,pent[i].grank,pent[i].ctxt_max);
#endif

// bail out if not a member of the new communicator
    if (col == MPI_UNDEFINED) {
        *pcomm = MPI_COMM_NULL;

        free(pent);

        return MPI_SUCCESS;
    }

// otherwise prepare for communicator creation
    if (_mpi_comm_max == _MPI_COMM_MAX) {
        free(pent);

        return MPI_ERR_INTERN;
```

```
    }

// count members of the same color (must be at least oneself)
    for (ncol = 0,i = 0; i < _mpi_comm[comm].size; i++)
        if (pent[i].col == col)
            ncol++;

    assert(ncol > 0);

// create a singleton communicator for oneself if alone
    if (ncol == 1) {
        _mpi_comm[_mpi_comm_max].prank = &_mpi_rank;

        _mpi_comm[_mpi_comm_max].size = 1;
        _mpi_comm[_mpi_comm_max].rank = 0;
        _mpi_comm[_mpi_comm_max].ctxt = _mpi_ctxt_max;
        _mpi_comm[_mpi_comm_max].errh = _mpi_comm[comm].errh;

        *pcomm = _mpi_comm_max++;
        _mpi_ctxt_max += _MPI_CTXT_INC;

        free(pent);

        return MPI_SUCCESS;
    }

// otherwise fill out the handle array and find out the max free context
    if ((hent = malloc(ncol*sizeof(struct _mpi_comm_split_s **))) == NULL) {
        free(pent);

        return MPI_ERR_OTHER;
    }

    for (n = 0,i = 0; i < _mpi_comm[comm].size; i++)
        if (pent[i].col == col) {
            if (ctxt_max < pent[i].ctxt_max)
                ctxt_max = pent[i].ctxt_max;

            hent[n++] = pent + i;
            if (n == ncol)
                break;
        }

// sort the handle array
    qsort(hent,ncol,sizeof(struct _mpi_comm_split_s **),(int (*)(const void *, const
void *))_mpi_comm_split_cmp);

// create a new communicator
    if ((_mpi_comm[_mpi_comm_max].prank = (int *)malloc(ncol*sizeof(int))) == (int *)
NULL) {
```

```
        free(hent);
        free(pent);

        return MPI_ERR_OTHER;
    }

    for (i = 0; i < ncol; i++)
        _mpi_comm[_mpi_comm_max].prank[i] = hent[i]->grank;

    _mpi_comm[_mpi_comm_max].size = ncol;
    for (i = 0; i < ncol; i++)
        if (_mpi_comm[comm].rank == hent[i]->rank) {
            _mpi_comm[_mpi_comm_max].rank = i;
            break;
        }

    _mpi_comm[_mpi_comm_max].ctxt = ctxt_max;
    _mpi_comm[_mpi_comm_max].errh = _mpi_comm[comm].errh;

    *pcomm = _mpi_comm_max++;
    _mpi_ctxt_max = ctxt_max + _MPI_CTXT_INC;

    free(hent);
    free(pent);

    return MPI_SUCCESS;
}
```

First of all, we spread the information available across all processes involved. The normal way of doing this is the MPI_Allgather(). We could have used a series of tricky MPI_Allreduce() operations to the same end, but this would make things unnecessarily messy and, if implemented naively, cost us several collective calls (and they are expensive). In fact, I considered this way before deciding that implementing the gather operations was a better option.

Then we deal with the degenerate cases in which the process does not belong to any new communicator and thus returns the MPI_COMM_NULL to the caller, or is an isolated singleton, in which case we basically create yet another MPI_COMM_SELF equivalent for it, inheriting the error handler from the old communicator.

Otherwise we proceed to create a non-trivial new communicator that needs to have its ranks sorted by the key, with the ties broken by the rank in the old communicator. Instead of sorting the original data directly, we sort an array of handles that point to the pertinent parts of the data associated with the respective color. This allows the use of the qsort(3) function. The respective comparison function looks as follows (see Listing 4.40):

Listing 4.40: Extended subset MPI: communicator splitting, comparison function (excerpt, see ext/v3/src/comm.c)

```
int _mpi_comm_split_cmp(const struct _mpi_comm_split_s **h1,const struct
_mpi_comm_split_s **h2)
{
    int keydiff = (*h1)->key - (*h2)->key;

    if (keydiff)
        return keydiff;

    return (*h1)->rank - (*h2)->rank;
}
```

We rely on the field rank to break the ties. However, we could have just as well, here and in the internal function, where we restore the correspondence between the process data entries and the ranks in the new communicator, saved on this data item and used the addresses contained in the respective handles themselves or, in the latter case, even the global ranks, respectively.

Another possible optimization is the local calculation of the global ranks rather than spreading this information in the MPI_Allgather() call. Whether this makes sense is a good question. I decided to go the lazy way, because here we simply send out the data we readily have on each process, and sending one more integer should not make a bit of difference from the performance point of view. The more natural resulting layout of the main data structure comes in as a bonus.

Finally, note that we describe a struct using a sequence of elementary datatypes. Although not strictly following the MPI standard, this should work in any reasonable C compiler.

Does this function look horrible? A little bit, yes. However, as we will find out later on, it works. Once there, we can revisit it and beautify it by adding some nice looking internal calls and more structure. There are several methods of addressing hard work. Mine is to make things work first, and make things look better once there, as necessary. Many books on programming will tell you a different story about design and forethought and structure and patterns and so on, but believe me, most of the time people do what I have just done. They only present the end result rather than the intermediate steps, for whatever reason.

4.5.3 Collective Operations Revisited

Well, we are almost done now. What remains is to extend the collectives to cover more than two processes, and not explode the package limited by at most one posted send, one posted receive, and one unexpected receive request at a time.

4.5.3.1 Design Considerations

There is at least one communication pattern that uses just as many pending transfer operations as we can support: namely, the *ring*. They are good at handling very large process numbers, especially if latency is not an issue and if the communication can be *pipelined* (cf. double buffering). In this chapter, we will stick to the rings and extend our knowledge of the collective algorithms substantially in Chapter 5.

Let's go through the operations provided by the current MPI subset one by one in the order in which I extended them. This seemed to be the most logical way, and you will soon see why.

4.5.3.2 Broadcast

This operation consists of only the fan-out from the root to the leaves. If a ring pattern is used, root rank 0 sends its buffer contents to the first leaf, that in turn addresses the second leaf, and so on, until the last leaf with MPI rank N − 1 needs to synchronize with the root, so that the latter does not run too far away while the leaves are busy digesting its message. We need to do this only because we have so few requests in the queues. Otherwise we could have allowed the root to run away and start another operation, be it point-to-point or collective, like yet another MPI_Bcast(), which would be typical of benchmarking, for example.

Here's what it looks like (see Listing 4.41, cf. Listing 4.25):

Listing 4.41: Extended subset MPI: broadcast (excerpt, see ext/v4/src/coll.c)

```
int MPI_Bcast(void *buf,int cnt,MPI_Datatype dtype,int root,MPI_Comm comm)
{
    int rank,size,rc = MPI_SUCCESS;
    MPI_Status stat;

#ifdef DEBUG
    fprintf(stderr,"%d/%d: MPI_Bcast(%p,%d,%d,%d,%d)\n",_mpi_rank,_mpi_size,buf,cnt,
dtype,root,comm);
#endif

    if ((rc = MPI_Comm_size(comm,&size)) != MPI_SUCCESS ||
        (rc = MPI_Comm_rank(comm,&rank)) != MPI_SUCCESS)
        return rc;

    _MPI_CHECK(root >= 0 && root < size,comm,MPI_ERR_ROOT);

    if (cnt == 0) {
        _MPI_CHECK(buf != NULL,comm,MPI_ERR_BUFFER);
        _MPI_CHECK_TYPE(dtype,comm);

#ifdef DEBUG
        fprintf(stderr,"%d/%d: -> success\n",_mpi_rank,_mpi_size);
```

```
#endif
        return MPI_SUCCESS;
    }

    if (size == 1) {
        _MPI_CHECK(buf != NULL,comm,MPI_ERR_BUFFER);
        _MPI_CHECK(cnt >= 0,comm,MPI_ERR_COUNT);
        _MPI_CHECK_TYPE(dtype,comm);

#ifdef DEBUG
        fprintf(stderr,"%d/%d: -> success\n",_mpi_rank,_mpi_size);
#endif
        return MPI_SUCCESS;
    }

    _mpi_inc_ctxt(comm);

    if (size == 2) {
        if (rank == root)
            rc = MPI_Send(buf,cnt,dtype,!root,_MPI_TAG_BCAST,comm);
        else
            rc = MPI_Recv(buf,cnt,dtype,root,_MPI_TAG_BCAST,comm,&stat);
    }
    else {
        if (rank == root) {
            if ((rc = MPI_Send(buf,cnt,dtype,(rank + 1)%size,_MPI_TAG_BCAST,comm)) ==
MPI_SUCCESS)
                rc = MPI_Recv(buf,cnt,dtype,(size + rank - 1)%size,_MPI_TAG_BCAST,
comm,&stat);
        }
        else {
            if ((rc = MPI_Recv(buf,cnt,dtype,(size + rank - 1)%size,_MPI_TAG_BCAST,
comm,&stat)) == MPI_SUCCESS)
                rc = MPI_Send(buf,cnt,dtype,(rank + 1)%size,_MPI_TAG_BCAST,comm);
        }
    }

    _mpi_dec_ctxt(comm);

    return rc;
}
```

Apart from a bit of code reshuffling to handle the argument checking and the corner cases of the one- and two-process communicators in a slightly stronger manner, the primary addition is contained at the end of the function body. To not lose the work done before, we just keep the one- and two-process specific code. What changes in the multiprocess case is the way in which we address the next and the previous MPI ranks. We will reuse these formulae throughout this section. We could have even embedded them in macros, but upon some contemplation I decided against this, since the result would be less instructive.

One interesting point here is that for simplicity, we just send the resulting buffer back to the root once the ring has been gone through. In principle, this is superfluous and runs a bit against the mandate of the standard that expects the buffer to remain untouched at root. We can also send only a zero-sized message here, or even add more requests to the queue and then bang on closing the ring, as mentioned before.

Note that the ring is built in such a way that there likely are not so many unexpected messages: first the processes other than the root get into the MPI_Recv(), then the root kickstarts the exchange by the first MPI_Send() operation, and hangs itself on the receive side until the data token passes through the ring. In other words, this is a self-synchronizing structure that gets in sync latest on the second operation, and can also be used as a barrier of sorts. Collective operations—apart from the MPI_Barrier(), of course—do not have to be synchronizing by the MPI standard, by the way. We will revisit this interesting topic when talking about the barrier (see Section 4.5.3.4).

4.5.3.3 Reduce and Allreduce

Reduction looks just like the broadcast described above, with the addition of the reduction function call (see Listing 4.42, cf. Listing 4.26):

Listing 4.42: Extended subset MPI: reduction (excerpt, see ext/v4/src/coll.c)

```
int MPI_Reduce(const void *sbuf,void *rbuf,int cnt,MPI_Datatype dtype,MPI_Op op,int
root,MPI_Comm comm)
{
    int rank,size,rc;
    MPI_Status stat;

#ifdef DEBUG
    fprintf(stderr,"%d/%d: MPI_Reduce(%p,%p,%d,%d,%d,%d,%d)\n",_mpi_rank,_mpi_size,
sbuf,rbuf,cnt,dtype,op,root,comm);
#endif

    if ((rc = MPI_Comm_size(comm,&size)) != MPI_SUCCESS ||
        (rc = MPI_Comm_rank(comm,&rank)) != MPI_SUCCESS)
        return rc;

    _MPI_CHECK_OP(op,comm);
    _MPI_CHECK(root >= 0 && root < size,comm,MPI_ERR_ROOT);

    if (cnt == 0) {
        _MPI_CHECK(sbuf != NULL,comm,MPI_ERR_BUFFFR);
        if (rank == root)
            _MPI_CHECK(rbuf != NULL,comm,MPI_ERR_BUFFER);
        _MPI_CHECK_TYPE(dtype,comm);
```

```
#ifdef DEBUG
        fprintf(stderr,"%d/%d: -> success\n",_mpi_rank,_mpi_size);
#endif
        return MPI_SUCCESS;
    }

    if (size == 1) {
        _MPI_CHECK(sbuf != NULL,comm,MPI_ERR_BUFFER);
        if (rank == root)
            _MPI_CHECK(rbuf != NULL,comm,MPI_ERR_BUFFER);
        _MPI_CHECK(cnt >= 0,comm,MPI_ERR_COUNT);
        _MPI_CHECK_TYPE(dtype,comm);

        if (_mpi_type_copy(rbuf,(void *)sbuf,cnt,dtype) == NULL) {
            MPI_Comm_call_errhandler(comm,MPI_ERR_INTERN);
            return MPI_ERR_INTERN;
        }

#ifdef DEBUG
        fprintf(stderr,"%d/%d: -> success\n",_mpi_rank,_mpi_size);
#endif
        return MPI_SUCCESS;
    }

    _mpi_inc_ctxt(comm);

    if (size == 2) {
        if (rank == root) {
            if ((rc = MPI_Recv(rbuf,cnt,dtype,!root,_MPI_TAG_REDUCE,comm,&stat)) ==
MPI_SUCCESS)
                (*_mpi_op[op])((void *)sbuf,rbuf,&cnt,&dtype);
        }
        else
            rc = MPI_Send(sbuf,cnt,dtype,root,_MPI_TAG_REDUCE,comm);
    }
    else {
        if (rank == root) {
            if ((rc = MPI_Send(sbuf,cnt,dtype,(rank + 1)%size,_MPI_TAG_REDUCE,comm))
== MPI_SUCCESS)
                rc = MPI_Recv(rbuf,cnt,(size + rank - 1)%size,_MPI_TAG_REDUCE,
comm,&stat);
        }
        else
        {
            int ext;

            if ((rc = MPI_Type_size(dtype,&ext)) != MPI_SUCCESS) {
                _mpi_dec_ctxt(comm);
                return rc;
            }
```

```
        if ((rbuf = malloc(ext*cnt)) == NULL) {
            _mpi_dec_ctxt(comm);
            MPI_Comm_call_errhandler(comm,MPI_ERR_INTERN);
            return MPI_ERR_INTERN;
        }

        if ((rc = MPI_Recv(rbuf,cnt,dtype,(size + rank - 1)%size,_MPI_TAG_REDUCE,
comm,&stat)) == MPI_SUCCESS) {
            (*_mpi_op[op])((void *)sbuf,rbuf,&cnt,&dtype);

            rc = MPI_Send(rbuf,cnt,dtype,(rank + 1)%size,_MPI_TAG_REDUCE,comm);
        }

        free(rbuf);
    }
}

_mpi_dec_ctxt(comm);

return rc;
}
```

The only non-trivial thing here is not to forget that the reduction function should be called only N—1 times when N processes are involved. Unlike the MPI_Bcast(), the ring must be closed here for the end result to get back to the root.

Note that our use of the function MPI_Type_size() is not completely correct, since data described by the derived MPI datatypes may in principle have some padding on either side of the actual data, or both, and potentially have some holes inside, too. In our case, however, it is easier to use this function on the contiguous datatypes we provide than to open Pandora's box of noncontiguous datatype machinery, including the old, convenient but deprecated call MPI_Type_extent(), and the new, better thought through but a little awkwardly formulated call MPI_Type_get_extent().

The MPI_Allreduce() operation does not change (see Listing 4.29). This is of course suboptimal, because inside MPI_Reduce() we need to allocate an intermediate buffer on all non-root processes, where the receive buffer is not meaningful according to the standard. If we knew that the MPI_Reduce() was called from the MPI_Allreduce(), we could have allocated the buffer up there, and probably even managed to do this only once per message size.

However, it is even more suboptimal to use an MPI_Bcast() to spread around information that is partially available on the processes thanks to the ring structure of the exchange pattern, right?

4.5.3.4 Barrier
This done, `MPI_Barrier()` is a snap (see Listing 4.43, cf. Listing 4.23):

Listing 4.43: Extended subset MPI: barrier (excerpt, see ext/v4/src/coll.c)

```
int MPI_Barrier(MPI_Comm comm)
{
    int rank,size,rc;
    char buf[1];
    MPI_Status stat;

#ifdef DEBUG
    fprintf(stderr,"%d/%d: MPI_Barrier(%d)\n",_mpi_rank,_mpi_size,comm);
#endif

    if ((rc = MPI_Comm_size(comm,&size)) != MPI_SUCCESS ||
        (rc = MPI_Comm_rank(comm,&rank)) != MPI_SUCCESS)
        return rc;

    if (size == 1) {
#ifdef DEBUG
        fprintf(stderr,"%d/%d: -> success\n",_mpi_rank,_mpi_size);
#endif
        return MPI_SUCCESS;
    }

    _mpi_inc_ctxt(comm);

    if (size == 2) {
        if (rank == 0) {
            if ((rc = MPI_Send(buf,0,MPI_BYTE,1,_MPI_TAG_BARRIER,comm)) ==
MPI_SUCCESS)
                rc = MPI_Recv(buf,0,MPI_BYTE,1,_MPI_TAG_BARRIER,comm,&stat);
        }
        else {
            if ((rc = MPI_Recv(buf,0,MPI_BYTE,0,_MPI_TAG_BARRIER,comm,&stat)) ==
MPI_SUCCESS)
                rc = MPI_Send(buf,0,MPI_BYTE,0,_MPI_TAG_BARRIER,comm);
        }
    }
    else {
        if (rank == 0) {
            if ((rc = MPI_Send(buf,0,MPI_BYTE,(rank + 1)%size,_MPI_TAG_BARRIER,comm))
== MPI_SUCCESS)
                rc = MPI_Recv(buf,0,MPI_BYTE,(size + rank - 1)%size,_MPI_TAG_BARRIER,
comm,&stat);
        }
        else {
```

```
        if ((rc = MPI_Recv(buf,0,MPI_BYTE,(size + rank - 1)%size,_MPI_TAG_BARRIER,
comm,&stat)) == MPI_SUCCESS)
            rc = MPI_Send(buf,0,MPI_BYTE,(rank + 1)%size,_MPI_TAG_BARRIER,comm);
    }
  }

  _mpi_dec_ctxt(comm);

  return rc;
}
```

Now, is this barrier really synchronizing? Probably not. At least, not on the first pass. But if you parse the ring in the reverse direction or even go around in the same direction for the second time, it will be.

Let's pause here and see what difference this will make (see Figure 4.1):

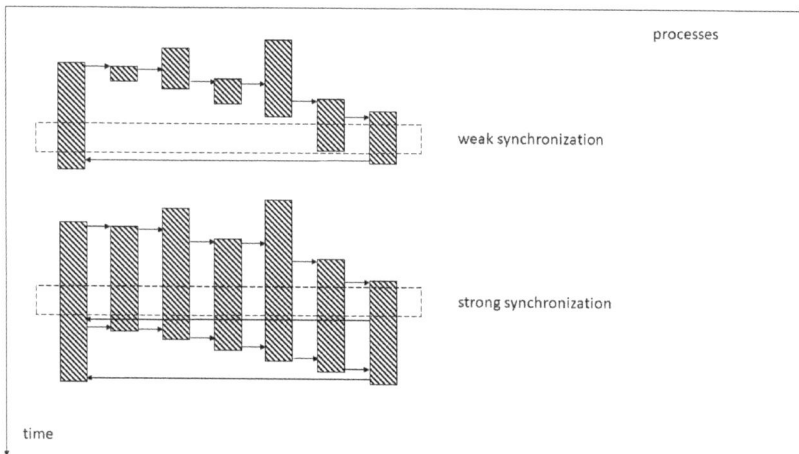

Figure 4.1: Weak and strong synchronization

We show here two state diagrams. The time axis runs vertically and down, the process axis runs horizontally and to the right. The striped rectangles designate the time spent by the corresponding process inside the barrier. Arrows show the messages sent. The root (rank 0) is on the left.

In order to be *strongly synchronizing*, a barrier must ensure that all MPI processes involved sit inside the MPI_Barrier() operation at a certain moment of (global) time. Our current barrier implementation does not guarantee that. It is synchronizing the processes involved, albeit *weakly*. Most likely, they will form a sort of ladder, with the root process enjoying the longest time inside the barrier. Whether all other processes will at a certain time all sit inside the MPI_Barrier() call is unclear. Most likely, they won't.

If, however, we add a backflow ring pass or a second forward flow ring pass to the operation (as shown), we will make it strongly synchronizing, but this will certainly increase the communication overhead and thus the time taken by the barrier. Whether this is always warranted, depends on what the MPI user expects. So, it may make sense to introduce strong synchronization as an option, preferably as one controllable at run time.

4.5.3.5 Gather and Allgather

Now we get to the real queen of the night. As you may recall, MPI_Gather() passes a lot more data around than any other collective call we have seen earlier in this section, except for the MPI_Allgather(), of course.

The added complexity lies in the fact that the root may have a nonzero MPI rank. So, just passing data around is not going to help—we need to accumulate the parts while we traverse the ring. To keep things simple and avoid messing around with buffer parts, let's simply pass around the whole of the receive array, gradually adding to it the contents of the send buffers of each process. Of course, this is a very raw solution, but it will work all right as long as the buffers do not get overly long. And as it happens, they normally won't, because the MPI_Allgather() is most often used to get pretty limited amount of data sent to the root.

Here's how this can be done in a quick and dirty way (see Listing 4.44, cf. Listing 4.30:

Listing 4.44: Extended subset MPI: gather (excerpt, see ext/v4/src/coll.c)

```
int MPI_Gather(const void *sbuf,int scnt,MPI_Datatype sdtype,void *rbuf,int rcnt,
MPI_Datatype rdtype,int root,MPI_Comm comm)
{
    int rank,size,rc;
    MPI_Status stat;

#ifdef DEBUG
    fprintf(stderr,"%d/%d: MPI_Gather(%p,%d,%d,%p,%d,%d,%d,%d)\n",_mpi_rank,_mpi_size,
sbuf,scnt,sdtype,rbuf,rcnt,rdtype,root,comm);
#endif

    if ((rc = MPI_Comm_size(comm,&size)) != MPI_SUCCESS ||
        (rc = MPI_Comm_rank(comm,&rank)) != MPI_SUCCESS)
        return rc;

    _MPI_CHECK(root >= 0 && root < size,comm,MPI_ERR_ROOT);

    if (scnt == 0) {
```

```
        _MPI_CHECK(sbuf != NULL,comm,MPI_ERR_BUFFER);
        _MPI_CHECK_TYPE(sdtype,comm);
        if (rank == root) {
            _MPI_CHECK(rbuf != NULL,comm,MPI_ERR_BUFFER);
            _MPI_CHECK(rcnt >= 0,comm,MPI_ERR_COUNT);
            _MPI_CHECK_TYPE(rdtype,comm);
        }

#ifdef DEBUG
        fprintf(stderr,"%d/%d: -> success\n",_mpi_rank,_mpi_size);
#endif
        return MPI_SUCCESS;
    }

    if (size == 1) {
        _MPI_CHECK(sbuf != NULL,comm,MPI_ERR_BUFFER);
        _MPI_CHECK(scnt >= 0,comm,MPI_ERR_COUNT);
        _MPI_CHECK_TYPE(sdtype,comm);
        if (rank == root) {
            _MPI_CHECK(rbuf != NULL,comm,MPI_ERR_BUFFER);
            _MPI_CHECK(rcnt >= 0,comm,MPI_ERR_COUNT);
            _MPI_CHECK_TYPE(rdtype,comm);
        }

        if (_mpi_type_copy2(rbuf,rcnt,rdtype,(void *)sbuf,scnt,sdtype) == NULL) {
            MPI_Comm_call_errhandler(comm,MPI_ERR_INTERN);
            return MPI_ERR_INTERN;
        }

#ifdef DEBUG
        fprintf(stderr,"%d/%d: -> success\n",_mpi_rank,_mpi_size);
#endif
        return MPI_SUCCESS;
    }

    _mpi_inc_ctxt(comm);

    if (size == 2) {
        if (rank == root)
        {
            int rext;

            if ((rc = MPI_Type_size(rdtype,&rext)) != MPI_SUCCESS) {
                _mpi_dec_ctxt(comm);
                return rc;
            }

            rext *= rcnt;
```

```
            if ((rc = MPI_Recv((char *)rbuf + rext*(!root),rcnt,rdtype,!root,
_MPI_TAG_GATHER,comm,&stat)) == MPI_SUCCESS) {
                if (_mpi_type_copy2((char *)rbuf + rext*root,rcnt,rdtype,(void *)
sbuf,scnt,sdtype) == NULL) {
                    _mpi_dec_ctxt(comm);
                    MPI_Comm_call_errhandler(comm,MPI_ERR_INTERN);
                    return MPI_ERR_INTERN;
                }
            }
        }
        else
            rc = MPI_Send(sbuf,scnt,sdtype,root,_MPI_TAG_GATHER,comm);
    }
    else {
        if (rank == root)
        {
            int rext;

            if ((rc = MPI_Type_size(rdtype,&rext)) != MPI_SUCCESS) {
                _mpi_dec_ctxt(comm);
                return rc;
            }

            rext *= rcnt;

            if (_mpi_type_copy2((char *)rbuf + rext*rank,rcnt,rdtype,(void *)
sbuf,scnt,sdtype) == NULL) {
                _mpi_dec_ctxt(comm);
                MPI_Comm_call_errhandler(comm,MPI_ERR_INTERN);
                return MPI_ERR_INTERN;
            }

            if ((rc = MPI_Send(rbuf,rcnt*size,rdtype,(rank + 1)%size,
_MPI_TAG_GATHER,comm)) == MPI_SUCCESS)
                rc = MPI_Recv(rbuf,rcnt*size,rdtype,(size + rank - 1)%size,
_MPI_TAG_GATHER,comm,&stat);
        }
        else
        {
            int sext;

            if ((rc = MPI_Type_size(sdtype,&sext)) != MPI_SUCCESS) {
                _mpi_dec_ctxt(comm);
                return rc;
            }

            sext *= scnt;

            if ((rbuf = malloc(sext*size)) == NULL) {
                _mpi_dec_ctxt(comm);
```

```
                    MPI_Comm_call_errhandler(comm,MPI_ERR_INTERN);
                    return MPI_ERR_INTERN;
              }

          if ((rc = MPI_Recv(rbuf,scnt*size,sdtype,(size + rank - 1)%size,
_MPI_TAG_GATHER,comm,&stat)) == MPI_SUCCESS) {
                    if (_mpi_type_copy((char *)rbuf + sext*rank,(void *)sbuf,scnt,sdtype)
== NULL) {
                        _mpi_dec_ctxt(comm);
                        MPI_Comm_call_errhandler(comm,MPI_ERR_INTERN);
                        return MPI_ERR_INTERN;
                    }
                    rc = MPI_Send(rbuf,scnt*size,sdtype,(rank + 1)%size,_MPI_TAG_GATHER,
comm);
              }

          free(rbuf);
      }

  }

  _mpi_dec_ctxt(comm);

  return rc;
}
```

It's a bit of heavy lifting, but in essence it is rather simple: the send buffers get copied over into the appropriate part of the receive buffer, which is allocated on non-root processes for this purpose, and then passed along the ring as a whole. This is of course rather memory consuming, and allocation itself is going to cost dearly.

An extra kink comes from the fact that the receive buffer and its description are not meaningful on the non-root process, just as in the case of MPI_Reduce(), so that we cannot rely on them being available and have to use the send buffer description instead. Is there a cure for this? Of course there is, in case we first propagate the receive buffer description across the ring. This will add some latency equivalent to an MPI_Bcast(), so it's up to you whether you want to go to this trouble.

The MPI_Allgather() is not changed (see Listing 4.32), at least for now. Optimization considerations comparable to those related to the MPI_Allreduce()apply here as well (see Section 4.5.3.3).

4.5.4 Testing

Woof, this was quite an adventure. Testing the result is relatively easy, though: we just need to add test runs for more than two MPI processes, which is a trivial undertaking.

4.5.5 Benchmarking

Benchmarking done on one, two, three, and four processes produced the following results (see Table 4.2, cf. Table 4.1):

Table 4.2: Two-, three-, and four-process benchmarks: minimum latency at 16 bytes, 16 KiB, 128 KiB, and 4 MiB (seconds)

Pattern	16 bytes	16 KiB	128 KiB	4 MiB
Ping-pong	2.30e-06	4.23e-06	1.36e-05	0.0006
Ping-ping	3.00e-06	4.95e-06	1.83e-05	0.0015
Pong-pong	3.28e-06	5.19e-06	1.89e-05	0.0012
Barrier*	4.68e-06			
2-process	5.11e-06			
3-process	8.72e-06			
4-process	11.69e-06			
Bcast	2.18e-06	3.64e-06	1.75e-05	0.0005
2-process	**0.50e-06**	4.31e-06	1.74e-05	0.0005
3-process	8.57e-06	12.42e-06	4.67e-05	0.0020
4-process	12.98e-06	16.88e-06	6.59e-05	0.0026
Reduce	**0.65e-06**	7.12e-06	4.05e-05	0.0017
2-process	2.27e-06	7.16e-06	4.118e-05	0.0016
3-process	8.82e-06	18.59e-06	9.35e-05	0.0036
4-process	11.77e-06	25.98e-06	13.61e-05	0.0051
Allreduce	4.64e-06	11.09e-06	5.09e-05	0.0022
2-process	4.74e-06	11.17e-06	5.12e-05	0.0022
3-process	16.57e-06	30.30e-06	13.97e-05	0.0058
4-process	22.73e-06	41.93e-06	19.96e-05	0.0077
Gather				
2-process	2.80e-06	3.20e-06	2.28e-05	0.0010
3-process	8.728e-06	22.69e-06	11.42e-05	0.0069
4-process	13.46e-06	42.13e-06	19.34e-05	0.0114
Allgather				
2-process	4.69e-06	8.02e-06	4.10e-05	0.0023
3-process	16.50e-06	43.96e-06	21.89e-05	0.0126
4-process	22.84e-06	84.72e-06	38.17e-05	0.0213

* Barrier is special: it's shown as a 16-byte result even though it uses zero-sized messages

We provide data from Table 4.1 for comparison in the first row of data for each pattern. After that there follow, if available, data items for 2-, 3-, and 4-process runs done using software developed in this section. The outliers that apparently occurred due to the process binding anomaly are in **bold**. Apart from those, the timings show expected behavior: grow with the number of processes and message size, and add up properly when a superimposed operation like MPI_Allreduce() is compared to its components. It is also noticeable that the MPI_Gather() and MPI_Allgather() timings for larger messages slowly become astronomical, just as we thought they would due to a very naïve implementation.

4.5.6 Exercises

Exercise 4.5 Correct the MPI_Bcast() by using a zero-size message to close the ring (see Section 4.5.3.2).

Exercise 4.6 Implement the MPI_Allreduce() directly using the ring pattern.

Exercise 4.7 Make the MPI_Barrier() strongly synchronizing this way or another (see Listing 4.43). Consider controlling the manner in which this is achieved, if at all, at run time.

Exercise 4.8 Improve the MPI_Gather() by sending and receiving only the required portions of the buffer along the ring (see Section 4.5.3.5).

Exercise 4.9 Implement the MPI_Allgather() directly using the ring pattern.

4.6 Subset Completion: Language Bindings

What is HPC without Fortran? A bad joke. So we have to add this binding to our library if it is to be called "MPI." Period.

4.6.1 Language Bindings

Let's add the following features to the underlying MPI subset:
- Fortran language binding based on the INCLUDE 'mpif.h' method for the simplest MPI subset defined so far

More complicated Fortran bindings can be considered later if such a need ever arises. Bindings for other languages are not described by the MPI standard and therefore are also out of scope of this chapter.

4.6.1.1 Design Considerations

There are several approaches to the language bindings:

- Most often, one binding (normally, the C language binding, but rarely the Fortran language binding) is taken as the basis, and all other bindings are formulated as so-called thunk layers on top of it either:
  - o Natively (i.e., in terms of the target language, possibly using standard features or vendor extensions, like Fortran, or a native method interface, like in Java)
  - o In terms of the basic implementation language (like C)
- Alternatively, all or at least some bindings are formulated in terms of a hidden, internal interface that is not accessible for the MPI user.

According to the 10/90 principle, we will use the first approach and express our Fortran binding using C language.

4.6.1.2 MPI Header File

The Fortran 77 header file mpif.h contains only constants equivalent to those present in the mpi.h header (see Listing 4.45, cf. Listing 4.1):

Listing 4.45: Extended subset MPI: Fortran 77 header file (see ext/v4/include/mpif.h)

```
INTEGER MPI_UNDEFINED
PARAMETER (MPI_UNDEFINED=-1)

INTEGER MPI_COMM_NULL,MPI_COMM_SELF,MPI_COMM_WORLD
PARAMETER (MPI_COMM_NULL=MPI_UNDEFINED)
PARAMETER (MPI_COMM_SELF=0,MPI_COMM_WORLD=1)

INTEGER MPI_BYTE,MPI_INTEGER,MPI_DOUBLE_PRECISION
PARAMETER (MPI_BYTE=1,MPI_INTEGER=4,MPI_DOUBLE_PRECISION=8)

INTEGER MPI_MAX,MPI_MIN,MPI_SUM
PARAMETER (MPI_MAX=0,MPI_MIN=1,MPI_SUM=2)

INTEGER MPI_STATUS_SIZE
PARAMETER (MPI_STATUS_SIZE=4)

INTEGER MPI_ERRORS_ARE_FATAL,MPI_ERRORS_RETURN
PARAMETER (MPI_ERRORS_ARE_FATAL=0,MPI_ERRORS_RETURN=1)

INTEGER MPI_MAX_ERROR_STRING
PARAMETER (MPI_MAX_ERROR_STRING=32)

INTEGER MPI_SUCCESS
PARAMETER (MPI_SUCCESS=0)
```

```
INTEGER MPI_ERR_BUFFER,MPI_ERR_COUNT,MPI_ERR_TYPE
INTEGER MPI_ERR_TAG,MPI_ERR_COMM,MPI_ERR_RANK
INTEGER MPI_ERR_REQUEST,MPI_ERR_ROOT,MPI_ERR_OP
INTEGER MPI_ERR_ARG,MPI_ERR_UNKNOWN,MPI_ERR_TRUNCATE
INTEGER MPI_ERR_OTHER,MPI_ERR_INTERN,MPI_ERR_LASTCODE
PARAMETER (MPI_ERR_BUFFER=1,MPI_ERR_COUNT=2,MPI_ERR_TYPE=3)
PARAMETER (MPI_ERR_TAG=4,MPI_ERR_COMM=5,MPI_ERR_RANK=6)
PARAMETER (MPI_ERR_REQUEST=7,MPI_ERR_ROOT=8,MPI_ERR_OP=9)
PARAMETER (MPI_ERR_ARG=10,MPI_ERR_UNKNOWN=11,MPI_ERR_TRUNCATE=12)
PARAMETER (MPI_ERR_OTHER=13,MPI_ERR_INTERN=14,MPI_ERR_LASTCODE=14)

DOUBLE PRECISION MPI_WTIME
```

I type Fortran code in capital letters so that it stands out compared to C. Fortran IV was my third programming language after BASIC and Algol-60. Man, some 35 years on, I still love it: its succinct beauty, its demanding nature, its intrinsic power. It's like Latin among programming languages, albeit still spoken today, with all the modern quirks and later additions that, to my taste, have badly spoiled this noble language of the real programmers of old.

One thing to notice here on the less spiritual side is that we define different MPI datatypes—different in name, but similar in nature, memory layout, and internal processing. And yes, the very last round bracket in the file occupies position number 72. This number, by the way, rather than 42, is the right answer to the question of life, universe, and everything.

4.6.1.3 Thunk Layer
Adding the actual implementation requires some research and experience. I will not bore you with theoretical details and just explain the things that I did instead (see Listing 4.46):

Listing 4.46: Extended subset MPI: Fortran 77 bindings (see ext/v4/src/mpif.c)

```
#include "mpi.h"

void mpi_init_(int *pierr)
{
    int argc = 1;
    char *argv[] = { "a.out"};

    *pierr = MPI_Init(&argc,(char ***)&argv);
}

void mpi_finalize_(int *pierr)
{
    *pierr = MPI_Finalize();
}
```

```
void mpi_comm_size_(int *pcomm,int *psize,int *pierr)
{
    *pierr = MPI_Comm_size(*pcomm,psize);
}

void mpi_comm_rank_(int *pcomm,int *prank,int *pierr)
{
    *pierr = MPI_Comm_rank(*pcomm,prank);
}

void mpi_send_(void *buf,int *pcnt,int *pdtype,
               int *pdest,int *ptag,int *pcomm,int *pierr)
{
    *pierr = MPI_Send(buf,*pcnt,*pdtype,*pdest,*ptag,*pcomm);
}

void mpi_recv_(void *buf,int *pcnt,int *pdtype,
               int *psrc,int *ptag,int *pcomm,int *pstat,int *pierr)
{
    *pierr = MPI_Recv(buf,*pcnt,*pdtype,*psrc,*ptag,*pcomm,(MPI_Status *)pstat);
}

double mpi_wtime_(void)
{
    return MPI_Wtime();
}
```

The naming convention used for the Fortran subroutine and function names represented in C depends on the compiler. The GNU Fortran compiler f77 normally expects them to be in small letters, with an underscore appended.

All Fortran 77 arguments are passed by reference, which means a pointer in C. We have already seen this convention used in the internal reduction and error handling routines (see Sections 4.3.4.3 and 4.4.7.5, respectively).

Since we do not have access to the main() program arguments here, and do not want to, actually, we fake them for now to make the MPI_Init() happy.

Finally, a seemingly risky conversion between (int *) and (MPI_Status *) takes care of the MPI_Status passing. Indeed, four integers here and there. Who cares?

That's it. If you want to extend this limited implementation to cover the whole of the current MPI subset, good luck with Fortran strings and the MPI_Request! In particular, remember the 10/90 principle before you start adding a general hash table to this simple thunk layer.

4.6.2 Testing

Instead of inventing something completely new, we will just take the existing test programs written in C and translate them by hand (oh!) into very-very old looking Fortran. First, the "Hello world" (see Listing 4.47, cf. Listing 2.17):

Listing 4.47: Extended subset MPI: simple process creation test (see ext/v4/test/t0.f)

```
INCLUDE 'mpif.h'

CALL MPI_INIT(IERR)

PRINT *,'HELLO!'

CALL MPI_FINALIZE(IERR)

END
```

I do use an implicitly defined variable IERR and output formatted so that there can be no mistake as to who is speaking.

After this we can take on the more demanding task of building the test. Here's how this can be done (see Listing 4.48):

Listing 4.48: Extended subset MPI: test Makefile (excerpt, see ext/v4/test/Makefile)

```
...

FFLAGS=-O2
#FFLAGS=-g

...

t0f: t0.f $(INCDIR)/mpif.h
        f77 $(FFLAGS) -I$(INCDIR) -o t0f t0.f -lfmpi $(LDFLAGS)

...
```

So we simply use the GNU Fortran compiler as we earlier used the GNU C compiler, and add the Fortran bindings library, called libfmpi.a and located in the lib directory alongside the main MPI library libmpi.a, to the compiler run string. The Fortran names are resolved in the Fortran binding library, while the functions called from inside the binding come from the main MPI library as usual.

Running the resulting program t0f does not differ from running a program t0 written in C. The Fortran library is also built like the C library. Note that by keeping them separate we pursue the modular way. It is possible to do this differently, but for these simple test programs, this would be an overkill.

Now, let's take on the more demanding test case (see Listing 4.49, cf. Listing 2.26):

Listing 4.49: Extended subset MPI: simple toast test program (see ext/v4/test/t1.f)

```
      INCLUDE 'mpif.h'

      INTEGER ISBUF(2),IRBUF(2),ISTAT(MPI_STATUS_SIZE)
      DATA ISBUF / 4HSALU,3HTE! /
      DATA IRBUF / 4HCHEE,3HRS! /

      CALL MPI_INIT(IERR)

      CALL MPI_COMM_SIZE(MPI_COMM_WORLD,ISIZE,IERR)

      IF(ISIZE .GT. 1) THEN
         CALL MPI_COMM_RANK(MPI_COMM_WORLD,IRANK,IERR)
         IF(IRANK .EQ. 0) THEN
            PRINT 1001,IRANK,ISIZE,ISBUF
            CALL MPI_SEND(ISBUF,2,MPI_INTEGER,1,0,
     &                    MPI_COMM_WORLD,IERR)
         ELSE
            IF(IRANK .EQ. 1)
     &          CALL MPI_RECV(IRBUF,2,MPI_INTEGER,0,0,
     &                    MPI_COMM_WORLD,ISTAT,IERR)
            PRINT 1001,IRANK,ISIZE,ISBUF
         ENDIF
      ENDIF

      CALL MPI_FINALIZE(IERR)

 1001 FORMAT(' ',I1,'/',I1,': ',2A4)

      END
```

The only really remarkable part here is the use of Hollerith constants to initialize integer buffers to human readable and printable contents. The modern GNU compiler honors this feat of nostalgia with multiple warnings, as if an old skeleton were trying to get out of the closet. Just to make it know I mean it, I also use an old-fashioned logical IF statement to receive data. I think it gets the message.

The output of both Fortran test programs is predictably similar to that of their C equivalents (not shown).

4.6.3 Benchmarking

It would be relatively easy to translate our many C benchmarks into Fortran. However, what's the point? If you want to measure the negligible overhead of the MPI thunk layer, you can do this.

4.6.4 Exercises

Exercise 4.10. Translate the unified point-to-point benchmark b.c into Fortran and compare the performance of both incarnations.

4.7 Conclusions

Well, this was easy, right? It took me a couple of weeks to get everything done, and I didn't spend all day (or night) coding, mind you. Now, however, we have a pretty well-rounded overstructure that warrants further optimization of the underlying communication machinery. This is what we are going to do in the next chapter.

In principle, this cycle of semantic extension up there followed by supporting technical extension down below permeates this entire book, both as a whole and in all its parts. This process is somewhat akin to the bootstrapping used to start an operating system, from simple loader and control program through single-user privileged mode to the full-blown multiprocess environment possibly spanning more than one machine.

Now, more statistics for sweets (see Table 4.3, cf. Table 3.1):

Table 4.3: MPI subsets vs. full MPI implementations

Subset	Entities	Incl. Functions	LOC
Shared memory	13	7	343
Sockets	18	11	852
Extensions	60	27	2137
MPI-1	244	128	~100000
MPI-3.1	855	450	~350000

We keep getting excellent results at a negligible fraction of the cost, don't we? Will optimization change anything here? If so, how? Next chapter will show this. *Avanti!*

Chapter 5
Optimization

In this chapter you will learn how to optimize MPI internally using advanced implementation techniques, available special hardware. We discussed optimization targets
briefly in Chapter 1 (see Section 1.4.7) and occasionally practiced the art of optimization across the previous three chapters. Here we will drive everything together
and review the entire field, noting those things that we have done before simply in
passing, and focusing chiefly on performance as the optimization target.

Due to the sheer volume of this topic, we will start programming less and looking
more into what other people have done, still providing an example or two of the technologies discussed in application to the MPI subset developed in Chapter 4. We will
be optimizing an existing MPI subset, extending it marginally once in a while when
the need arises, so the style of the narration will change to fit the new task.

5.1 Unifabric Optimization

Unifabric optimization is dedicated to the increase in performance or other useful
characteristics of any given communication medium—such as shared memory,
sockets, or another network. It should be differentiated from tuning. Optimization
presumes source code changes. Tuning, on the contrary, is restricted to proper selection of the existing code paths and thresholds, either beforehand or depending on the
operating environment. We will consider tuning in Chapter 6.

5.1.1 Design Decisions

We will consider all parts of the MPI library below—process start-up, connection
establishment, point-to-point communication, and collective operations—in order to
identify possible optimization opportunities and to indicate and sometimes exem-
plify their exploitation. In each of these topics, we will follow the top-down approach
mentioned in Section 1.4.7 in order to introduce optimizations at the most effective
level first, starting with algorithms and going down to the wire only once we have
exhausted all other possibilities above it.

DOI 10.1515/9781501506871-005

5.1.2 Code Refactoring

There are a few things that we need to do to the code before we proceed to the optimization per se. Consider this as a clean-up of the work bank before starting a new
project.

    The first thing to do is the separation of the device layer from the rest of the MPI
library proper. So far, we have been keeping all functions related to the communication in one file (called xfer.c, see Section 4.3.5). This was okay as long as we wanted
to learn a particular fabric or extend an MPI subset working over it.

    Now we are going to work on the fabric itself, so it will possibly make sense to
have several variants of the fabric interface implementation. This implementation is
normally called an *MPI device*, or an *MPI device layer*. For brevity, in this book we will
be calling it simply a device layer, since we are talking about MPI, right?

5.1.2.1 Design Considerations

There are several ways of separating a device layer from the rest of the library. As
usual, there are at least two schools of thought as to where a device should live with
respect to the main source code of the MPI library—above or alongside it. We will
follow the most trivial approach for now, just taking all calls that relate to the actual
byte pushing into a subdirectory src/dev. This way we will keep all things nicely
tacked in one place. We will do the transformation step by step, so that the structure
we use is well suited for the optimization kind we are going to perform, and so that
this structure evolves naturally until we get it completely correct and flexible by the
end of this chapter.

5.1.2.2 Device Layer

The actual code manipulation is rather simple. We will only show what calls (and
their dependents) are moved up a notch (see Listing 5.1):

Listing 5.1: Optimized subset MPI: simple device layer header (see opt/v1/src/dev/dev.h)

```
int _mpi_dev_init(int *pargc,char ***pargv);
int _mpi_dev_fini(void);
int _mpi_progress(void);
```

Yes, that's it! We need to initialize the device and terminate it, and then kick its
progress engine. If you look into the rest of the MPI communication implemented in
Chapter 4, you will see that (quite accidentally) all the MPI library does is enqueue
message requests inside the MPI_Isend() and MPI_Irecv() calls, and then complete
them inside the MPI_Test() call (see Section 4.3.5). The rest of the library uses these
three high-level MPI functions to express all the remaining MPI subset functionality,

including the `MPI_Send()`, `MPI_Recv()`, `MPI_Wait()`, and the collective operations. Of course, this is only one of the possible solutions, but it will do for now.

The contents of the corresponding source file `dev.c` are trivial (not shown).

5.1.2.3 Queue Extension

Another issue to address is the queue length limitation that we have been carrying along since Chapter 3. One message request per queue (posted send, posted receive, and unexpected receive) was enough for simple tests and ring-based collectives. Of course, we need to break out of this jail if we are to do more complicated exchanges, both point-to-point and collective.

Now, where should we put the queue implementation? Naturally, it should straddle both the high-level (MPI library) and the low-level (MPI device) communication functions. Putting it alongside the high-level source code files, even as a separate one, would be a little unwise because the queue structure may (and sometimes must) depend on the nature of the MPI device used. So it looks like we should put the request management calls alongside the device calls, using a separate source code file for that.

What we need to move there is shown below (see Listing 5.2):

Listing 5.2: Optimized subset MPI: simple request queue management header (see `opt/v1/src/dev/req.h`)

```
#define _MPI_QREQ_MAX     10

struct _mpi_q_s {
    struct _mpi_request_s req[_MPI_QREQ_MAX];
    int flag[_MPI_QREQ_MAX];
    int n,head,tail;
};

extern struct _mpi_q_s _mpi_rq,_mpi_xq,_mpi_sq;

struct _mpi_q_s *_mpi_findq(MPI_Request req);
MPI_Request _mpi_newreq(struct _mpi_q_s *pq);
MPI_Request _mpi_findreq(struct _mpi_q_s *pq,int rank,int tag);
void _mpi_freereq(struct _mpi_q_s *pq,MPI_Request *preq);
MPI_Request _mpi_firstreq(struct _mpi_q_s *pq,MPI_Request *preq);
MPI_Request _mpi_nextreq(struct _mpi_q_s *pq,MPI_Request *preq);
```

Again, all these calls are known to you. They will just live elsewhere. Another matter to address is whether we should expose the queues in the way shown here. In principle, this will do for now, so there's no need to overcomplicate things right away. Of course, we could have added an initialization and a termination call for the message queue subsystem, a variable number of requests per queue, and even extension of the request queue should its size prove to be insufficient at run time. We could have even hidden the queues completely behind the interface. But why bother?

One interesting detail is that instead of using some extra field or playing with the bits of the message request structure, we move all we need for the queue implementation into a parallel array flag[]. This way we separate the queue implementation and the message request per se, which will pay off as soon as we decide to change the queue implementation.

As to the implementation, we will do a very simple extension for now (see Listing 5.3):

Listing 5.3: Optimized subset MPI: simple request queue management (see opt/v1/src/dev/req.c)

```c
#include <stdio.h>          // NULL, fprintf(3), etc.

#include "mpi.h"
#include "imp.h"
#include "req.h"

#include <assert.h>         // assert(3)

struct _mpi_q_s _mpi_rq,_mpi_xq,_mpi_sq;

struct _mpi_q_s *_mpi_findq(MPI_Request req)
{
    if (req >= _mpi_rq.req && req < _mpi_rq.req + _MPI_QREQ_MAX)
        return &_mpi_rq;
    else if (req >= _mpi_xq.req && req < _mpi_xq.req + _MPI_QREQ_MAX)
        return &_mpi_xq;
    else if (req >= _mpi_sq.req && req < _mpi_sq.req + _MPI_QREQ_MAX)
        return &_mpi_sq;

    return NULL;
}

MPI_Request _mpi_newreq(struct _mpi_q_s *pq)
{
    if (pq->n == 0) {
        pq->flag[(pq->head = pq->tail = 0)] = 1;
        return pq->req + (pq->n++);
    }
    else if (pq->n < _MPI_QREQ_MAX) {
        pq->flag [(pq->tail = (pq->tail + 1)%_MPI_QREQ_MAX)];
        pq->n++;
        return pq->req + pq->tail;
    }

    return NULL;
}
```

```
MPI_Request _mpi_findreq(struct _mpi_q_s *pq,int rank,int tag)
{
    if (pq->n > 0)
    {
        int i;

        for (i = pq->head;; i = (i + 1)%_MPI_QREQ_MAX) {
            if (pq->flag[i])
            {
                MPI_Request req = pq->req + i;

                if (req->rank == rank && req->tag == tag)
                    return req;
            }

            if (i == pq->tail)
                break;
        }
    }

    return NULL;
}

void _mpi_freereq(struct _mpi_q_s *pq,MPI_Request *preq)
{
    if (pq->n > 0)
    {
        int i = *preq - pq->req;

        if (pq->flag[i]) {
            pq->flag[i] = 0;

            if (i == pq->head)
                for (i = (i + 1)%_MPI_QREQ_MAX;; i = (i + 1)%_MPI_QREQ_MAX) {
                    if (pq->flag[i]) {
                        pq->head = i;
                        break;
                    }

                    if (i == pq->tail)
                        break;
                }
            else if (i == pq->tail)
                for (i = (_MPI_QREQ_MAX + i - 1)%_MPI_QREQ_MAX;; i = (_MPI_QREQ_MAX +
i - 1)%_MPI_QREQ_MAX) {
                    if (pq->flag[i]) {
                        pq->tail = i;
                        break;
                    }
                }
```

```
                        if (i == pq->head)
                            break;
                    }

                *preq = NULL;
                pq->n--;
            }
        }
    }

    MPI_Request _mpi_firstreq(struct _mpi_q_s *pq,MPI_Request *preq)
    {
        if (pq->n > 0)
            return (*preq = pq->req + pq->head);

        return NULL;
    }

    MPI_Request _mpi_nextreq(struct _mpi_q_s *pq,MPI_Request *preq)
    {
        if (pq->n > 1)
        {
            int i = *preq - pq->req;

            for (i = (i + 1)%_MPI_QREQ_MAX;; i = (i + 1)%_MPI_QREQ_MAX) {
                if (pq->flag[i])
                    return (*preq = pq->req + i);

                if (i == pq->tail)
                    break;
            }
        }

        return (*preq = NULL);
    }
```

Linear search in the circular list of a predefined length is all we need. The head of the list is indexed by the field head, the tail is appropriately indexed by the field tail. We use indices rather than pointers because we need to do quite a bit of queue browsing, which is better done using integer variables. If this implementation ever becomes a bottleneck, we will be able to address it quite scientifically.

Of course, we could have also defined several functions for setting the message request fields. However, since we have a very limited set of high-level MPI calls to manage, it will be overkill.

One more fine point: you see that I do not mention in our header files other headers that they are dependent upon. Again, there are at least two schools of thought here. One is including right into the header file involved all headers it needs, so that every header is self-contained, whatever happens. This, however, requires that the headers be protected from accidental multiple inclusion. Not a big deal, but still, some extra fuss. The other school of thought just mentions in the header what has to be there; that is, only the new things it declares. The inclusion of the proper dependent header files is then delegated to the respective source code file. This way is more flexible but requires a bit more attention. I feel like a hero this morning, so we take the latter, hero path.

5.1.2.4 Testing and Benchmarking

Unavoidable testing and benchmarking shows no regressions. As an expected bonus for the queue extension, we get the ping and pong patterns re-enabled, and open up the possibility of using more complicated collective algorithms that may generate more than one pending send and/or receive request at a time.

5.1.3 Start-up and Termination

Process start-up and termination are often overlooked where MPI optimization is concerned. Indeed, for small jobs and fast networks, the overall contribution of the initial and final stages of the execution to the total wall clock time is normally very low. It grows rather quickly with the job scale in particular, however.

Termination is important both from the execution time and clean-up points of view. You need to recall the TIME_WAIT issue to grasp the clean-up aspect. Sometimes, this task becomes so formidable that an MPI implementation provides an additional command to locate and free the system resources that accidentally remain registered to the finished MPI jobs.

5.1.3.1 Design Considerations

Process start-up assumes that the respective executable image is available for loading into the memory. Sometimes this assumption is true, say, on a local node, or across the cluster, when the Network File System (Wikipedia.org, 2018e) or its equivalent has been set up correctly. If, however, this assumption is false, the executable must be copied over either by the MPI library, or implicitly by the MPI *process manager* (like an extended mpiexec), or even explicitly by the user. Now, once the executable has been made accessible to a particular node, this way or another, it must be started there. Again, there are more ways to do this than meets the eye.

5.1.3.2 Alternative Networks

This way is trivially accessible via sockets. In order to exploit it, just use IP addresses assigned to the respective alternative network interface, say, IPoIB, and the rest will work as it did over the Ethernet, but faster. If your NFS has been set up to use this alternative network, the executable will reach its destination(s) faster as well. That's it.

5.1.3.3 Built-in Facilities

On some systems, the lower interfaces may provide special methods for starting processes. Of the ways I have personally used, there are remote fork(2) and remote exec(2) system calls, and/or system libraries that provide access to these or comparable facilities. Once you see one in the respective system description, you will know what to try. However, pay attention to the actual speed of the respective implementation: sometimes they may disappoint you and your users.

5.1.3.4 Process Manager Revisited

We have defined and refined our simple MPI job launcher mpiexec earlier (see Section 3.3.2.2.1). This was a first step toward a full-blown MPI process manager. If you remember, we had one issue with this implementation, namely, the recurring TIME_WAIT matter. In order to work around it, we would modify the port range to use via an environment variable.

However, there is a better way. Imagine that in each node we have a remote agent that we can connect to using a well-known address. Once connected, we can instruct this remote server process to start an MPI process or two on our behalf. Their addresses will then be relayed back to us by the server that started them.

Naturally, this server process will have to be started somehow, so that we know we can connect to it before we start an MPI job. Such a server component is normally implemented by a *daemon*—a program that runs without being connected to any particular terminal or session. In operating systems other than UNIX, these daemons can be called *services*, etc.

There are two schools of thought here. One argues that the server should be a special cut one, developed for the MPI. The other tries to rely on the already existing services such as ssh(1) or the like. Whatever the backend is, the frontend interface of the mpiexec command is specified by the MPI standard.

5.1.3.5 Job Managers

Big clusters can rarely be used directly, due to the high demand and high cost of their maintenance. Most of the time, access to them will be isolated by a so-called *job manager*. This system component accepts job execution requests (not just MPI ones) from the users, schedules them, allocates resources, launches the job in due time, and then takes care of its execution, timely termination (if requested), and proper clean-up. Sometimes a job manager is split into a scheduler and a launcher, but this is not important to this book.

In any case, a job manager normally provides an interface or two that are to be used. This may be a set of commands and/or special library. Your process manager can use them explicitly. Alternatively, the user can allocate a partition using a job manager, and then run the MPI job inside this partition. The latter step can be done manually, by logging in onto one of the allocated nodes and forming the list of node addresses by hand, or semi-automatically, if the MPI process manager recognizes that it has been started within an allocated partition and uses available information to build the respective node address list.

Every known job manager exposes information about the partition details in some manner, be that a set of special environment variables or a configuration file. The start-up is then done by the job manager for you, in the hope that it knows how to do this quickly. Again, it may make sense to benchmark this, and fall back upon the socket-based start-up in case it is faster. Note that if the job manager is good, it may make no sense to start MPI's own process manager daemons—their job can be handled directly by the job manager daemons or services.

Note that job managers are normally quite good at resource usage monitoring, including maximum execution time allowed, maximum number of processes and nodes allocated, memory consumptions, and many other accounting aspects. This also helps in handling termination and the associated clean-up rather efficiently and tidily.

5.1.4 Connection Establishment

Just like process start-up and termination, connection establishment is often overlooked during optimization. It is surprising to learn that only advanced MPI users know about this stage at all, to say nothing about the ways in which this can be done, if this is indeed necessary.

5.1.4.1 Design Considerations
As soon as processes are up and running, they will want to communicate with each other. For that they need to know who to talk to, what credentials to present, and then what connections to establish, if any. Let's consider these aspects in more detail.

5.1.4.2 Address Propagation
We have been solving this problem by defining certain environment variables that were then interpreted by the MPI processes. This is not the only way, of course. Assuming the file system is shared by all the nodes involved, a configuration file or two would work just as well. When, however, one cannot rely on this *out-of-band communication* channel, one has to establish another channel.

This is where an extended process manager comes in handy. All its daemons have to communicate with each other anyway. This side communication channel is normally established when the process manager infrastructure is set up—either by the user or by the system—but it can also be established on demand, especially if a system service like sshd(8) is used under the hood. Once this infrastructure gets an application start-up request from the mpiexec command or its equivalent, these resident daemons spawn a set of additional processes dedicated to handling the new task. If the resident daemons run under root privileges, for security reasons, this new set of processes normally assumes the identity of the user who started the job. Depending on the structure of the process manager, another set of processes may be spawned to handle a particular job on those nodes that actually run it. It is this last level of management processes that actually spawn the MPI processes the user wanted to start. The MPI processes get their data and can use a special interface, like PMI (Balaji, et al., 2012), to communicate with each other via the respective management processes. This communication can be used, among other things, for propagating the information necessary for connection establishment between the MPI processes.

It is quite likely that the process management infrastructure uses sockets for its out-of-band communication. If the sockets run over some kind of Ethernet, all typical latencies and bandwidths apply. However, the sockets can be mapped upon Infini-Band or another fast network, in which case we can expect substantially higher performance.

There is one more issue to keep in mind in this context: system setup errors. In practice, it is not impossible to stumble upon an installation that has incomplete, incorrect, or simply broken address tables, so that different nodes have different views of their surroundings. For example, symbolic names of the nodes may not be equal on all nodes, and an attempt to resolve them into IP addresses may either fail or result in different MPI processes receiving conflicting information. This, of course, will most likely lead to the MPI job suspension or even failure at the later connection establishment phase. There is an elegant way to resolve this issue, however (US Patent No. 7,644,130, 2010): the head node on which the mpiexec command is run must have correct information about all the nodes it is attempting to start the MPI job upon. This information can be propagated to all the MPI processes, so that they get correct IP addresses, and simultaneously select the right network in case there are more than one of them (say, Ethernet and InfiniBand with IPoIB enabled).

5.1.4.3 Lazy Connection Establishment

So far, we have always established all connections upfront. This may make sense until the MPI job size becomes relatively high, say, greater than 32 or 64 processes. After that, one may gain quite a bit both in speed and in scalability by establishing only those connections that are really going to be used by the MPI job. It is normally difficult to predict what connections will be needed, so this detective work is delegated to the MPI job itself. Once it requires a certain new connection, the MPI library should oblige.

The necessary change goes right to the heart of the MPI, namely, the progress engine. In this way or another, a request for connection establishment reaches it. In the case of sockets, this may be a connection request that hits the respective listening socket. This request has to be detected and then served by the progress engine. Since high scalability is not targeted in this book, we are not going to implement this stuff. I will instead indicate how this can be done:

- Introduce a new queue for connection handling (say, _mpi_cq).
- When you want to send a message to a process that has not yet been connected to, put an appropriately formed request for connection into said queue.
- Inside the progress engine, process the request indicated above before all other requests. Use a state machine to govern the processing as necessary. Note that a file descriptor for a connected socket will never be equal to zero.
- On the receiving side, monitor the listening socket that we wisely added to the _mpi_fds array. Take care of the connection requests possibly crossing, e.g., when both sides want to send messages to each other (cf. ping-ping pattern). Use MPI ranks or another value to break the eventual ties.

Taking care of the incoming and outgoing connection requests will therefore add an extra activity to the progress engine. This will make the whole a little bit more expensive, so it will make sense to count the established connections and switch the connection request processing as soon as all possible connections have been established. Moreover, sometimes it makes sense to start establishing connection in the lazy mode only if the job size exceeds a certain threshold. Below this size, all connections can be established upfront statically, which saves a lot of extra hassle down the road.

Also, since during lazy connection establishment respective requests will most likely be coming relatively less often as the job progresses, it may make sense to consider some kind of a *back-off strategy* that will also improve the latency of the message transfer. One idea is to introduce a limit counter that will guide the frequency of the connection queue polling compared to the message queue polling. Starting with this limit counter in a relatively low value area, one can increase the value as time passes and more connections are established, in the hope that the more of them are already provided, the less will be needed in the future.

Another idea to consider is the use of the select(2) or poll(2) system calls, or the more modern epoll(7) subsystem in order to avoid polling all file descriptors. This may benefit the message transfer as well, by the way, especially at high process counts.

5.1.4.4 Connectionless Communication

Another opportunity to save on the connection establishment time is the use of connectionless communication. This comes at a cost, because you will most likely have to take care of the message segmentation, data loss, and retransmission yourself.

However, you will have less fuss with the connection establishment in return. At a certain scale, this may actually be the only way to go.

5.1.4.5 Adaptive Connections

In the end, what you want is to send data as fast as possible, spending as few resources for this as necessary. Hence the idea of combining all the techniques mentioned in this section, in order to use the fastest, but also the heaviest, connection-oriented protocols for the most frequently used destinations and rely on the connectionless protocols for all other messages.

There are various approaches as to what destination to choose for this preferential treatment. One can start, for example, with the connections to the immediate neighbors in the MPI rank sequence, in the assumption that they should in principle communicate most intensively with each other. Another approach is to start in connectionless mode everywhere and count messages or bytes sent, in order to "promote" a certain hot connection when a particular threshold has been met. As usual, the possibilities are infinite.

5.1.5 Point-to-Point Communication

Point-to-point communication is the basis of most if not all other MPI functionality. Unsurprisingly, then, is that optimization of this part of MPI gets paid a lot of attention. Only optimization of the collective operations can overtake in significance the point-to-point communication, and that mostly at scale.

5.1.5.1 Design Considerations

In this book, we take a somewhat unusual approach to point-to-point optimization. You can find any number of articles out there describing this or that interface of a particular network or a whole family of networking products in their application to the MPI internals. Here, we consider them all inside the point-to-point optimization, with a possible excursion into the process start-up, connection establishment, and collectives, if they are covered at all by the respective interfaces.

To separate the essence of the matter from its external appearance, we distinguish between interfaces and protocols. Protocols determine in what particular way data is sent over a particular communication medium. Interfaces define how these protocols can be used in your program.

There are *high-level* and *low-level protocols*. The former are somewhat independent of the medium used. The latter are very medium dependent, up to and including a completely different paradigm and application interface compared to what we have seen so far.

5.1.5.2 High-Level Protocols
There are several popular high-level protocols. We will consider two of them in this book: eager and rendezvous.

5.1.5.2.1 Eager Protocol
Remember what we did every time we got a message to send? Right: we either pushed it away immediately because this was all we could do then (see Chapter 2), or added a message request to the pending send queue and let the progress engine handle the transfer (see Chapter 3). In any case, we have always tried to push the data out as soon as possible. This way of handling messages is called *eager protocol*. This protocol works well if we can take the burden of an accidental extra copy in case the message arrives unexpectedly. For short and medium-size messages this is normally a safe bet.

5.1.5.2.2 Rendezvous Protocol
However, as the message length increases, so does the cost of allocating, managing, and filling in the respective intermediate buffer. At a certain point, all savings provided by sending the message away immediately are more than covered by the aforementioned losses on the receiving side. In this case it may be more profitable to let the sending and the receiving sides synchronize first and make sure that the message contents can be delivered right into the user receive buffer. One way of doing this is provided by the *rendezvous protocol* that normally works as follows:
- On the sending side, once a message is ready, a special control message is sent to the receiving side. This message is usually referred to as "ready-to-send."
- On the receiving side, this "ready-to-send" message make the MPI implementation browse its posted receive queue in an attempt to find a matching receive operation. If such an operation is found, a special control message is returned to the sender. This message is usually called "ready-to-receive." If a matching receive cannot be located, the sender is not notified, and it has to assume that the data delivery has been postponed, and act accordingly.
- Once the sender gets the return "ready-to-receive" message back, the actual data transfer is started. Since the receiving side has already located the matching receive operation, the data is put directly into the destination buffer.

You will immediately see that the rendezvous protocol could help us prevent the 8 MiB issue we have experienced (and tactfully hid) in our socket communication. The choice of the respective threshold depends on the medium used: the extra overhead created by the additional back-and-forth should be offset by the bandwidth gained by skipping the extra copying of the unexpected messages.

5.1.5.3 Low-Level Protocols

We have so far considered only two fundamental media: shared memory and sockets. They do not cover the whole of the available fabrics and protocols, and even within them we have only scratched the surface, being limited by this book's scope. Let's consider the most prominent options available.

5.1.5.3.1 Special Memory

The easiest thing to do is to put the shared memory segment into a special memory if it is available on the platform. There are sometimes so-called scratchpads, or high-bandwidth memory areas, or special caches, or special modes of operation in the memory, sometimes with a stronger alignment requirement attached to them. Without being exhaustive, let me point out that most platforms these days provide a special memcpy(3) function that can bypass the caches. This prevents cache pollution on one hand and may speed up bulk data transfer on the other. If a special memory area is involved, it can be allocated using a special set of calls like memalign(3) and friends, or by setting additional flags that control and modify the behavior of the well-known malloc(3) and its many relations.

5.1.5.3.2 Direct Memory Access

Another idea is to eliminate copying user data into and out of the shared memory segment, and to go for the so-called *Direct Memory Access* (DMA). This feature can be provided in software and/or may be supported by hardware. In software form, it is usually served up by a particular kernel module that may have to be loaded in addition to the standard ones. In hardware form, this feature may rely on a special piece of hardware called DMA engine or such that may in fact even sit on the networking card. Since copying data from one process memory directly to another process' memory eliminates the need for extra copying, these protocols are also referred to as *zero-copy* protocols.

The main point here is to overcome the process memory protection scheme. Ironically, what was introduced to keep data safe and sound by preventing accidental (or, nowadays, intentionally malicious) interprocess data access, has become a bottleneck in the case of HPC. There are several variants of software DMA like XPMEM (Cray, 2014), CMA (Nazarewicz, 2012), and KNEM (Goglin & Moreaud, 2013). Although their specific interfaces differ from each other, one task is common to all of them: the sending process has to somehow learn the destination address to pass data to. For this purpose, the existing shared memory communication mechanism can be used as a side channel.

All three mechanisms mentioned above have been compared to each other in the context of the Open MPI implementation, and the XPMEM was found to be superior (Squyres, 2014). So, if you want to give DMA a try, better start with the XPMEM.

5.1.5.3.3 Remote Direct Memory Access

The idea of DMA can be extended to span more than one node. In this case it is called *Remote Direct Memory Access* (RDMA). In fact, the whole InfiniBand hype was built around this idea. There are several variants of the underlying interfaces, from the historical InfiniBand verbs that were vendor-dependent, through user-level Direct Access Programming Library (uDAPL) that tried to unify them with some degree of success, to the relatively modern OpenFabrics Enterprise Distribution (Openfabrics. org, 2017).

The RDMA interfaces normally have a couple of kinks that need to be addressed by the MPI implementor. One of them is the need to lock (or register) the respective memory areas so that they are protected from remapping during context switches and especially swapping. Most old interfaces required this registration to be done explicitly, which took time. Hence, the need to strike the right balance between the performance of the RDMA operation per se on one hand, and the need to register the memory on the other. This led to the creation of registration caching and other schemes that tried to minimize the need for repetitive registering of the same memory areas in case the MPI application involved sent data back and forth using a relatively static set of user buffers, as they normally would.

The other big issue is the detection of the message transfer completion. In principle, all these interfaces provide a clean way of monitoring the data transfer using special request queues connected to hardware. However, despite the so-called kernel bypass technology, working with these queues incurs extra overhead that basically kills the latency. Due to this, MPI implementors reverted to dirty tricks, detecting transfer of a certain pattern placed in a well-known place, like the buffer end or the expected message end (the latter would save one RDMA write operation). If the underlying data transfer mechanism guaranteed ordered data delivery, as it often did, this scheme worked well, replacing the request queue handling by polling on a certain memory location, which was considerably faster.

You can gain better understanding of the complexity of these interfaces by reading papers dedicated to *MVAPICH* and inspecting the source code of this MPI distribution (The Ohio State University, 2018). The main idea addressing the aforementioned issues is as follows:

- For each pair of processes, a circular set of buffers is established on either side, and their credentials are shared by the two processes involved, so that every time a message can be sent from one process to the other using the current buffer that sort of revolve with each message sent.
- When messages fit within these pre-allocated and preregistered revolving buffers, data transfer is done using them directly. This, of course, incurs extra copying on both the sending and the receiving side. When messages get bigger, the revolving buffers are used to pass control information between the processes, so that a sort of low-level rendezvous occurs:

  o The sending process notifies the receiving process of the wish to send a certain amount of data with a specific message envelope

  o The receiving process responds with the address of the receive buffer as soon as this address is known

  o The sending process initiates the RDMA write operation with the user data going directly from the send buffer to the receive buffer

There is also the possibility of the receiving process pulling data from the send process, in which case an RDMA read operation is performed by the receiving process once it learns the address of the respective send buffer on the other side. What operation to choose depends on what works faster: RDMA write or RDMA read. This needs to be found experimentally beforehand, because sometimes one of the operations is layered on top of the other, with the understandable extra overhead added.

5.1.5.3.4 Higher-Level Interfaces

Of course, RDMA was such a low-level protocol that it did not satisfy the MPI implementors or the hardware vendors completely. The former had to fight against multiple intricacies and the fragility of the interfaces provided. The latter were locked into a very low-level paradigm that imposed severe restrictions upon what could be done to speed up the transfer, and made the vendors compete on raw performance only.

As a natural reaction to that on both sides, higher-level interfaces emerged, such as Portals (Sandia National Laboratories, 2017), *Messaging Accelerator (MXM)* (Mellanox Technologies, 2017a), and others. This way or another, tag matching capability was added to present semantics that looked closer to MPI's. Now, instead of doing the RDMA operations, the MPI implementors could use the more convenient send and receive operations, with the added bonus of doing the tag matching behind the scenes, which in turn simplified the corresponding part of the MPI library.

This "crawling up the stack", as I call it, continued and was temporarily crowned by the interfaces that basically replaced MPI to the point where a corresponding MPI call such as MPI_Isend() or MPI_Irecv() could be mapped upon one underlying "lower-level" call. A good example of this are the OFA libfabrics (OFI Working Group, 2017) and Mellanox' *Unified Communication X (UCX) Framework* (Mellanox Technologies, 2017b) interfaces. This tendency held true not only for the InfiniBand but also for Ethernet.

The downside of this move is that alongside the higher-level operations, these libraries keep offering lower-level interfaces as well, which makes a thorough MPI implementor wonder (and hence benchmark) relative performance of the provided interfaces on all target platforms. This in turn may lead to the need to support several execution paths inside the MPI library, depending on what interface performs better in a particular situation found out there in the field.

Another possible issue is that in their push for generality, these interfaces also offer shared memory connectivity, the quality of which may be lacking in comparison to the hand-crafted implementation specific to a particular platform. However, this is the usual price one pays for convenience, right?

5.1.5.3.5 Multirail

Finally, with the advent of relatively cheap networking, it became possible to send data along several paths, e.g., wires or virtual connections, which led to the corresponding increase in bandwidth. This approach is called *multirail*. Depending on the underlying interface used, and normally those are vendor-specific ones, details may differ, but the idea remains unchanged: messages are either sent as a whole across different wires, to be reordered in case later messages happen to overtake the earlier ones, or split into pieces and multiplexed, to be appropriately reassembled on the receiving end. This means some overhead, so the expected gain should more than compensate for this. Again, the task for synchronizing all that is happening across the wires can be approached by using one path for transferring control messages, thus simplifying the sorting of what belongs where.

5.1.6 Collective Operations

We looked into collective operations before (see Sections 4.4.5 and 4.5.3). Here we will try to improve their implementation in selected cases using various techniques, from better algorithms to special hardware.

5.1.6.1 Design Considerations

Just like point-to-point communication, optimization of collective operations can be done in at least two levels: up there in the MPI space; and down there, right on the wire. The high-level optimizations normally include the use of more advanced collective algorithms. The low-level optimizations use specific capabilities of the underlying medium involved. It is not unusual that both ways coexist and interoperate, especially since only a part of the processes may be amenable to a particular low-level optimization, by virtue of sitting inside a particular network subdomain, using a particular transport layer, sharing a common memory segment, and so on. In this case, the low-level optimizations may be associated with the respective MPI communicator, say, by way of a function table pointing to the optimized, fabric-specific versions of certain collective operations.

5.1.6.2 Algorithms

Algorithms, if you remember, are the most powerful tool in your hand. Instead of feeding you with endless cost estimates and such, although you can certainly find and enjoy them elsewhere (Pješivac-Grbović, et al., 2007), I decided to show their value in practice. Let's take the MPI_Bcast() operation and see what we can do to it at the algorithmic level.

5.1.6.2.1 Tree

A *tree pattern* is good for higher process counts because it takes a logarithm of the number of processes involved into a collective operation. We have seen a binary tree in action when we were starting MPI processes in our first, naïve shared memory MPI subset (see Section 2.3.2.3.2). This same approach could be applied to the MPI_Bcast(), but let's go a different way and do an iterative rather than a recursive implementation this time (see Listing 5.4):

Listing 5.4: Optimized subset MPI: tree-based broadcast (excerpt, see opt/v1/src/coll.c)

```
#ifdef TREE_BCAST
        int n = 4,diff = (size + rank - root)%size;

        while (n <= size)
           n <<= 1;

        n >>= 1;

        for ( ; n > 0; n >>= 1)
           if (rank == root) {
               if (diff + n < size)
                   if ((rc = MPI_Send(buf,cnt,dtype,(rank + n)%size,_MPI_TAG_
BCAST,comm)) != MPI_SUCCESS)
                       break;
           }
           else if (rank == (root + n)%size) {
               if ((rc = MPI_Recv(buf,cnt,dtype,root,_MPI_TAG_BCAST,comm,&stat)) !=
MPI_SUCCESS)
                   break;
               root = rank;
           }
           else if ((size + rank - root)%size > n)
               root = (root + n)%size;

...

#endif
```

According to our custom, we introduce an internal macro TREE_BCAST in order to experiment with a new technology but keep the escape hatch open. You will probably want to take a sheet of paper and verify this implementation. It took me a bit of time to figure out how to do this in a more or less elegant manner. First, I find the power of two just below the number of processes involved (there are better algorithms for this than my simple loop, for sure). Then I go down the tree, halving the step each time, and figuring out whether the current process, being the root, has to send data, or rather receive it first and then possibly send it farther down the tree as the new root.

Now is it time to do some benchmarking? Almost. Remember that we closed the ring in our first MPI_Bcast() implementation (see Listing 4.41). This was caused by our worry about the root process running away and possibly exhausting the very limited request queues of the time. Now we are free to break that last link and see what happens if a ring (or an *O-ring*) becomes a *horseshoe* (or a *C-ring*) (see Listing 5.5):

Listing 5.5: Optimized subset MPI: horseshoe broadcast (excerpt, see opt/v1/src/coll.c)

```
#ifdef SHOE_BCAST
        if (rank == root)
            rc = MPI_Send(buf,cnt,dtype,(rank + 1)%size,_MPI_TAG_BCAST,comm);
        else {
            if ((rc = MPI_Recv(buf,cnt,dtype,(size + rank - 1)%size,_MPI_TAG_BCAST,
comm,&stat)) == MPI_SUCCESS)
                if ((rank + 1)%size != root)
                    rc = MPI_Send(buf,cnt,dtype,(rank + 1)%size,_MPI_TAG_BCAST,comm);
        }
...
#endif
```

We also keep the old implementation in the code, and thus can directly compare the timings of all three algorithms (see Table 5.1):

Table 5.1: Ring, horseshoe, and tree broadcast benchmarks: minimum latency at 16 bytes, 16 KiB, 128 KiB, and 4 MiB (seconds)

Pattern	16 bytes	16 KiB	128 KiB	4 MiB
Bcast (old)	2.18e-06	3.64e-06	1.75e-05	0.0005
2-process	**0.50e-06**	4.31e-06	1.74e-05	0.0005
3-process	8.57e-06	12.42e-06	4.67e-05	0.0020
4-process	12.98e-06	16.88e-06	6.59e-05	0.0026
Bcast (ring)				
2-process	2.0972e-06	2.8976e-06	1.01626e-05	0.000513673
3-process	9.75759e-06	1.36192e-05	4.82165e-05	0.00198621
4-process	1.33993e-05	1.84357e-05	6.53723e-05	0.00261324
5-process	1.80148e-05	2.42678e-05	8.69835e-05	0.00345519
6-process	2.39478e-05	3.05862e-05	0.00011059	0.00442529
7-process	2.96729e-05	3.61028e-05	0.000133896	0.00529108
8-process	3.4496e-05	4.2645e-05	0.000160117	0.0120734
Bcast (horseshoe)				
2-process	2.51231e-06	3.21641e-06	1.04047e-05	0.000515331
3-process	**9.29594e-07**	7.23488e-06	2.77629e-05	0.001514
4-process	1.20029e-06	9.74891e-06	3.71019e-05	0.00190903
5-process	**7.98798e-07**	9.8923e-06	3.58547e-05	0.00289177
6-process	1.05e-06	1.21267e-05	5.20773e-05	0.00360915
7-process	1.20001e-06	1.42925e-05	5.91439e-05	0.0041519
8-process	1.15678e-06	1.73864e-05	5.26181e-05	0.00518266
Bcast (tree)				
2-process	2.495e-06	3.53379e-06	1.07501e-05	0.00052972
3-process	3.2872e-06	6.2109e-06	1.93553e-05	0.00104628
4-process	4.07491e-06	6.42481e-06	2.10101e-05	0.00191234
5-process	5.21481e-06	8.97582e-06	3.04928e-05	0.0024729
6-process	5.7889e-06	1.19654e-05	3.87143e-05	0.00322537
7-process	6.73702e-06	1.40779e-05	5.32739e-05	0.00374505
8-process	6.4362e-06	1.27699e-05	4.49109e-05	0.00443978

Since I had the so-called "Spectre" and "Meltdown" bug fixes applied, withdrawn, and then again reapplied to my computer sometime between writing Chapters 4 and 5, I keep the old (pre-fix) data from Table 4.2 here for comparison. I kept the special treatment of the 1- and 2-process configurations unchanged; the 2-process data is directly comparable across all the patterns, while the 3- and 4-process data is directly comparable across the old and the new ring patterns. I again consistently took the timing measured on MPI rank 0 (the root). As usual, the outliers are highlighted in **bold**.

This time, however, when collecting new timings, I went up to 8 MPI processes hoping to see the power of the logarithm. But where the hell is it? The tree algorithm definitely outperforms the original ring one, especially at higher process counts, but the horseshoe pattern works better than the tree most of the time!

Well, this is a mirage. What you see in the table are *local timings*. Naturally, the horseshoe outperforms most of the rest in this metric, while the timing taken on a process just reflects how long this process is busy transferring its part of the data. And since processes forming the horseshoe naturally synchronize with each other, they almost immediately build a *pipeline*. Hence, the overall time spent by all processes inside a particular instance of the horseshoe broadcast is longer, and it is equal to the number of processes minus one times the latency of each process involved.

In the case of the ring and the tree, however, what you see in the table is the *global timing* of the whole operation! In the case of an 8-process ring-based broadcast, the root process actually waits for the token to go across the whole ring. In the case of an 8-process tree-based broadcast, MPI process rank 0 (the root) sends its data to process rank 4, then 2, and then 1, thus doing three transfers, which is the maximum expected at this process count. This is certainly lower than 7 times the latency which is the overall timing of the horseshoe, and 8 times the latency of the original ring.

In principle, the trap that we have just nearly fallen into is but a shadow of the holy war between the proponents of local and global timings. Both camps have good arguments for their preference and even better ones against the other choice. However, what matters is what you personally want to measure, and what you say about the numbers so obtained.

One more notice is warranted here: the *pipelining algorithm* mentioned above is a very powerful one, and it can be applied not only between the similar operations but also inside them. Imagine, for example, an MPI_Bcast() that splits the data to be sent into smaller chunks, and starts pushing them one by one, possibly receiving the incoming contributions, if any, on the other end of the ring or, for that matter, any other exchange pattern like the tree. This way, one can lower the operation start-up overhead, use the optimal message size to achieve the highest possible bandwidth (remember the shared memory cache effects, for one), and also avoid the saturation of the channels at very high message sizes. One can also extend the MPI subset by introducing the MPI_Sendrecv() call that is ideally suited for this sort of communication arrangement. And once you think about multithreading or multiple cores, the pipelining idea becomes very promising indeed.

5.1.6.2.2 Hypercube

The collective algorithms we have considered so far—the ring, closed or not, and the tree—work well for all process counts. There are however algorithms specific to a particular kind of job size, the most prominent being those for the powers of two. For good or for bad, application programmers tend to love the powers of two, so that it may make sense to have an algorithm or two in the library specifically suited for this situation.

Hypercube is one of those special patterns. In one dimension, a hypercube is a segment connecting two vertices. In two dimensions, this is a square. In three—well, a common cube. And then you can go into four, five, six dimensions, and so on, still keeping the pattern exemplified above. This pattern is very appealing for operations that have to distribute data to everyone, like the MPI_Allreduce() or MPI_Allgather(). Indeed, exchanges between the processes involved always go along one direction or the hypercube edges, thus preventing any interference and keeping the network traffic relatively well structured and low. At the same time, the nature of the hypercube again takes a logarithm of the problem dimension, and reduces, for example, an operation for eight processes to three exchanges per process—and uniformly, unlike the tree. In other words, during a hypercube driven exchange, all processes really do useful work all the time, rather than idly wait for the data to come or overwork themselves sending the data out.

5.1.6.2.3 Advanced Algorithms

There are many substantially more sophisticated collective algorithms. In fact, one could write a whole separate book or a great number of articles dedicated to this topic alone. For the sake of brevity, what I can recommend you at this point is to go and learn more about this exciting topic using a good starting point, like "Performance analysis of MPI collective operations" (Pješivac-Grbović, et al., 2007). If after this overview you still want to deal with this matter, pursue the references from that paper, then have a look in another paper coming from the competing camp (Thakur, Rabenseifner, & Gropp, 2004) and note both the names of the authors and the references they cite. Once done, you will know most of those involved and will have learned more than an average MPI implementor ever had to, but you can still add some on top by searching the Internet for more recent additions. Note however that some of the algorithms used out there may have never been described, mainly because of intellectual property considerations.

5.1.6.3 Native Implementation

Our collectives have so far been mapped upon the point-to-point exchanges. This is a reasonable approach when the implementation of the operations involved needs to stay general and portable to all the targeted fabrics. If, however, the targeted fabrics are known beforehand, one can use the respective fabric-specific expressive methods to gain extra performance.

5.1.6.3.1 Lock-Free Collectives

One medium where the mapping upon point-to-point operations looks particularly wasteful is the shared memory. Indeed, having a shared memory segment or even better mechanisms (see Section 5.1.4.2.2), one wastes a lot of time going through the MPI_Send() and friends if, for example, a few processes have to be synchronized. There are multiple well-known methods based upon the synchronization primitives such as mutexes, semaphores, and conditional variables that can do this faster and better.

There is however one more possibility that was described in a pretty old paper of mine (Supalov A. , 2003). You can look the matter up in this article that did this relatively naïvely, targeting a low number of SMP processes typical of the time. The main idea was to dedicate a special part of the shared memory segment to the shared memory collective operations, and then to apply the double buffering scheme discussed earlier to the data to be transferred. Note that by the virtue of the MPI-1 standard, any MPI process can be involved in just one collective operation at a time across all communicators it participates in. Alas, this is no longer true in MPI-3 due to the advent of the nonblocking collective operations. Once again, there is extra stress put upon the device level by any extra semantics introduced up there.

Of course, the progress has not stopped then. We will revisit this topic in Section 5.2.8 while dealing with multifabric optimization.

5.1.6.3.2 Multicast

Networks like Ethernet and InfiniBand have built-in capabilities for multicast, that is, sending a message to all or a group of nodes, the latter determined by the subnet mask, a virtual process group, or another mechanism. In my experience, using these mechanisms is normally cumbersome, while the expected profit is rather low.

5.1.6.3.3 Collective Extensions

Higher-level interfaces often provide additional native interfaces for doing the collective communication. Mellanox' *Fabric Collective Accelerator (FCA)* (Mellanox Technologies, 2017c) as well as OpenFabrics libfabric have them, too. In the past, there were also other interfaces, but they are irrelevant now, like collective extensions to the uDAPL interface (Intel Corporation, 2011). Here, however, the expected profit can be rather noticeable, so it makes sense to invest in benchmarking and, in case of success, use these extensions as prescribed by the respective specifications.

5.1.6.3.4 Special Registers

On some platforms, hardware supports collective operation through special registers and the like. One popular mechanism is the global synchronization lock that is provided by hardware. Unfortunately, it is often restricted to the whole of the machine, without the possibility of splitting off parts of it for an MPI job, to say nothing about a communicator inside it.

5.1.6.3.5 Extra Networks
Really expensive custom-built machines come with more than one network installed, not counting the Ethernet and the InfiniBand that to some extent co-exist in big clusters, the former being normally the control network, the latter serving the application data exchanges and possibly the NFS file transfers, thus assisting the MPI process start-up. What I mean here are real special-purpose additional networks that also have a special topology dedicated to a particular task, be that the data transfer or the interprocess synchronization. A typical configuration would include a low latency, high throughput primary network of some nature, mostly organized as a torus, a butterfly, or some other fancy pattern. A secondary network organized as a tree spanning all the nodes would be dedicated to the internode synchronization. However, either of these networks could also be (mis)used for other purposes.

5.1.6.3.6 Accelerators
Accelerators are now everywhere. Certainly, an MPI implementor will never skip yet another method for making the MPI library faster, and so accelerators have their heyday now. By their very nature, they are seldom suitable for running fully-blown MPI jobs, and rather, serve as a helping vehicle, especially if access to their memory range is restricted or indirected, which was very often the case in the not so distant past. One point where they can be very useful indeed, even under these conditions, are the reduction operations. In some sense, graphics cards can be considered poor man's vector units, provided they are programmed properly.

5.1.7 Exercises

Exercise 5.1 Implement the MPI_Sendrecv() call. What other call(s) would you add as well?

Exercise 5.2 Optimize the MPI_Bcast() using the horseshoe pattern and the pipelining idea (see Section 5.1.6.2.1).

Exercise 5.3 Implement the MPI_Reduce() using the tree pattern.

Exercise 5.4 Implement the MPI_Allreduce() using the hypercube pattern and the MPI_Sendrecv() call.

Exercise 5.5 Describe the lock-free shared memory barrier.

5.2 Multifabric Optimization

Multifabric optimization is built upon combining or selecting particular fabrics for particular parts of the MPI library activity. This advanced area of HPC is densely infested with software patents, so the following description will have to be substantially oblique in order not to lead you into a trap. I will highlight those patents I know—since I have also contributed to the infestation, *mea culpa, mea ultima culpa!*— and for the rest of it, you have been warned.

5.2.1 Design Decisions

In order to optimize anything using the multifabric capability, one has to have this capability in the first place. We have developed two (rudimentary) MPI devices so far—one for shared memory and one for sockets. We can combine them now in several ways (see Figure 5.1):

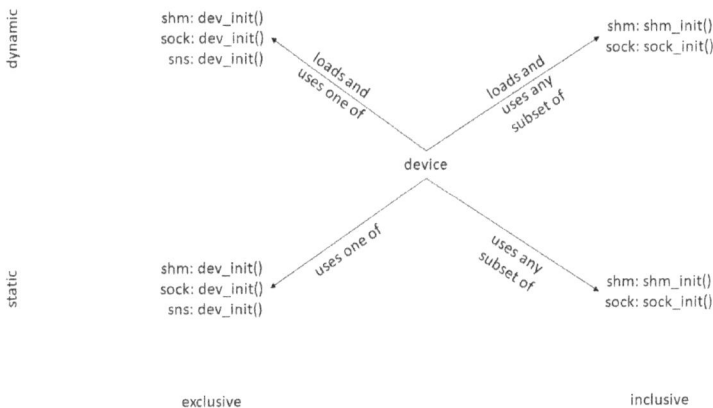

Figure 5.1: Multifabric capabilities

- Targeting static, exclusive configuration (lower left corner), we can build our subset MPI in such a way that it will use the shared memory, or the sockets, or a combined shared memory/sockets device (called here and elsewhere "sns" for "shared memory and sockets"). Of course, multifabric capability is only present during the run time if the sns multifabric device has been configured and built inside the current subset MPI. Note that in this case every device can provide a standard interface prefixed, say, by the dev_ abbreviation, since we do not need to differentiate between them at run time.

- We can do this dynamically as well (upper left corner), in which case the upper MPI device layer will load and use one of the shared memory, socket, or multi-fabric devices. The ability to load a device at run time can be rather convenient to the user who can then choose a precise configuration necessary for a particular run. However, this extends the development effort because all three devices (in this case) will have to be created, tested, benchmarked, and then maintained by the MPI implementor. Moreover, this adds a tiny bit of latency because using a dynamically linked library may be marginally more expensive, at least on some legacy platforms.

- If we however decide to use more than one device at a time—say, shared memory and sockets—we need to call the device functions in some fabric-specific way. This can be done statically (as in the lower right corner) or dynamically (as in the upper right corner). In this case there is no real need for a combined device, but in that case the individual devices for shared memory and sockets need to behave themselves. We will consider this matter in more detail in Section 5.2.4.5. Of course, even in this case, there may still be a need for a combined device for performance reasons, just because shared memory is so much faster than the sockets, and every little bit of fat in the shared memory critical path or on the way there will make itself clearly felt.

In this book, we will use the static, exclusive approach (lower left corner in Figure 5.1) because this is the easiest way, and it will let us get all the necessary details and performance right with the least possible effort. The 10/90 principle, as usual.

5.2.2 Code Refactoring

Well, it is going to be a long hunt. So far, we have been dealing with one fabric at a time and made them work as far as our test programs were concerned. In order not to reinvent the wheel, we will now reuse most of the code created so far, simultaneously highlighting those transformations that are necessary to make the former unifabric code capable of functioning in the multifabric environment.

5.2.2.1 Design Considerations

In order to go through this process in a structured fashion, we will follow this sequence of steps, basing on the extended code optimized earlier in this chapter:

- Separate the socket-specific device code
- Add the shared memory device from Chapter 2, throwing in a simple progress engine

–   Unify the process management infrastructure of the shared memory and socket devices

We will have to do testing along the way to make sure that nothing gets broken or degrades performance. By now you ought to have become so accustomed to this, that we will not make big fuss about details, all right?

5.2.2.2 Socket Device

Well, we have to bump the socket code up a notch again, adding the directory sock in the src/dev hierarchy and moving socket-specific code there. Our request management code is fabric-independent, and thus it stays put. All of the fabric-specific layers will provide the same set of capabilities—initialization, termination, and progress—so that the device header shown in Listing 5.1 remains unchanged.

We will select a device to use at build time for now, so that, as mentioned above, we do not need any device-specific function names. The Makefile structure will change accordingly (not shown).

5.2.2.3 Shared Memory Device

Fortunately, we do not have to recode everything. Taking the modified socket device as a template, we fill it out using the shared memory code from Chapter 2.

5.2.2.3.1 Device Initialization

Device initialization is a snap (see Listing 5.6):

Listing 5.6: Optimized subset MPI: shared memory device initialization and termination (excerpt, see opt/v2/src/dev/shm/dev.c)

```
...

int _mpi_dev_init(int *pargc,char ***pargv)
{
    int rc;

    assert(_mpi_init == 0);
    assert(pargc != NULL && *pargc > 0);
    assert(pargv != NULL);

    if ((rc = _mpi_getargs(pargc,pargv)) != MPI_SUCCESS ||
        (rc = _mpi_allocate()) != MPI_SUCCESS   ||
        (rc = _mpi_split(0,_mpi_size)) != MPI_SUCCESS) {
#ifdef DEBUG
        fprintf(stderr,"%d/%d: -> failure\n",_mpi_rank,_mpi_size);
#endif
```

```
        return rc;
    }

    return MPI_SUCCESS;
}

int _mpi_dev_fini(void)
{
    assert(_mpi_init);

    return MPI_SUCCESS;
}
```

You may have divined that we kept the simple process management infrastructure unchanged, so that the shared memory device still expects and accepts the -n and -np program options. Device termination is trivial (not shown).

5.2.2.3.2 Progress Engine

And when it comes to the progress engine, the proven structure developed for the sockets is combined with the data transfer snippets that worked well for shared memory (see Listing 5.7):

Listing 5.7: Optimized subset MPI: shared memory device progress engine (excerpt, see opt/v2/src/dev/shm/dev.c)

```
static int _mpi_progress_send(MPI_Request req)
{
    struct _mpi_shm_s *pshm;
    struct _mpi_shm_cell_s *pcell;
    char *buf;
    int cnt;

    if (req->flag)
        return MPI_SUCCESS;

    pshm = _mpi_shm + _mpi_rank*_mpi_size + req->rank;
    pcell = pshm->cell + pshm->nsend;

    while (pcell->flag)
        ;

    for (buf = req->buf,cnt = req->cnt,pcell->tag = req->tag,pcell->len = cnt; cnt >
BUFLEN; buf += BUFLEN,cnt -= BUFLEN) {
        memcpy(pcell->buf,buf,BUFLEN);
        pcell->flag = 1;

        pcell = pshm->cell + (pshm->nsend = !pshm->nsend);
```

```
            while (pcell->flag)
                ;
    }

    memcpy(pcell->buf,buf,cnt);
    pcell->flag = 1;
    pshm->nsend = !pshm->nsend;

    req->stat.MPI_SOURCE = req->rank;
    req->stat.MPI_TAG = req->tag;
    req->stat.cnt = req->cnt;
    req->flag = 1;

    return MPI_SUCCESS;
}

static int _mpi_progress_recv(MPI_Request req)
{
    struct _mpi_shm_s *pshm;
    struct _mpi_shm_cell_s *pcell;
    char *buf;
    int cnt,unexp = 0;

    if (req->flag)
        return MPI_SUCCESS;

    pshm = _mpi_shm + req->rank*_mpi_size + _mpi_rank;
    pcell = pshm->cell + pshm->nrecv;

    while (!pcell->flag)
        ;

    if (req->tag != pcell->tag)
    {
        MPI_Request reqx;

        if ((reqx = _mpi_newreq(&_mpi_xq)) != NULL) {
            reqx->rank = req->rank;
            reqx->tag = pcell->tag;
            reqx->cnt = pcell->len;
#ifdef DEBUG
            fprintf(stderr,"%d/%d: unexpected message, tag %d != %d, cnt = %d\n",_mpi_
rank,_mpi_size,req->tag,reqx->tag,reqx->cnt);
#endif
            req = reqx;
            unexp = 1;
        }
        else {
#ifdef DEBUG
```

```
            fprintf(stderr,"%d/%d: -> failure, tag %d != %d\n",_mpi_rank,_mpi_size,
req->tag,pcell->tag);
#endif
            return MPI_ERR_INTERN;
        }
    }

    if (unexp) {
        if ((req->buf = malloc(req->cnt)) == NULL) {
#ifdef DEBUG
            fprintf(stderr,"%d/%d: malloc buf: %s\n",_mpi_rank,_mpi_size,
strerror(errno));
#endif
            return MPI_ERR_OTHER;
        }
    }
    else if (req->cnt < pcell->len) {
#ifdef DEBUG
        fprintf(stderr,"%d/%d: -> failure, cnt %d < %d\n",_mpi_rank,_mpi_size,
req->cnt, pcell->len);
#endif
        return MPI_ERR_TRUNCATE;
    }
    else
        req->cnt = pcell->len;

    for (buf = req->buf,cnt = req->cnt; cnt > BUFLEN; buf += BUFLEN,cnt -= BUFLEN) {
        memcpy(buf,pcell->buf,BUFLEN);
        pcell->flag = 0;

        pcell = pshm->cell + (pshm->nrecv = !pshm->nrecv);
        while (!pcell->flag)
            ;
    }

    memcpy(buf,pcell->buf,cnt);
    pcell->flag = 0;
    pshm->nrecv = !pshm->nrecv;

    req->stat.MPI_SOURCE = req->rank;
    req->stat.MPI_TAG = req->tag;
    req->stat.cnt = req->cnt;
    req->flag = 1;

    return MPI_SUCCESS;
}

...
```

Now that we have a fairly straightforward shared memory progress engine, we can attend to the question of what happens when there are more MPI processes than processors, cores, and hardware threads to actually run them. In this case (called *processor oversubscription*), one process occupying the hardware thread would block it until the work it had was done, or until the kernel decided that this process had had enough of the CPU attention and pushed it out for others' sake. Once in a while this could mean that something would block indefinitely. In this case, calling the sched_yield(2) system call every so often may eliminate this issue. However, forcing a context switch, which is what this system call usually does, is expensive. So, it hardly makes sense to use this method in case of *undersubscription*. Whether or not this call should be issued when the number of MPI processes is exactly equal to the number of CPU units to run them, which is often desired unless threading has to be used by the MPI application, is open to debate and generally depends on the level of additional system activity you are willing to pay for in terms of lost performance versus the remote possibility of an MPI application suspension.

Note that any MPI device that uses system calls in the critical path, like the socket device we have been dealing with, provides enough opportunities for the kernel to preempt an overly zealous process, since every system call potentially can be used as a context switch point between processes managed by the operating system.

5.2.2.4 Start-up Unification
We pass the process management options to the MPI library from a shared memory specific mpiexec.shm command (see Listing 5.8):

Listing 5.8: Optimized subset MPI: shared memory mpiexec command (see opt/v2/bin/ mpiexec.shm)

```
#!/bin/sh
launchLocal () {
#    echo "launchLocal $*"
    local size=$1
    shift
    local prog=$1
    shift

#    echo $prog $size: "$*" "[wait]"
    $prog -np $size $*
}

if [ $# -gt 0 ]
then
#    echo "$*"

    if [ "$1" = "-n" -o "$1" = "-np" ]
```

```
      then
          shift
#             echo "$*"

          if [ $# -gt 0 ]
          then
              size=$1
              shift
#               echo "$size x $*"

              if [ $# -gt 0 -a $size -gt 0 ]
              then
                  launchLocal $size $*

                  exit $?
              fi
          fi
      fi
fi

echo "Usage: mpiexec -n[p] number_of_processes program [program_arguments]"
exit 1
```

The socket-specific mpiexec becomes mpiexec.sock, but it does not change at all other-wise (see Listing 3.7). A symbolic link mpiexec points to the "right" mpiexec.* variety. Setting it up correctly will be left to you. This is a minor nuisance that we will get rid of soon. The same holds for the option parsing machinery that drags along here even though comparable functionality also exists in the socket device code (see Listing 5.6). We consistently follow the main programmer's maxim: first make things work, then optimize!

5.2.2.5 Testing and Benchmarking
Testing and benchmarking confirm that none of the functionality or performance was lost in the process. Hurray!

5.2.3 Multifabric Device

Now that we have both unifabric devices following the same pattern, and also have both process management parts of the package unified, we can proceed to creating a multifabric device that will combine both shared memory and socket capabilities under one roof. Just to be on the safe side, we will not remove the existing old devices yet, because we want to make sure that the new combined device works correctly and performs well—hopefully, just as well as the old ones. Should this be the case, we will have the option of removing the unifabric devices and thus reducing further develop-ment and maintenance effort by a factor of three.

5.2.3.1 Design Considerations

We can and should use this chance to unify more of the paraphernalia common to all three devices. This way we can get rid of the code duplication mentioned above and have only one set of functions that interpret the contents of the options and configu-ration files. They are inherited from the socket device (see Listing 3.9) and provide the following interface (see Listing 5.9):

```
int _mpi_getint(char *var,int val);
char **_mpi_getarr(char *var,int len);
```

Listing 5.9: Listing 5.9. Optimized subset MPI: environment query functions (excerpt, see opt/v3/ src/dev/env.h)

5.2.3.2 Device Initialization

For the device initialization, we basically smash the shared memory and socket codes into each other, taking care of the possible need to use only shared memory or only sockets, or both of them in one job (see Listing 5.10):

Listing 5.10: Optimized subset MPI: multifabric device initialization (excerpt, see opt/v3/src/ dev/sns/dev.c)

```
...

static int _mpi_shmon = 1,_mpi_sockon = 1;

int _mpi_dev_init(int *pargc,char ***pargv)
{
    int rc;

    assert(_mpi_init == 0);
    assert(pargc != NULL && *pargc > 0);
    assert(pargv != NULL);

    _mpi_size = _mpi_getint("MPI_SIZE",_mpi_size);
    assert(_mpi_size > 0);
    _mpi_rank = _mpi_getint("MPI_RANK",_mpi_rank);
    assert(_mpi_rank >= 0 && _mpi_rank < _mpi_size);
    _mpi_root = _mpi_rank;

    _mpi_hosts = _mpi_getarr("MPI_HOSTS",_mpi_size);
    assert(_mpi_hosts != (char **)NULL);
    _mpi_ports = _mpi_getarr("MPI_PORTS",_mpi_size);
```

```
    assert(_mpi_ports != (char **)NULL);

    _mpi_shmon = _mpi_getint("MPI_SHM",_mpi_shmon);
    _mpi_sockon = _mpi_getint("MPI_SOCK",_mpi_sockon);
    assert(_mpi_shmon || _mpi_sockon);

    _mpi_lsize = _mpi_getint("MPI_LSIZE",_mpi_lsize);
    assert(_mpi_lsize > 0);

    if (_mpi_shmon) {
        if ((rc = _mpi_allocate(_mpi_lsize)) != MPI_SUCCESS) {
#ifdef DEBUG
            fprintf(stderr,"%d/%d: -> mmap failure\n",_mpi_rank,_mpi_size);
#endif
            return rc;
        }
    }

    if (_mpi_lsize > 1) {
        if ((rc = _mpi_split(0,_mpi_lsize)) != MPI_SUCCESS)
#ifdef DEBUG
            fprintf(stderr,"%d/%d: -> fork failure\n",_mpi_rank,_mpi_size);
#endif
        return rc;
    }

    if (_mpi_sockon)
        if ((rc = _mpi_connect_all()) != MPI_SUCCESS ||
            (rc = _mpi_nonblock_all()) != MPI_SUCCESS) {
#ifdef DEBUG
            fprintf(stderr,"%d/%d: -> sock failure\n",_mpi_rank,_mpi_size);
#endif
            return rc;
        }

    return MPI_SUCCESS;
}
```

The environment variable MPI_LSIZE tells the MPI library how many processes are to be started locally using the fork(2) system call. The process start-up for the first process of a local group—normally called *group leader*—is managed via the ssh(1) or directly, as it used to be for the socket device (see Listing 3.7). Thus, we introduce a so called *heterogeneous start-up* that uses different mechanisms for creating different MPI processes in the fastest and most appropriate manner, contrary to the earlier used *homogeneous start-up* that favored simplicity over speed.

Local connections use an anonymous shared memory segment, remote connections use sockets as before. Note that by spawning the local processes first, we also quite naturally open sockets between all processes, including the local ones. We do not use these local sockets just yet, although in principle we could. Had we done the latter, we would introduce *heterogeneous connections* as opposed to the currently used *homogeneous* connections. We will revisit both topics—heterogeneous start-up and heterogeneous connections—in more detail later in this chapter (see Sections 5.2.5 and 5.2.6).

5.2.3.3 Start-up Unification

We just keep on the practice of adding device-specific mpiexec commands (see Listing 5.11):

Listing 5.11: Optimized subset MPI: multifabric device mpiexec command (see opt/v3/bin/mpiexec.sns)

```
#!/bin/sh
shm=${MPI_SHM:-1}
sock=${MPI_SOCK:-1}

hostfile="mpi.hosts"
portbase=${MPI_PORTBASE:-10000}
rankstep=${MPI_RANKSTEP:-1}

launchLocal () {
    echo "launchLocal $*"
    local rank=$1
    shift
    local size=$1
    shift

    local lsize=$rankstep
    local nextrank='expr $rank + $rankstep'

    if [ $nextrank -ge $size ]
    then
        lsize='expr $size - $rank'
    else
        launchLocal $nextrank $size $*
    fi

    if [ $rank -eq 0 ]
    then
        echo $rank/$size: "$*" "[wait]"
        (export MPI_RANK=$rank; export MPI_SIZE=$size; export MPI_SHM=$shm; export
MPI_SOCK=$sock; export MPI_LSIZE=$lsize; export MPI_HOSTS="$hosts"; export
MPI_PORTS="$ports"; $*)
    else
        echo $rank/$size: "$*" "[nowait]"
```

```
        (export MPI_RANK=$rank; export MPI_SIZE=$size; export MPI_SHM=$shm; export
MPI_SOCK=$sock; export MPI_LSIZE=$lsize; export MPI_HOSTS="$hosts"; export
MPI_PORTS="$ports"; $*) &
    fi
}

launchRemote () {
#    echo "launchRemote $*"
    if [ $1 -lt $2 ]
    then
        local rank=$1
        shift
        local size=$1
        shift
        local host
        if ! read host
        then
            host="$last"
        fi

        launchRemote 'expr $rank + $rankstep' $size $*

        if [ $rank -eq 0 ]
        then
#            echo $rank/$size: ssh $host "$*" "[wait]"
            ssh $host "export MPI_RANK=$rank; export MPI_SIZE=$size; export
MPI_LSIZE=$rankstep; export MPI_HOSTS=\"$hosts\"; export MPI_PORTS=\"$ports\"; $*"
        else
#            echo $rank/$size: ssh $host "$*" "[nowait]"
            ssh $host "export MPI_RANK=$rank; export MPI_SIZE=$size; export
MPI_LSIZE=$rankstep; export MPI_HOSTS=\"$hosts\"; export MPI_PORTS=\"$ports\"; $*" &
        fi
    fi
}

if [ $# -gt 0 ]
then
#    echo "$*"

    if [ "$1" = "-n" -o "$1" = "-np" ]
    then
        shift
#        echo "$*"

        if [ $# -gt 0 ]
        then
            size=$1
            shift
#            echo "$size x $*"
```

```
                if [ $# -gt 0 -a $size -gt 0 ]
                then
                    if [ -r $hostfile ]
                    then
                        last="localhost"
                        rank=0
                        while [ $rank -lt $size ]
                        do
                            if read host
                            then
                                last="$host"
                            fi
                            hosts="$hosts $last"
                            ports="$ports 'expr $portbase + $rank'"
                            rank='expr $rank + 1'
                        done < $hostfile

                        launchRemote 0 $size $* < $hostfile
                    else
                        rank=0
                        while [ $rank -lt $size ]
                        do
                            hosts="$hosts localhost"
                            ports="$ports 'expr $portbase + $rank'"
                            rank='expr $rank + 1'
                        done

                        launchLocal 0 $size $*
                    fi

                    exit $?
                fi
            fi
        fi
fi

echo "Usage: mpiexec -n[p] number_of_processes program [program_arguments]"
exit 1
```

As you can see, the externally defined environment variable MPI_RANKSTEP, if set, determines how many MPI processes need to be started per each node of the MPI job. This variable is transformed into the earlier described internal environment variable MPI_LSIZE according to the number of processes that actually need to be started on a particular node. This number can be equal to MPI_RANKSTEP, but it can also be lower on the last node(s) of the group in case there are not enough processes left.

   This is a very simple interface that makes testing easier, because once set, the MPI_RANKSTEP influences the whole of the test runs. I considered other opportunities, including the well-known -ppn option that is used to the same end in many MPI implementations, but decided against that, since this would require extensive changes to the respective Makefile. If you do not like my way, you are free to implement the traditional method.

Arbitrarily sophisticated extensions, including but not limited to noncontiguous rank layouts, up to and including manual process placement on particular nodes, are also possible. Just imagine that ranks are assigned according to an array called, say, MPI_LRANKS instead of simply incrementing the group leader rank. This will, however, lead to certain complications of the adjacency matrix and result in additional code and overhead. Are they worth it? You decide.

5.2.3.4 Progress Engine

The progress engine deals now with two fabrics instead of one (see Listing 5.12), and nothing else changes in the data transfer machinery:

Listing 5.12: Optimized subset MPI: multifabric device progress engine (excerpt, see opt/v3/src/dev/sns/dev.c)

```
...

static int _mpi_progress_send(MPI_Request req)
{
    if (_mpi_shmon)
        if (req->rank >= _mpi_root && req->rank < _mpi_root + _mpi_lsize)
            return _mpi_shm_progress_send(req);

    if (_mpi_sockon)
        return _mpi_sock_progress_send(req);

    return MPI_SUCCESS;
}

static int _mpi_progress_recv(MPI_Request req)
{
    if (_mpi_shmon)
        if (req->rank >= _mpi_root && req->rank < _mpi_root + _mpi_lsize)
            return _mpi_shm_progress_recv(req);

    if (_mpi_sockon)
        return _mpi_sock_progress_recv(req);

    return MPI_SUCCESS;
}

...
```

Appropriately prefixed fabric-specific progress engines deal with the shared memory and socket parts of the task. Since we use a very simple process placement scheme (that could, however, be arbitrarily extended if need be), knowing what MPI rank starts the local process team and how many processes are in it, is sufficient to determine whether a certain message targets this or another node. Again, all this could be made super generic and extensible if need be—but mind the associated costs.

5.2.3.5 Testing and Benchmarking

Surprisingly enough, the new multifabric device works and shows performance matching that of the respective individual devices (not shown). However, there is a hitch here that spoils the celebration: we cannot get the ping-ping and pong-pong patterns running. Why? In order to answer this question, we need to look closer into what happens and how a real progress engine works.

5.2.4 Progress Engine Revisited

The ping-ping and pong-pong patterns represent so called *head-to-head* communication. In the ping-ping pattern both participating processes try to send messages to each other, in the pong-pong pattern, they try to receive messages from each other. Remember that we did not put the original simple shared memory subset MPI to this rigorous test and added the head-to-head communication for sockets only. Even there, we had to reduce the size of the buffer down to 8 MiB, otherwise something would hang, even during ping-pong. Now is the time to understand what and why.

5.2.4.1 Design Considerations

Let us look at the head-to-head communication first. For it to work, both participating processes must be able to simultaneously send and receive messages to and from each other without hanging if that can be avoided. This sounds counterintuitive, but this is a consequence of the so-called progress requirement of the MPI standard. At the same time, looking into our current progress engine, we notice that it is not suitable for handling this situation.

5.2.4.1.1 Blocking vs. Nonblocking Progress

First of all, a progress engine is a *blocking* one if once we enter it, we return only if there was a message to send or to receive, and we have processed it. Certainly, this is not what should be happening, because the progress engines on both sides will start to either send or receive data, and they will naturally deadlock. Note however that even though a blocking progress does have its limitations, it also has a very substantial advantage in the unifabric case: it may block once in a while. Due to this, it can do many clever and useful things, including waiting on interrupts or events, and backing the polling off up to and including putting the process to sleep.

5.2.4.1.2 Whole vs. Partial Message Delivery

Second, our progress engine can only handle *whole messages*: it's just all or nothing, the progress engine cannot pause at the middle of a message, put it aside for a while, and start dealing with another message. This would help in the case of head-to-head communication, for one, so that once a receive operation starts and finds no data, it could be put aside, and the send operation, that does have data, could be started. By jumping from the receive operation to the send operation, the progress engine would be able to deliver and get the message in parts, respectively.

These limitations become absolutely critical in the multifabric environment. Indeed, if a process is blocked, say, sending a message via shared memory while another process is blocked waiting for this process to send data over a socket connection, the second process gets blocked. And what if for some reason, including an application error, nothing ever comes through shared memory? Well, both processes will sit there waiting for something to come, and this will be a hopeless undertaking, also for those other processes that may be waiting on those two directly or indirectly.

5.2.4.1.3 Request- vs. Data-Driven Progress

There is one more hitch here that has not come to fruition just yet, but, by Murphy's Law, which actually is the main governing law of programming in general and of parallel programming in particular, it definitely will. Imagine a sequence of messages with tags that we denote A, B, A, B, going from one process to another. Note that on both sides the progress engines are request driven, that is, the sending process works through a list of pending send requests, so that it will emit messages in the aforementioned sequence. The receiving process also has a list of messages to process, which hopefully (just hopefully, but not necessarily) follow the same sequence, because in the presence of tag matching, MPI should be able to handle situations when certain messages overtake each other if the receiving process so decree and the tags so prescribe. However, even if the receiving process wants to receive messages in the said sequence, this is what may happen due to the processes working independently of one another:

- The receiving process tries to receive a message with tag A. However, this process runs ahead of the sending one, and there is nothing to receive yet. Our new nonblocking progress engine returns.
- Being driven by the pending receive message queue, the receiving process tries to receive a message with tag B. In the meantime, the sending process has caught up and started to send the first message with tag A. Our progress engine receives this message as an unexpected one, putting it into the unexpected message queue.

– Now, the receiving process goes to the next element in the pending receive request queue and tries to receive a second message with tag A—the third one in the queue. Well, the receiving process finds such a message in the unexpected message queue! And so, a message with tag A that was intended to match the first request in the pending receive request queue, is matched by the third message in this queue, also with tag A. In other words, two messages with tag A happen to be received in reverse order, thus violating the pairwise ordered message delivery requirement of the MPI standard. This is unacceptable.

We see that our progress engine needs a major rewrite. Instead of being *request driven* on the receiving side, it needs to be driven by the *data* it detects in the communication channel being served. In this, it needs to match the incoming messages to the elements of the pending receive queue, and if nothing matches there, put the incoming message into the unexpected receive queue. This element will then be matched once a receive request is queued that holds a matching signature; that is, both the matching source process and the matching message tag. In order to avoid unnecessary polling of the open connections, the engine may use facilities like poll(2), select(2), epoll(7), or equivalent fabric specific mechanisms, including sensing a cache line state change, and respond only to those connections that go live during this progress cycle.

5.2.4.1.4 Fair vs. Unfair Progress

However, this is not the end of the story, not by a long shot. Imagine that we get messages delivered via both types of connections, shared memory and sockets. What connections are to be served first? What is the cost of doing this? Will latency accumulated while serving one channel reflect itself upon the overall latency experienced by the MPI application?

Well, it is typical to serve the fastest connections first and, thanks to the fairness non-requirement of the MPI standard, be pretty relaxed about situations in which one of the channels may get little or even no attention from the progress engine for rather extended periods of time. This was actually the governing idea behind a patent that described a so-called multifabric *pragmatically truncated* progress engine (US Patent No. 7,567,557, 2009). This simple trick lead to almost pure shared memory latency in the shared memory channel and almost pure socket latency in the socket channel, respectively, because typical shared memory latency was negligible when added to the typical socket latency, and socket latency was not interfering with the shared memory due to immediate return from the progress engine as soon as at least one shared memory message was detected and its processing started.

If this simple approach is deemed too risky or is simply unable to handle all eventualities, including lazy connection establishment, one can control the frequency at which particular fabrics are revisited by the progress engine. One possible way of doing so is described in another patent (US Patent No. 8,245,240, 2012). Note, however, that the progress engine is probably the hottest spot in the MPI patent minefield I've mentioned before, so that you need to be very careful about handling this matter in public. For this very reason I will not be able to provide you with a full source code of a multifabric progress engine in this book. You can however always find a good example of it in the MPICH or Open MPI code. I hope that the general discussion presented here will help you make sense of what is happening there and why.

5.2.4.1.5 Requirements

To sum it up, we need to make our new multifabric progress engine generally *non-blocking* and *partial*; and, in addition, also *data driven* and possibly *unfair* on the receiving side. Of course, if done, this will benefit the unifabric communication as well, and if we later decide to drop the unifabric devices, we will have to develop such a progress engine only once. Accidentally, this will also resolve the inconvenience of having to maintain several device-specific mpiexec commands and to use the right one that matches the device involved.

It is actually a more general principle that reveals itself here: a correct design decision will always manifest itself later on in beneficial and often unforeseen impacts. Likewise, a wrong design decision will always create ever more repercussions down the road. It may be advisable to use this criterion alongside Occam's razor to verify all design work.

5.2.4.2 Synchronous Progress Engine

So far, we have been dealing exclusively with so called *synchronous progress engines*: to get them work, the MPI library had to explicitly "kick" them (as MPI implementors casually call the action of invoking the progress engine). The point at which this is done is very important.

First of all, not all MPI calls should kick the progress engine, only those directly or indirectly involved in communication. Otherwise one may get unpredictable behavior of seemingly innocuous and purely local calls like MPI Comm_rank().

Second, when the progress engine is invoked, one has to pay attention to the consequences of doing this too early or too late. For example, in our earlier blocking progress engines we always kicked the engine before enqueuing a new request, both on the sending and on the receiving side. This way we ensured that if we had any data in the channel on the sending side, we would push it out before adding more work. On the receiving side, if there was data in the channel, it would be handled as an unexpected message immediately, regardless of whether we have a matching request for it. We would pay double for handling the memory copying (once for the intermediate buffer, once for the user buffer), but we kept the channel as free as possible. Of course, it might be interesting to try kicking the progress engine after enqueuing a fresh send or receive request, to see how this influences performance and correctness of the program. This, however, may cause some rewrite of the code.

Note that our current progress engine is blocking, so that one has to pay attention not only to performance but also to the possibility of the test program suspension if the progress engine is used too actively. With the nonblocking progress engine we outlined in Section 5.2.4.1, there may be an additional issue of handling too many messages as unexpected if on the receiving side the progress engine is kicked before yet another receive request is enqueued.

Finally, an invocation of the progress engine changes the global state of the MPI library, because it affects both the communication channel and the request queues. Whatever decisions have been made beforehand on the basis of this information, need to be reviewed upon the progress engine invocation, otherwise one may keep on going in a rather outdated and possibly wrong direction.

5.2.4.3 Asynchronous Progress Engine

Synchronous progress engines have a big intrinsic issue: for them to work, the MPI library needs to be invoked fairly often and at rather regular intervals. This happens most of the time, but there are a couple of peculiar yet completely correct situations when the MPI library may not be called for an arbitrarily long period of time. A completely synchronous progress engine will not be able to handle this situation, especially in presence of the lazy connection establishment: an unhandled connection request will eventually time out. The situation becomes even more acute if one-sided communication, especially the so called passive target variety of it, is involved: there, the target MPI process may not even know that the source MPI process is trying to put or get some data remotely!

This is why in some packages MPI implementors experimented with more involved progress engines. One of them the is the so-called *asynchronous progress engine*: basically, an additional thread that handles all of the requests in background. This requires the MPI library to be thread safe at least on the inside—a topic that we will consider in some detail in Chapter 7. Leaving the nitty-gritty to that future happy time, it is sufficient to say that, in this case, the main thread that executes the MPI application and hence the MPI library proper interacts with the progress thread via a set of request queues that are properly protected against potentially conflicting simultaneous actions of these two threads, be that either adding a new request, or modifying its state, or deleting it. This can be achieved in a number of ways that we will consider later on. Whatever the way, however, some overhead exists just because running two threads with access to a common set of data always creates some friction in the system, either directly, say, on the mutual exclusion primitives, or indirectly, say, in the cache hierarchy, thread management, processor time allocation, context switching, and the main memory interface.

Therefore, however attractive at first glance—one thread does the upper level MPI work and lets the progress thread handle the data pushing, which will look rather well especially in case of abundant access to separate hardware threads or even dedicated cores available in the current processors—the asynchronous progress engine frequently loses performance compared to the synchronous one. Here absolute correctness is traded for impaired performance of most applications. Is there a way out of this situation?

5.2.4.4 Hybrid Progress Engine

Yes, there is such a way: to do those things that can be done synchronously in the main thread right there, and to delegate the more involved tasks like connection establishment or passive target one-sided communication to a helper progress thread, either strictly ("you do this always, and you do that always") or flexibly ("oh, I have time right now, let me do this here for a change"). I heard about this approach for the first time when attending a vendor session at a conference where a package called WMPI II was presented (Christensen, Brito, & Silva, 2004). Whether this was the first case of a so-called *hybrid progress engine* or not, is open. In any case, it looks like a nifty idea, whether it's been patented or not, right? Of course, there is a lot of detail to be settled when such a complicated progress engine is built, so we leave this topic and ascend to the next metalevel.

5.2.4.5 Generic or Meta Progress Engine

Can one combine several independent MPI devices, and hence several independent progress engines inside a meta MPI device? This becomes pertinent when the constituent MPI devices each can handle a mutually exclusive subset of fabrics and do this so well that there is no desire, time, or indeed any need to redo this work.

Well, such a *generic progress engine*, sometimes also called a *meta progress engine*, is indeed feasible if all the constituent MPI devices behave themselves. First of all, they all need to be purely nonblocking—clear, since otherwise they will not be able to follow the progress rule. Second, they all need to be mutually protected against conflicting simultaneous access to the critical data structures like message queues and such (of course, all these data structures should ideally be shared between devices without undue duplication). And third, these devices all need to be keenly aware of each other and especially make no fuss if they see a request that they cannot handle themselves.

There is, however, one MPI feature that may be rather difficult to handle correctly in this situation—this is the MPI_ANY_SOURCE wildcard. Again, we have looked into this matter only in passing so far and will deal with it more specifically in Chapter 7. Let us revisit the meta progress engine there, once we understand all the implications of this very powerful and yet controversial feature of the MPI standard.

5.2.5 Start-up and Termination

Basing on what we have learned about process start-up in the unifabric case (see Section 5.1.3), the extension to the multifabric situation is going to be rather trivial and yet very effective.

5.2.5.1 Design Considerations
We have used heterogeneous (or hybrid) shared memory and socket-based start-up in Section 5.2.3.2. This was a specific example of a more general approach applied whenever there is more than one way to create MPI processes. We have seen the usual Linux system calls fork(2) and, implicitly, exec(2) in action so far, but certain parallel installations and operating systems may have additional facilities.

5.2.5.2 Heterogeneous Start-Up
One of these additional facilities is normally referred to as *remote fork*. It can be thought of as a system service that automatically clones the current process image upon another node using whatever mechanisms provided by the operating system or its parallel extension. Another possibility is the so-called *remote exec*. Here, the operating system manages loading an indicated process image upon another node, again using whatever internal mechanisms available.

It may sound counterintuitive, but whenever you meet such a facility out there, you need to measure thoroughly how it performs before opting for it in your MPI library. More than once in my experience a carefully hand-crafted implementation would outperform the respective system call. This is normally not a sign of sloppy programming on the part of the OS developers. They may have had a wider, more complicated scenario in mind when they formulated the product requirements and the respective design criteria. Just compare this situation to the subset MPI performance versus that of substantially wider full MPI implementations. Who did a better job here? Well, both, in their own way.

5.2.6 Connection Establishment

All of the chemistry developed for the unifabric devices naturally applies to the multifabric ones (see Section 5.1.4). However, the ability to use different connections separately or in parallel adds an extra level of complexity and power in the multifabric case.

5.2.6.1 Design Considerations
So far, we have been using very compact representation of the connections used by our devices. The shared memory used a shared memory array representing all the possible connections in their entirety, whether they were actually used or not. The sockets were described by a sole integer representing the respective file handle. This was a kind of ad-hoc approach that grew as we were adding features one by one.

Once we start thinking about other types of connections, especially those used in assorted combinations between the same pair of MPI processes, this simple representation may become a limitation. This is why mature MPI implementations introduce the concept of a *virtual connection*. This is normally a data structure that contains a description of the connection, including its type and the required additional information fields describing the connection state and specifics. These data structures become then the basis of formulating all operations, up to the point of defining a set of functions to be called whenever a connection is being used. These functions naturally depend on the connection type and use the aforementioned additional information fields in a connection-specific manner. The rest of the operation remains transparent to the code that uses this connection. This is somewhat comparable to the file interface of the operating system kernel: whatever the underlying device driver, its user can normally count on the read(2) and write(2) system calls to be available.

5.2.6.2 Heterogeneous Connections
There are essentially two ways of using the power of virtual connections:
- You can choose what type of connection to use between a particular pair of processes, either for the entire duration of the MPI job, or dynamically, depending on the traffic observed so far or predicted for the near future
- You can combine several types of connections, which is especially beneficial if there is some additional data transfer hardware or system services that yield better performance

Let us consider these varieties one by one.

5.2.6.2.1 Static Connections
We have used this scheme in our multifabric device (see Section 5.2.3.2): local processes would use shared memory exclusively, while remote processes would rely on sockets. This way we achieved low latency locally and acceptable bandwidth remotely.

5.2.6.2.2 Adaptive Connections
However, we could have applied a different scheme, especially for bigger jobs. For example, the shared memory segment could be reformulated so that only those connections that require lowest latency had a place in it that might be dynamically

assigned to a particular pair of processes. This assignment could potentially change in the course of the MPI run. The rest of the connections would be universally served by sockets. Comparing the memory consumption for the shared memory and sockets (in our case, double buffering structure versus one integer plus its counterpart structure in the kernel), you can get the idea of the scale of savings. If other mechanisms were applied, say, the RDMA or fancier interfaces, the total memory consumption could be kept under control even in very large jobs. The connectionless data transfer methods would further enhance this affect.

5.2.6.2.3 Parallel Connections

Even more interesting is the situation in which a pair of processes is connected using different types of connections acting in parallel—say, shared memory and sockets. In this case, one could, for example, use the shared memory for small messages, and add sockets or DMA for the bulk data transfer. This concept is described in substantial detail in yet another patent (US Patent No. 8,281,060, 2012).

A fascinating variety of parallel connection is the *heterogeneous multirail* arrangement. Here, several sorts of connections can be arranged as if they were multirail wires (see Section 5.1.5.3.5) in order to achieve higher performance. Of course, these "wires" may themselves be of the *homogeneous multirail* nature, described earlier.

5.2.7 Point-to-Point Communication

All that has been done before—the start-up, the connection establishment, all the tricks mentioned—were intended to speed up the job execution in general, and point-to-point communication in particular.

5.2.7.1 Design Considerations

From the design point of view, the less upper layers of the MPI library know about what is happening during the data transfer, the better. So, all details of how exactly the bytes are being pushed back and forth are normally totally isolated inside the MPI device, and even within the device—in the virtual channel mechanism.

5.2.7.2 Virtual Channels

When there is only one type of connection available between two given processes, there is nothing new here compared to the situation we were dealing with before. The interesting part starts when there are more connection types available. The biggest question that arises, then, is the ordering of the messages or their segments, in case multiplexing is done across the available channels. If you decide to use one of the connections for carrying control messages, you should be aware of at least one patent (US Patent No. 8,281,060, 2012).

5.2.8 Collective Operations

Collective operations readily inherit all the good and bad features of the point-to-point calls if the collectives are mapped upon them. However, what is important to understand is that the order in which these point-to-point operations are called may greatly influence the overall performance. This is also true if some collectives are supported directly by the respective devices on a particular subset of MPI processes (see Section 5.1.6.3).

5.2.8.1 Design Considerations
We have seen large differences between intranode performance, especially if shared memory was used locally, and internode performance, especially if sockets and Ethernet were used for remote connections. In this situation, it almost certainly makes sense to sort all connections into local and remote, and then use local connections as much as possible. Note that, in this case, the upper MPI layer should actually know pretty well what is happening under the hood, at least in terms of the expected performance.

5.2.8.2 Locality Exploitation
The principle of locality exploitation is straightforward: do all you can locally, do what you must remotely. Sometime, just skip doing anything remotely: for example, surprisingly enough, it may be more beneficial to perform a reduction of a full vector on each of the processes involved into the MPI_Allreduce() operation, even though you will have to do an MPI_Allgather() first to get all input data to all participating processes.

Special attention should be paid to the result reproducibility when a particular job is run on different sets of processors and nodes. While ignorable in the case of logical and integer data entities, a different order of floating point operations may lead to differing—and sometimes dramatically differing and even wrong—results, including occasional creation of numeric exceptions (Goldberg, 1991). Due to this, a high quality MPI library should provide a way to either control or eliminate the possible effects of the ordering of local and remote operations upon the resulting value.

This is true within the run, when different processes might, due to a careless or overly aggressive optimization, contribute their partial results to the total in a different order, depending on the relative timing of the contributions and other transient factors. This is, however, especially true for different runs that may be done using a different number of processes and/or their layout. In this case, the result reproducibility requirement may strictly limit the options available to the implementor of the respective collective operation, and sometimes impair their performance. In other words, reproducibility should probably not be a high priority goal or the default setting.

Apart from the reductions, all other collectives may and normally do profit greatly from the proper use of the process proximity as well as the use of respective special mechanisms. You will have no problem finding references that explain this approach and illustrate its effect, especially if you start with a good article (Li, Hu, Hoefler, & Snir, 2014).

5.2.9 Exercises

Exercise 5.6 Add sched_yield(2) into the shared memory path and measure the slowdown if this system call is issued on every iteration of the progress engine.

Exercise 5.7 Implement the -ppn option instead of the MPI_RANKSTEP environment variable in the mpiexec command. What takes precedence if both the variable and the option are used at the same time?

Exercise 5.8 Kick the progress upon enqueueing a new send request rather than before that. What happens to performance? Why?

Exercise 5.9 Like Exercise 5.8, but for the receive request. Same questions.

5.3 Conclusions

We have gone pretty much through everything that can be done on the optimization side. Further performance improvements will not require source code changes, at least as far as heavy-duty parts of the MPI library are concerned. Of course, some coding will be necessary, but it will deal with getting data in and out of the library in the process of tuning that we will consider in the next chapter.

Let us complete our complexity investigation now (see Table 5.2, cf. Table 4.3):

Table 5.2: MPI subsets vs. full MPI implementations

Subset	Entities	Incl. Functions	LOC
Shared memory	13	7	343
Sockets	18	11	852
Extensions	60	27	2137
Optimization	60	27	2745
MPI-1	244	128	~100000
MPI-3.1	855	450	~350000

As you can see, the complexity increment introduced by our optimization is rather small, even though we added the multifabric capability.

Chapter 6
Tuning

In this book, the term MPI *tuning* denotes MPI performance optimization done by proper selection of the already existing code paths and their parameters rather than by source code changes. Tuning attempts to make the product achieve the best possible performance on a particular system—out in the field, rather than in your own test lab.

Tuning seeks to improve latency or bandwidth of a certain communication pattern like ping-pong, of a set of collective operations, of a target application performance expressed in wall clock time or an application-specific rating, or of any sensible mixture thereof. However, tuning is only the last and rather expensive polish applied to an otherwise finished MPI library. If you have not carefully selected your algorithms, implemented your primitives, and optimized your fabrics, tuning will only be able to go as far as the basis allows. No amount of tuning will be able to fully offset bad design, sloppy programming, and careless benchmarking.

6.1 Overview

Tuning can be precomputed in advance (in anticipation of the expected target platforms) and then applied on the spot. Additional tuning can also be done individually prior to the run, during it, and even between runs. Let's first review the various forms of tuning in order to better understand how they work and how they interact with each other as well as the MPI library and the MPI users. We will review the respective techniques in more detail later on.

6.1.1 Default Settings

Tuning starts with selecting the most appropriate *default settings* of the MPI library. These settings are determined by the MPI implementor on the basis of the entire body of performance information collected during product development and its use, the latter done principally by other people called end users—those who use both the MPI library and MPI applications built on top of it. Sometimes they also develop or at least tune those applications, but this group of MPI application developers and tuners is rather small—bigger than that of the MPI implementors, but substantially smaller than the whole group of the MPI users.

For a particular performance-relevant setting to be accepted as a default, it needs to be beneficial for at least 90% of target platforms and applications. This is a rule of thumb, of course, so that reasonable flexibility on the part of the MPI implementor is

DOI 10.1515/9781501506871-006

welcome here. Proper default settings can make or break the product, because what people see during the first run of their favorite application often determines whether they have a second look at this particular MPI library at all. This is why extreme care and caution are exercised when default settings need to be changed in a new library release. This change should by no means disappoint the existing user base, but still cater to the new and potential users.

6.1.2 Static Settings

In some cases, even the most carefully selected default settings may be insufficient to make a particular application sing. In this case, either the MPI user or the MPI library can try to improve performance of the target application by applying further changes to the performance-relevant application- and platform-specific MPI parameters.

For this *static tuning* to work, the MPI user and/or the MPI library must be aware of the platform being used: the processor frequency, type, and model; the memory subsystem characteristics, starting with the number, size, and mutual relation of the cache levels; the available networks, their parameters and topology; and so on. Most of this data is inquired by the MPI library anyway using the many various services provided by any mature operating environment. The MPI library also knows enough about the MPI job itself: how many processes there are, where they are being executed, and how they are connected to each other. MPI users, to the contrary, are quite often blissfully unaware of all this, so that the MPI library must come to the rescue— its own rescue, by the way, since a poorly performing library is going to be dropped on the spot, as mentioned before.

The knowledge of the platform and job parameters mentioned above constitutes the basis for individual tuning. Using this information, the MPI user and/or the MPI library can select the most appropriate precomputed configuration of the tuning parameters to start with, can try to improve them further during the run, and can understand enough about the application at hand to further optimize its execution when it is invoked the next time. Let's consider these variants of tuning in more detail.

6.1.2.1 Precomputed Settings
The use of the precomputed settings assumes that some performance data has been collected and analyzed prior to the MPI library release in order to compose a body of interdependent settings deemed most suitable for a particular target application and/ or platform. We call this process MPI *pretuning*. It may affect the default MPI library settings mentioned above, in case certain settings common to the majority of the target applications and platforms qualify for this promotion. The main result of the pretuning process, however, is a multitude of precomputed sets of MPI parameters

that can be applied by the MPI user and/or library immediately, without additional performance analysis.

There are essentially two sorts of precomputed settings:
- *Platform-specific settings* that are considered beneficial for a particular platform as a whole, whatever the application. These settings are normally computed beforehand by the MPI implementor and affect all applications executed on a particular platform. Note that these settings may, and normally will, change with the product upgrade, so you should pay close attention to the regression testing and benchmarking here.
- *Application-specific settings* that are considered beneficial for a particular application, with or without platform dependency. These settings may have been computed by the MPI implementor or the MPI user. These settings normally override the platform-specific settings in case of a conflict—it is assumed that the application developer or at least the application user knows more about application performance and how to tune it.

6.1.2.1.1 Platform-Specific Settings
Platform-specific pretuning is intended to produce platform-specific settings that can be retrieved later on depending on the parameters of the actual MPI run—primarily the platform type, the number of MPI processes, their layout, and pinning. These settings are precomputed in a massive job that may take days and weeks to complete, since so many variants of the aforementioned settings need to be tried and verified in the process. Various techniques directed at saving resources during this process are exemplified by another patent (US Patent No. 9,448,863, 2016).

The result of this work is normally stored in a set of configuration files or a database that is queried by the MPI process manager whenever a new MPI job is started. Whether this happens or not depends on the way these settings are controlled by the user. In some cases, their use can be explicitly desired and explicitly requested. In other cases, when tuning is applied by default, it can be explicitly disabled.

6.1.2.1.2 Application-Specific Settings
The easiest way to obtain application-specific settings is to do a number of runs of the target application with various aspects of the MPI library tuned this way or that, to see what settings yield a better result. In order to make this tuning process a little less haphazard, you can also try to analyze the application algorithm and the way in which it uses the MPI library. The analysis is described in a number of books, e.g., (Supalov, Semin, Klemm, & Dahnken, 2014).

Once found, the application-specific settings can either be entered in the command line or, better still, put into a special application-specific configuration file to be used explicitly or by default whenever this particular application is launched. A good example of this approach is given by those applications that rely heavily upon certain collective operations that happen to function best when a certain algorithm or their combination is used. The same is true of extra settings that may improve certain point-to-point settings, be that the eager protocol threshold or the internal MPI buffer size.

6.1.2.2 Individual Settings

Sometimes the precomputed settings may need additional adjustment or may have to be scrapped altogether in favor of a very particular, normally application-dependent set of parameters that make a particular target application fly on a particular target platform and possibly workload. In this case, the results of performance analysis or even inspiration (yes, this happens, too!) are entered as control parameters for the respective MPI run in the hope that they will fix things.

These kinds of settings normally take the form of specific environment variables and run time options set for a particular run. When the parameters settle down, the respective values are often put into a configuration file to be used with a particular application on a particular platform, explicitly. With time, if the relationship between the MPI implementor and the MPI user are good, these *individual settings* may become a part of the respective MPI library distribution, or even make their way into the precomputed and default settings.

6.1.3 Dynamic Settings

With the most appropriate static settings identified, you can go a step further and try to improve MPI library performance on the fly. Doing this correctly is a bit tricky. First, the overhead associated with monitoring the MPI library performance, figuring out the required parameter change, and applying this change consistently across all the affected MPI processes must be justified by the expected saving of total execution time or another metric used as the cost function.

Second, the change itself should be synchronized globally or at least within the job segment that is affected by the change. This is a little bit similar to the dynamic rerouting done by some networks: if there are any routing tables and no per-message routing done according to transient and local routing information only, these tables need to be synchronized across the affected network segment.

Third, these changes should be stable in that they do not introduce a slowdown or a deadlock. The latter is very well possible if, for example, one process changes its eager protocol threshold while its partner remains unaware of this. Now imagine what happens if some MPI processes try to use different collective algorithms for the same collective operation.

Another aspect of these *dynamic settings* is the time scope. You can always start with some predefined settings and improve them inside the run, basically redoing the work done in the previous runs that used the same or a similar application work-load. This purely *intrajob tuning* looks a bit wasteful. What if, instead of starting with the same initial static settings, you use the result of the previous tuning session as the initial settings for the next run instead? Then, with due care, you may hope to improve the respective application-specific settings further and further in a kind of closed tuning loop. We call this process *interjob tuning*.

6.1.3.1 Intrajob Settings
When a certain application runs long enough for dynamic tuning to make sense, the results of it are used immediately. If no trace of this work remains and every time the tuning is going to be done anew, they can only be application-specific by intent, and platform-specific by nature.

6.1.3.2 Interjob Settings
The situation changes if the results of dynamic tuning are being stored for later reuse. The order in which the dynamic settings can be changed differs somewhat from the static case, too. There, we would start with platform-specific settings, and then refine them on a per application basis (see Section 6.1.2.1). Here, to the contrary, since the parameter changes occur during a particular application run, the settings changed first are by definition application-specific. Later on, if the changes tend to converge to a certain common pattern across some or all of the applications being run on a particular platform, there comes a time when application-specific settings may be elevated to become platform-specific. In this case, all applications will automatically inherit the modified platform settings that can still be overridden—in case they care, of course. Finally, with time, certain common settings may qualify for a promotion to the default.

6.1.4 Exercises

Exercise 6.1 Recall the double buffer size tuning from Section 2.3.6.4. Was this static or dynamic, platform- or application-specific tuning?

6.2 Static Tuning

Tuning is basically minimization (or maximization) of a certain cost function. Doing this in advance, in order to obtain precomputed settings considered to be good for a particular application and/or a particular platform, may take an arbitrarily long time and thus can be both very comprehensive and very clever. Whether this is done by a human operator or by a special program, depends on the circumstances.

6.2.1 Design Decisions

The process of tuning depends on the data being passed by the *tuner* to the MPI library and back. The "tuner" here stands for the MPI implementor or the tuning automation machinery exploited by the development team. On the forward path, parameters like algorithm selectors, threshold values, and other control data need to be passed to the MPI library. On the backward path, certain performance metrics—like timing, call frequency, buffer lengths and so on, up to and including the settings actually in effect—need to be collected and passed back to the tuner on request, implicit or not.

As usual, it makes sense to make the parameter passing paths sufficiently universal in expressive power and relatively simple in implementation. In addition to this, it is always pleasant and beneficial when the facilities defined for tuning can also be used to other useful purposes—for example, verifying the parameter settings in effect, outputting certain statistical data for postprocessing, producing debugging output for checking product correctness, gauging performance of certain internal suboperations for fine adjustments, and so on.

The results of tuning may be applied immediately or stored for later use. For the latter, some system needs to be developed that is capable of storing precomputed settings and recalling them when there is a need.

6.2.2 Parameter Control

Most of the MPI tuning is dedicated to finding proper settings of those parameters that influence MPI performance in the most noticeable and effective ways. Experience shows that the following settings normally need to be properly set up first and roughly in this order:

- *Fabric selection.* It makes no sense, for example, to tune TCP/IP performance if your MPI jobs will all fit within the node, nor does it make any sense to use TCP/IP if you have a faster network protocol available. Note that modern MPI libraries normally select the most appropriate fabrics automatically. This choice can, however, be overridden, within reasonable limits, and should be closely controlled and monitored during tuning, otherwise you can easily spend days and even weeks tuning a completely wrong type of connections.
- *Process layout.* Placing the most heavily intercommunicating processes so that they can use the best possible connection may single-handedly create a great performance result or completely ruin it.

- *Process binding.* Even if processes are mapped properly, they may still take wrong positions inside the node and suffer performance slowdown due to this or to undesired process migration. Good selection of adjacent hardware threads versus adjacent cores and sound accounting for the NUMA system configuration are all very important.
- *Point-to-point protocol selection thresholds.* Of these, the eager protocol threshold is probably the most important single parameter to be tuned upfront, once the fabrics mix, the process layout and their bindings are all fixed. In selecting the correct thresholds, the predominant point-to-point patterns should be considered as well. For example, settings for an application that does pairwise ping-pong exchange may be radically different from an application that fills the communication channel with multiple messages of the ping-ping style.
- *Collective algorithm selectors.* Depending on the number of processes in the MPI job, their mapping upon the nodes and processors, the binding of the MPI processes, and the fabrics involved, wildly different collective algorithms need to be used for achieving best performance. Note, for example, the difference between the odd and even process counts and pay special attention to the powers of two that are universally loved by MPI users. The most frequent operations like MPI_Barrier() and MPI_Allreduce() need to be tuned first.

You may note that we apply here the top-down approach outlined earlier in this book (see Section 1.4.7). If you want to learn more about tuning and performance analysis, you may want to read more about this topic (Supalov, Semin, Klemm, & Dahnken, 2014).

6.2.2.1 Design Considerations

Two basic parameter control mechanisms are available almost universally across all MPI implementations: environment variables and runtime options. They can be entered individually or read by the MPI implementation more or less transparently from special MPI configuration files.

Whether or not every runtime option should have its counterpart in the environment variable space and vice versa is an open question. Some MPI implementations make a point of having a complete equivalence between these two ways of passing parameters to the MPI library. Other implementations consider some of the settings so tactical as to warrant no extra environment variables when a certain option is made available anyway. I personally do not like unnecessary duplication, primarily because it may lead to interesting hiccups out in the field. One of the most frequent issues encountered in this case is a "forgotten" setting of the number of processes per node. This issue alone may easily spoil a whole series of benchmarking runs and go unnoticed until it's too late.

In any case, keeping everything as unified as possible helps avoid a confusing mixture of data input mechanisms. For example, if configuration files are supported, it is better to have them use those options and environment variables that are accessible in the mpiexec command line and at shell command prompt, respectively, without introducing additional features or fancy file formats.

6.2.2.2 Parameter Input

There are several ways in which data can be passed to the MPI library, most of which we have considered, used, or at least mentioned before. Let's make a more comprehensive list and see how these methods apply to the process of MPI tuning.

Environment variables are probably the most flexible but not always the most convenient way of affecting the MPI library—for example, in the case of selecting a particular threshold, algorithm, or another control parameter. Once set, these variables stay defined in the current session and may be accidentally inherited by the next session. This is why the less permanent control data is normally passed via runtime options like -np (for the total number of processes), -ppn (for the number of processes per node), and such.

Both environment variables and runtime options can be collected and stored in the configuration files, which spares the user repeating this information every time an MPI job is started. These files are normally organized in hierarchical fashion, so that there are system-wide, user-specific, and job-specific configuration files. They are put into a system-specific place (e.g., the /etc. directory under Linux), a user-specific place (e.g., the user home directory), and a job-specific place (e.g., the current directory), respectively. In addition to this, platform-specific configuration files may also be organized based on a different principle and live elsewhere, especially if they are automatically selected by the MPI library depending on the set of job parameters, whether that is the total number of processes, their layout, and so on.

These static data sources are normally ordered by priority, so that:
- Runtime options override
- Environment variables that override
- Job-specific configuration file that overrides
- User-specific configuration file that overrides
- System-wide configuration file(s), including those used for platform-specific tuning

This way, the MPI user may be inheriting a lot of settings from the lower priority levels without ever knowing it, and at the same time can always refit the job to his or her liking using selected user-level controls of various operational agility.

6.2.2.3 Parameter Output

The most basic performance characteristic needed for tuning the MPI library need not be output by the library itself. This is wall clock time that can be measured externally, either by the time(1) command or by many other means. In our benchmarking, for example, we took the timings using the MPI_Wtime() call in order to stay as system independent as possible.

A well-evolved MPI library will normally have a built-in statistic gathering module. The data collected by the library on the number of calls, data buffer lengths, their usage frequency, and so on, directly or via the PMPI profiling interface, can all be used for tuning. In some cases, the statistics are gathered all the time and output on request. In others, the statistics are gathered and output only on request, possibly using a special configuration of the MPI library and associated tools. The advantage of the built-in statistics module is that it can go much deeper and cause less over-head than any external tool, even if the novel MPIT tool interface is involved (Message Passing Interface Forum, 2015). In addition, the internal statistical data is readily available at any moment.

The MPI library can output a lot of information, both on the current settings and on their effect on the library, including but not limited to the aforementioned statisti-cal data in some format. This data can be then read in by the offline tuner or analyzer.

Finally, some applications try to measure their performance using applica-tion-specific methods, whether that is timing akin to ours or calculation of a certain rating that may need minimization or maximization, depending on its nature. This data can also be used as the cost function for the parameter tuning process.

6.2.2.4 Parameter Tuning

With the data input and output paths clarified, what remains is the application of the usual minimization or maximization methods driven by the respective cost function. Some ideas of how to do this can be found in another patent (US Patent No. 7,769,856, 2010). This art covers automatic tuning, but the general principles and specific approaches work for manual tuning as well. Comparing this information to your own experience in tuning certain aspects of the subset MPI performance gained so far, you should get a pretty good idea of the process and the expected results.

The biggest issue in tuning is that by looking for a local extremum (be that the minimum or maximum of the cost function), you may easily miss another, substan-tially better parameter setting if it is separated from the current local extremum by a deep valley or a high ridge (as seen in the respective numeric space). All the tricks that help avoid this deplorable situation should in principle be applied to the MPI tuning, as are all accelerator methods known from linear algebra and functional analysis.

Another matter to consider is the sheer volume of the space that needs scouting for best results. Do not forget the possible and sometimes even clear interdependency between the many parameters that influence MPI performance either. The number of processes and their layout are the most obvious ones. However, point-to-point proto-col thresholds, collective algorithm selection criteria, and even the sequence of oper-ations used for benchmarking all may influence the result and its applicability to the general case.

One practical issue is the total time needed for tuning. Running a full applica-tion or a full-blown benchmark may be prohibitively expensive, especially if the parameter space is sufficiently large. Note, however, that not only a benchmark or an MPI application, but also an isolated or synthetically generated kernel (US Patent No. 7,856,348, 2010) or even an analytical model (US Patent No. 9,448,863, 2016) can be used as the optimization object or a facilitating technology just as well. Provided the interdependencies mentioned above are respected, using a truncated optimiza-tion run may greatly help in saving time and achieving a better result.

6.2.2.5 Parameter Storage and Retrieval

Once settings are computed, they need to be stored somewhere for later retrieval. In a minimalistic design, it is sufficient to store them in configuration files, either platform-specific or application-specific, in plain format normal for these files. Then an appropriate file can be selected either automatically by the MPI process manager during the job start-up, as soon as the MPI job characteristics like the number of pro-cesses and so on become known, or explicitly by the user who knows best, hopefully, what settings suit the MPI application at hand.

Certainly, this is just the beginning of it. One can easily imagine an involved hier-archical database that stores all this information in some internal, preprocessed form and does a lot of magic to select proper values and interpolate between them if such a need arises, since not all possible job configurations can be tested during an exten-sive precomputation. The introduction of such a database becomes rather pertinent when nontrivial actions like closed loop optimization are considered—at least for the future of the product. Whether to go to this trouble up front or not depends on the time and funding available. We will revisit this exciting topic in Section 6.3.2.4.

6.2.3 Process Remapping

Although process layout control can be considered as a kind of parameter control, because it basically redefines the mapping of the MPI processes upon the available processors and nodes, it stands a little outside of the normal tuning. It requires inter-action between the process management system and the MPI library *per se*. This is why we consider this topic in a separate section.

6.2.3.1 Design Considerations

Putting processes upon the nodes is a lovely game. Recalling the effects of the locality upon performance, making sure that the most intensively communicating processes have the lowest latency and highest bandwidth available to them may result in extreme variations in the MPI performance and hence the application run time.

There are three basic ways to control the process layout:
- From inside the MPI application
- From inside the MPI library
- From outside of the MPI job

In order to use the best possible layout, an MPI user may decide to use the virtual process topologies with the attending calls like `MPI_Comm_cart_create()` and `MPI_Graph_create()` in the hope that MPI implementors did a good job actually optimizing things there. Frankly, this is rarely the case, because MPI users are assumed to be smart enough not to shoot themselves in the foot, and so are the modern networks that use dynamic routing and optimization internally. Some corrective actions may indeed be undertaken and the biggest mistakes corrected by the MPI library when a proper virtual topology is requested, but counting on this happening all the time and to the best possible performance is probably a little naïve. In some cases, however, the very imperfection of the virtual topology mechanism can be exploited to astounding results. I remember a case, for example, when merely swapping Cartesian dimensions led to a 30% performance hike of a certain application because heavy messaging that had been taking place internode before this change, went intranode and profited from the shared memory bandwidth.

The MPI library itself may try to optimize the process layout by monitoring performance and migrating processes during the MPI run. This is a very advanced feature called process migration that we will discuss in Section 6.3.3. Again, counting on it being available and always doing a great job may be futile.

What practically remains is the careful placement of the MPI processes when the MPI job is started, or their dynamic remapping if a performance issue is detected. This can be done by the MPI library and its process management, within limits, but this can also be done during so called load balancing that is affected by system components like job schedulers and job managers. Let's leave the latter two to their own devices, for the sake of keeping the book focused upon the MPI, and see what can be done in the MPI library and its process management infrastructure to help MPI users get better performance under otherwise equal conditions by manipulating the process layout from outside the MPI library.

6.2.3.2 Static Remapping

The easiest way to remap processes statically, before launching an MPI job, is to change the host list of the mpiexec command. However, this will not be the most convenient, because once in a while the user may want to reorder processes inside the nodes, as well. In addition, changing the host list may lead to undesired host file proliferation.

Introducing one more level of indirection works best in this case. For example, in our unified sns device, we simply incremented the MPI ranks starting from the local group leader but could have assigned the ranks according to a remapping list called, say, MPI_LRANKS of size MPI_LSIZE for local MPI processes (see Section 5.2.3.3) or MPI_RANKS of size MPI_SIZE for all of them. Setting such a list up might require additional parameters in the external interface, and would certainly take some fuss internally, but this would put the whole process placement onto universal footing.

In this case, whatever the external interface was—a rank remapping file, a machine file with sophisticated description of the process layout, or just the -ppn option or an equivalent setting of an environment variable—it would not be seen by the library at all. Everything would be expressed in terms of the aforementioned two rank lists. This is probably the most elegant way of introducing this feature unless extreme scalability is targeted by the MPI library.

Regarding the cost function involved in the creation of the rank remapping file, it is normally driven by observed latencies and bandwidth across individual connections. Moving intensively communicating processes closer, in the sense of this cost function, and letting other processes get their share as they can will do the trick. The connections can be monitored by the library itself or via an additional tool that gets data via a trace file created using the PMPI profiling interface. The remapping can be computed using one of many graph optimization algorithms and packages available out there.

6.2.4 Exercises

Exercise 6.2 Inspect the implementation of the MPI_Cart_create() and MPI_Graph_create() calls in your favorite MPI implementation. Do they reorder process ranks for better performance?

6.3 Dynamic Tuning

Depending on when you do the tuning—beforehand, on the fly, or between the runs—the general approach remains unchanged: this is minimization (or maximization) of a certain cost function. Doing this beforehand, in order to obtain precomputed settings considered to be good for a particular application and/or a particular platform, may take an arbitrarily long time and thus can be both very comprehensive and very clever. Tuning on the fly, to the contrary, must be agile and quick, otherwise the time spent computing the settings may not be compensated by the observed gain in performance. Tuning done between runs, in a sort of closed loop optimization, may again be a little more relaxed, but need not be as comprehensive as tuning done beforehand.

6.3.1 Design Decisions

The biggest decision to be made here is whether the dynamic tuner will be a part of the MPI library itself or an outside agent that is capable of tuning more than one MPI implementation. The latter is possible in principle, now that the MPI standard defines a so-called MPIT tools interface. However, this interface determines only the mechanics of the parameter query and control rather than the semantics of the entities at hand. Thus, the tool must still be aware of the specifics of the underlying MPI library.

In my experience, everything done from inside the library is more compact and more focused than anything done from outside. If you have to account for the library specifics anyway, you can just as well do everything from inside this library. However, if you want to infuse a good deal of thought into the tuning framework, wish to target more than one MPI implementation and possibly other communication mechanisms as well, and are prepared to deal with the unavoidable inefficiencies that will arise in the process, your mileage may differ. We will consider dynamic tuning in terms applicable to both methods for the sake of generality below.

6.3.2 Parameter Control

Typical MPI applications consist of the preprocessing, computational, and postprocessing stages. During preprocessing, data to be processed is read in or generated and/or distributed by the application. During the postprocessing stage, data computed earlier is collected and/or output for further analysis by the application user. You can hardly optimize these stages dynamically, because these are one-off activities. All you can do is detect them this way or another and possibly apply certain precomputed parameters that make the application as fast as it would go here. Bulk messages and one-to-many or many-to-one collective operations are most probable at these stages. The computational stage is the one that is most amenable to dynamic tuning. This stage normally has an iterative character and it should take the biggest slice of the execution time in a properly designed application. It is here that you should try to detect, analyze, and optimize dynamically. Relatively short messages and neighbor exchange patterns, as well as respective collective operations done over restricted communicators are most likely at this stage.

6.3.2.1 Design Considerations

Detection of iterative computations is not too complicated, and so is detection of the iteration boundaries. With that in hand, you can time a couple iterations to find out what kind of messages, patterns, and collective operations are being used repetitively. Sorting the findings by the presumed influence, you can start tuning the parameters, either one by one or in batches. Applying a certain precomputed set of parameters right away is also possible and sometimes quite beneficial. In the end you may end up with three parameter sets: one for the preprocessing, one for the computational, and one for the postprocessing stages. You can easily grasp why this is possible if you recall our tuning experiment from Section 2.3.6.4, where we found that double buffer length of 16 KiB was best for messages of 4 bytes, while 64 KiB to 256 KiB were noticeably better for messages of 4 MiB (see Table 2.4).

6.3.2.2 Parameter Input and Output

In addition to the static data sources considered in Section 6.2.2.2, there are also dynamic ways of passing information to the MPI library: from binary patching of the executable on the fly using a symbolic debugger or another runtime agent, possibly dedicated solely to the MPI library control; through reading and writing a file or two, up to and including talking to a special socket open by the MPI library for accepting tuning information and reporting performance metrics back. In fact, MPI communication can also be used to control the MPI library. Why not?

The same considerations apply to the parameter output. You would not make the tuner capture screen images of the endless application output scrolling by, right? Let Hollywood directors do that. They apparently love it.

6.3.2.3 Parameter Tuning

Given the time constraints imposed upon dynamic tuning, the search algorithm applied needs to be simple and quick. Nothing else changes compared to the static case.

6.3.2.4 Parameter Storage and Retrieval

The use of an external database is more likely during dynamic tuning than in the static case. Since applications behave differently depending upon the workload and job configuration, it will be insufficient to key the parameters by the application name or path alone. With time, the database will contain enough entries to cover most of the commonly used applications and workloads, as well as the respective job configurations to make tuning loop closing possible. However, there are also other, less sophisticated approaches possible (see Section 6.3.4).

6.3.3 Process Migration

This is probably the most advanced of all automated optimizations possible within an MPI job. The decision in favor of process migration should not be taken lightly, because it involves really huge data volumes and substantial administrative overhead.

6.3.3.1 Design Considerations

The easiest way to implement process migration is via the backup/restore mechanism provided by certain operating environments and third-party packages. In this case, processes go into backup from certain, presumably less fortunately laid out nodes, and are restored to a better layout afterward. Since many big, long-running applications do use backup and restore in order to minimize the chance of irretrievable data loss due to a component failure, these time points provide very convenient moments for the process migration. Note that restore will of necessity include a new connection establishment phase, since it is quite unlikely that system resources like sockets and shared memory segment, to mention those that we have actually used so far, will be captured and stored by the backup machinery.

Of course, migration can also be done in the piecemeal fashion, up to and including pairwise process swaps done intranode or even internode. It can be synchronized with the likewise piecemeal restore/backup methods, or it may be completely independent of that. In this case, whatever the mechanism, some global or at least local synchronization will be necessary, to make sure that the changed layout is accounted for by all the processes that communicate with the reconfigured process subset.

6.3.3.2 Dynamic Remapping

Dynamic remapping uses the same cost functions as the static one. However, calculations of the cost function should not be too expensive or involve too many global operations. Once per iteration, as detected by the dynamic machinery anyway, is probably just right. For example, once the heaviest exchanges are detected, and their mapping is considered suboptimal, you can make the decision to remap the processes either all at once or pairwise.

6.3.4 Closed Loop Tuning

If an MPI application or the MPI library outputs sufficient information about its settings and performance observed during the respective MPI job execution, and this information is associated permanently with a particular set of the MPI run parameters like the application involved, the workload, the number of processes and so on, the MPI tuning loop can be closed in a way akin to the compiler profile-guided optimization.

6.3.4.1 Design Considerations

Of course, artificial intelligence just begs to be added to the product at this point. What else can comprehend the multitude of parameters that need to be taken into account while selecting the right slot for the most appropriate MPI settings?

In fact, a part of this complicated task can be put upon the shoulders of the MPI user. What closed loop tuning implies is that dynamic tuning is started not from the default or platform-specific settings selected at the job start, but from the application-specific initial settings that depend on all those job parameters. Putting them all into a configuration file and making the MPI library read this file in upon initialization and write a new file out at termination time is all that is necessary. The rest will be taken over by the dynamic tuning machinery discussed above.

6.3.4.2 Profile Guided Tuning

A comprehensive system that will automatically detect the necessary set of the initial parameters, and then propagate and elevate them across the system—from application-specific to user-specific and then to system-wide settings is also possible, including the tuning of the operating system, too—is very well possible. How far you want to go depends only upon the available time and allocated budget, as well as the level of control you exercise of the operational environment.

6.3.5 Exercises

Exercise 6.3 Implement statistics gathering mode that will collect data on the operations and buffer sizes used per process involved and output this information on request.

6.4 Conclusions

Tuning is the last but very powerful polish to be applied to the MPI library. Now we are ready to review the rest of the MPI functionality that can be introduced step by step in the cyclic extension-optimization-tuning manner described above.

Chapter 7
And the Rest

In this chapter we will focus mainly on those remaining MPI features that incur substantial rework of the library internals as compared to the extended MPI subset defined in Chapter 4. Most of the technical work related to adding new MPI features will be addressed only in passing. We will not be programming much, we will rather be discussing the general direction and referring to the full-scale MPI implementations instead. The reason is that if we were to go into any serious programming here, we would have to add some 90% of the MPI functionality, which would naturally take at least nine times as long as the work done so far. And this would go against the guiding 10/90 principle. So what's the point, when this work has already been done so many times? The purpose and scope of this book are intentionally different.

The material will be presented according to the appearance of the features in the MPI standards, with the necessary forward and backward references, so that you can make sense of what came when, why, and what changes had to be done to the standard and the underlying implementation(s). This way, we will go through the MPI-1.1, MPI-2.3, and MPI-3.1 standards, including all additions of the respective minor standard versions. Each time, major changes that justified the major increment of the standard version will be taken care of first, while smaller but notable additions will be dealt with in bunches. If the remaining features are mentioned, it is only in passing. It's not because I did not do all of it once or that I didn't hold a candle and stand close enough to see what was happening, it's the sheer volume of the MPI standard and relative triviality or obscurity of the matters involved.

Apart from the technical perspective from inside the MPI implementation, I will give you my personal point of view on the usability of the MPI features under review. In order to understand that a little better, you may want to know that the MPI standard, as any big piece of creative work, was driven by the individual ambitions of the participants, their desire to do good for themselves, their user base, and human-kind as a whole, and the voting rules that governed the formal side of the standard extension process. This way, the arguments of interested persons or parties had to be strong enough to convince the majority of the MPI Forum in a rather structured and consistent fashion. Of course, there were still some hiccups and casualties in the process, but most of them were either detected and fixed, or deprecated and ultimately removed, or left to die their natural death by lack of use.

7.1 MPI-1 Features

We have intentionally covered the most important basics of the MPI-1 (meaning MPI-1, MPI-1.1, MPI-1.2, and MPI-1.3) standard in previous chapters. As far the total

DOI 10.1515/9781501506871-007

MPI functionality is concerned, we have covered about 10% of it so far. As far as the pure MPI-1 functionality is concerned, we probably stayed at about 25% of the total. Indeed, if you recall Table 1.1 in Chapter 1, you will see that MPI-1.3 has about 255 entities by the counting method we used there, while we have introduced about 60 entities so far, counting different MPI error codes. This relates to about 855 entities in the full MPI-3.1 standard. If you ask me, these 10% (or 25%, for that matter) cover about 90% of all MPI application needs. In other words, if we were to represent the work done versus its effectiveness as a graph, we would be close to the "knee" of this graph (see Figure 7.1).

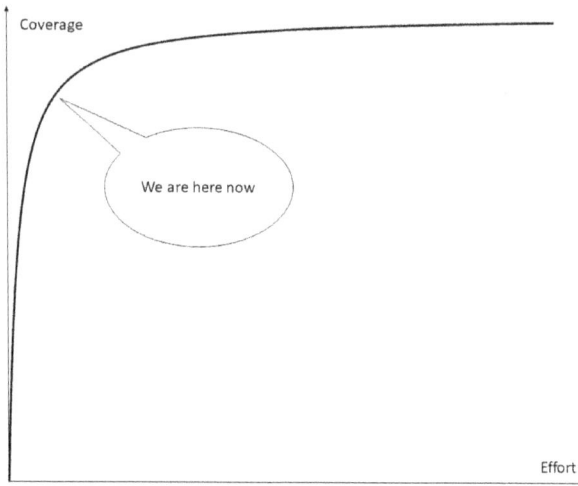

Figure 7.1: Effort expended vs application coverage

Beyond this point, extra effort spent yields smaller and smaller result as far as usefulness is concerned. True, for some very advanced applications, every ounce of extra functionality may mean a couple of seconds shaven off the total wall clock time of several hours. However, from the MPI implementor's point of view, this work becomes less and less effective, and thus more and more tedious. You will see this particularly well once we advance into MPI-2 and even more so into the MPI-3 domain.

7.1.1 Communication Modes

So far, we have been implementing and using only the standard point-to-point communication calls of the blocking and nonblocking varieties. If you look into the MPI standard, however, you will find that there are many more communication modes, and the standard mode we have been focusing on until now is actually a combination of some other behaviors that can be called upon by the MPI user at will.

Let's list all communication modes, together with some representative MPI calls (note that all modes are determined by the sending side; the corresponding MPI_Recv() and friends need to differentiate between the modes automagically on the receiving end):

- *Ready mode:* MPI_Rsend() and friends assume that a matching receive call has been posted on the receiving side before the respective send call is issued by the sending side. The basic performance relevant assumption here is that inside the sending call, the MPI can start pushing out the bytes without bothering to buffer them away or doing any additional work common to the standard mode send operation.
- *Buffered mode:* MPI_Bsend() and friends may simply buffer the user data away into a system buffer (possibly affected by the calls of the MPI_Buffer_attach() group) and return control back to the caller as soon as possible. The actual data transfer can take place later on, which is exactly what a busy application needs.
- *Synchronous mode:* MPI_Ssend() and friends make sure that the sending side and the receiving side synchronize before the data is transferred. In a sense, this looks like the ready mode with an added assurance that the matching receive operation has indeed been posted by the receiving side. Of course, this may add some latency but possibly save transfer time by avoiding extra handling of the unexpected receives.
- *Standard mode:* MPI_Send() and friends are a combination of the above plus some salt. Not necessarily, but with a high degree of confidence, the standard mode will work as follows:
  o For very short messages, a standard send operation may work as a ready mode send operation, assuming that the system will be able to handle a short buffer transfer somehow—say, by using the shared memory buffer that we relied on in Chapter 2. If this buffer is empty, we can simply put data into it and let the progress engine do the rest in due course.
  o For larger messages that do not fit into the system buffer mentioned above, the MPI may decide to buffer the data away into some intermediate storage on the sending side. Again, the actual transfer will go through the usual progress engine and message queue mechanisms. Thus, in this situation, the standard mode will look like the buffered mode.
  o For very large messages that would unduly stress system resources on either or both the sending and receiving sides, the standard send mode may force a synchronous transfer, in which both sides will first synchronize and then, being sure of the availability of the appropriate receive buffer, start transferring data right into its proper place. Of course, in this case the standard mode will look like the synchronous mode.

The selection of the respective switchover points is normally controlled by the MPI implementation itself based on a set of static and dynamic criteria, like buffer size and the available buffer area, with the help of extra hints that may be passed to the implementation by the user.

7.1.2 Wildcards

We have intentionally avoided a serious look at the wildcards, i.e., the MPI_ANY_SOURCE and MPI_ANY_TAG constants so far. Why? Well, I personally feel that if you need to use them, you do not really understand what is happening in your MPI program. There are very few situations, with the master-slave configuration (that is not at all well suited for the MPI, actually) being the biggest of them, where these wildcards may indeed make sense once in a while. For the remaining applications, these two methods of receiving messages from any source process (MPI_ANY_SOURCE) and/or of any tag (MPI_ANY_TAG) make things more complicated both to the MPI user and to the MPI implementor, with the MPI_ANY_SOURCE in particular being also a performance hog. In other words, these are typical "and the rest" features and may look nice in principle but cost dear in real life.

7.1.2.1 Matching Any Tag

Why is this so? Let's look at the MPI_ANY_TAG first. Its typical representation is -1, that is, all bits set to 1 in the 2's complement arithmetic. This observation leads to the following easy modification to the comparisons that involve the tag inside the MPI_Recv() call of the shared memory MPI subset from Chapter 2 (see Listing 7.1, cf. Listing 2.25):

Listing 7.1: Modified message tag handling (excerpt, see shm/mpi/v4/mpi.c)

```
...

#ifdef MPI_ANY_TAG
    assert(tag == MPI_ANY_TAG || (tag >= 0 && tag <= 32767));
#else
    assert(tag >= 0 && tag <= 32767);
#endif

...

#ifdef MPI_ANY_TAG
    if (tag == MPI_ANY_TAG)
        tag = pcell->tag;
    else
#endif
```

```
    if (tag != pcell->tag) {
#ifdef DEBUG
        fprintf(stderr,"%d/%d: -> failure, tag %d != %d\n",_mpi_rank,_mpi_size,tag,
pcell->tag);
#endif
        return !MPI_SUCCESS;
    }

...


    pstat->MPI_TAG = tag;

...
```

You can see here that we have to adapt the initial assertion, then the comparison in the main loop, and finally (albeit implicitly) the final assignment, because we do not want to return MPI_ANY_TAG as the value of the message tag that matched the selector. This is handled by changing the value of the tag selector when the comparison on the MPI_ANY_TAG matches it. All this complexity is guarded by the conditional compilation constructs that unfold differently depending on the presence or absence of the MPI_ANY_TAG definition in the mpi.h header file. This way we can create two different libraries from one source code, and then compare their performance.

7.1.2.2 Matching Any Source Rank

Now to the MPI_ANY_SOURCE that leads to substantially more trouble internally. If you go through all our code examples, you will see that at every point where we bravely used the source rank as provided by the respective MPI call for calculating the respective connection data structure, be that a shared memory segment cell or a file descriptor, we will have to now take additional precautions for this source rank being possibly the MPI_ANY_SOURCE constant. Thus, instead of using this value directly, we will first have to figure out first what connection we are going to work with.

Again, in the case of the shared memory subset, the changes are relatively obvious (see Listing 7.2, cf. Listing 2.25):

Listing 7.2: Modified source rank handling (excerpt, see shm/mpi/v4/mpi.c)

```
...


#ifdef MPI_ANY_SOURCE
    assert(src == MPI_ANY_SOURCE || (src >= 0 && src < _mpi_size && src !=
_mpi_rank));
#else
    assert(src >= 0 && src < _mpi_size && src != _mpi_rank);
#endif
```

...

```
#ifdef MPI_ANY_SOURCE
    if (src == MPI_ANY_SOURCE)
        for (;;)
            for (src = 0; src < _mpi_size; src++) {
                if (src == _mpi_rank)
                    continue;

                pshm = _mpi_shm + src*_mpi_size + _mpi_rank;
                pcell = pshm->cell + pshm->nrecv;
                if (pcell->flag)
                    goto found;
            }
    else {
#endif
    pshm = _mpi_shm + src*_mpi_size + _mpi_rank;
    pcell = pshm->cell + pshm->nrecv;

    while (!pcell->flag)
        ;
#ifdef MPI_ANY_SOURCE
    }
found:;
#endif
```

...

```
    pstat->MPI_SOURCE = src;
```

...

The handling of the MPI_ANY_SOURCE in the case of the sockets subset from Chapter 3 is not too difficult either. However, the situation becomes substantially more complicated once we have more than one fabric to serve. The point here is that when a message request with the MPI_ANY_SOURCE wildcard is registered with the MPI library, every underlying fabric becomes an open ear: every message coming from every process, provided its tag matches the request tag selector (including the MPI_ANY_TAG if present), becomes a potential hit. Whether this is a hit or not depends upon whether another fabric has already started processing of one of its messages as the matching one. In other words, fabrics become interdependent as far as matching the incoming messages is concerned. There are several ways of dealing with this issue, e.g., this patent (US Patent No. 7,539,995, 2009).

7.1.3 Derived Datatypes and Heterogeneity

Truly heterogeneous HPC platforms with vastly different data representations are rather rare these days. Most often you will find homogeneous clusters made up of the same of at least fully compatible pieces of hardware. Still, you never know, so that it makes sense to be careful about the datatypes used in the program.

MPI applications describe the buffer contents both on the sending and the receiving side. This way, the data conversion, if necessary, can be done on either side of the communication, or on both of them:

– In the former case, the sender may transform the data into the format expected by the receiver, or the receiver may transform the data from the format of the sender.
   o The first way may be better suited for point-to-point communication, when the sender knows for sure what process it is sending the data to.
   o The second way is better for collectives, where the receiving process knows where it got the data from, and since all receivers may in principle have different data representations, the sender is off-loaded of the burden of doing many data conversions.
– Finally, it is possible, albeit very rare, that the data conversion is done by both sides. In this case, the data is transformed into some universal representation (such as external32 specified by the MPI-2 standard) on the sending side, transferred in this univeral representation, and then transformed back into the specific representation needed by the receiving side. Of course, this means double transformation which costs time and space. This was what actually killed PVM once, if you remember.

Whatever the actual method, the transformation is either performed transparently by all the usual MPI communication calls or done explicitly by the packing and unpacking calls MPI_Pack(), MPI_Unpack() and their friends and relatives. Down under the hood, any datatype is represented by a series of memcpy(3) or internal data conversion calls that copy or transform and copy user data from the input buffer to the output buffer. Since a sequence of, say, memcpy(3) calls dealing with pieces of a bigger contiguous buffer is typically slower than a big memcpy(3) call on the whole buffer, the internal datatype description may be optimized during the MPI_Type_commit() call. The rest of the datatype handling is basically tedious list management that is interesting in principle but irrelevant practically for the reason that is mentioned at the very beginning of this section.

How efficient is this stuff? Normally, if well optimized, it works about as fast as a carefully handcrafted corresponding routine, but there may be exceptions. The main problem here is that this part of the MPI code is exercised relatively rarely, with the majority of applications preferring to work with contiguous buffers. The situation may change radically if a datatype system is well implemented and especially supported in hardware. Good optimization will detect contiguous data pieces even if they are described as noncontiguous ones. In the absence of data conversion, this will lead to fewer memcpy(3) calls and, generally, better performance. Hardware support will map the MPI datatype descriptions upon the underlying gather/scatter capabilities, and then the copying will be done by an agent different from the central processor (e.g., by the networking card itself). So, every time you want to decide what data structure to use, you should give them a try on both usability and performance first.

7.1.4 Profiling Interface

Every MPI function with the exception of potential macros like MPI_Wtime() exists in two incarnations: once as a user called MPI_ prefixed call, and once as a half hidden PMPI_ prefixed call. In practice these days, the MPI_ prefixed calls are represented with weak symbols pointing to their PMPI_ prefixed counterparts. This way, any redefinition of an MPI_ prefixed call that comes first in the linkage sequence overtakes the default weak symbol definition, and thus becomes the new MPI_ prefixed call visible to the MPI application. Typically, such a redefinition is done en masse in order to instrument the respective MPI_ prefixed calls by, say, capturing their entry and exit points, and recoding their arguments for further analysis, checking for correctness, or for statistical purposes. As an interesting alternative, such a redefinition may also be used to try out new collective algorithms if the default ones are deemed too slow, especially when the source code of them is unavailable. There are also a few others, more esoteric applications of this versatile interface, like those described in patents (US Patent No. 7,856,348, 2010) and (US Patent No. 7,966,624, 2011).

Technically, a typical piece of code implementing the profiling interface may look as follows (see Listing 7.3):

Listing 7.3: Typical profiling interface implementation via weak symbols

```
#pragma weak MPI_Finalize  =        PMPI_Finalize

int PMPI_Finalize(int code)
{
    ...
}
```

There may still be platforms around that do not support weak symbols, in which case a couple of different approaches can be used to implement the profiling interface. In that case you will define trivial MPI_ prefixed functions that call PMPI_ prefixed ones and let the MPI user supersede the MPI_ prefixed functions as they please by putting their library with their MPI_ prefixed alternatives first in the MPI application linkage command line.

Another matter to discuss is the profiling of the Fortran language binding. You can try to do this natively or delegate the profiling to the C binding only. Both ways have been implemented. The benefit of the native Fortran profiling is that you can try to write replacement functions using Fortran, of course. The benefit of the C-only solution is simplicity. Which to choose depends upon your prospective user base and their language predilection.

7.1.5 Other Features

Here is an assortment of the MPI-1 features that did not make my cut to be named major ones, but that have implementation complexity or rather unexpected repercussions across the rest of the library.

7.1.5.1 Communicator and Group Management

There are many calls dedicated to the group and communicator management—too many, one may say. Essentially, all of them are dedicated to manipulation with the rank lists. These lists may be plain, like the ones we have used in this book, or complicated, with some packing scheme for extreme scalability.

Since communicators are based upon groups semantically, it may seem natural to use references to the groups inside communicators. However, the extra indirection related to the decoding of the group handle or of the pointer to the group may be too expensive in the eyes of some MPI implementors, so that sometimes (like in MPICH) rank lists are cached on the communicators, too.

All opaque objects in MPI are reference counted, that is, they are stored only once, but the occurrences of their usage are kept track of by incrementing and decrementing a so-called reference count. This way, when a certain object is marked for deletion, it may still exist as long as other entities refer to it. This begs for a general internal mechanism that is often provided.

7.1.5.2 Intercommunicators

Intercommunicators are communicators that have two groups—one local and one remote—attached to them. Normal intracommunicators have only one local group. This is the only difference between them. The intercommunicators remained mostly useless until dynamic process management and extended collectives were introduced in MPI-2 (see Sections 7.2.1 and 7.2.5.1, respectively).

7.1.5.3 Communicator Attributes

Communicators get passed to almost all MPI communication calls, and as such they provide a very convenient vehicle for getting user data into libraries built on top of the MPI without introducing extra arguments. In addition to this, predefined attributes carry information about certain features of the MPI job and MPI implementation, such as the highest supported value of the MPI tag, biggest possible size of the MPI universe, and so on. While useful in principle, attributes are very rarely used in practice, partly due to the very limited nature of the values that can be stored in them (basically, an integer or a pointer), and partly due to the cumbersome mechanism introduced for creating and managing a new attribute.

7.1.5.4 Virtual Process Topologies

An interesting application of the communicator attributes was introduced right at the start. In addition to the linear rank sequence intrinsic to any MPI group and communicator, a virtual Cartesian topology created by the MPI_Cart_create() call provides a view of the new communicator that has so many dimensions of a certain size, possibly circularly closed. The ranks of the new communicator may be reordered to better suit the underlying interconnect topology. Likewise, virtual graph topology created by the MPI_Graph_create() call provides a view of the new communicator whose ranks, possibly reordered, represent graph vertices. In addition to this, there are calls that permit conversion from the virtual topology coordinates to the linear MPI ranks and back. Originally intended to not only simplify the life of the MPI application programmer but also to let the MPI library optimize the process layout, these calls are seldom properly optimized.

Note that there is an intrinsic discrepancy between the Cartesian and graph virtual topologies in the MPI-1 standard. The former is defined quite tersely and locally in terms of the overall grid dimensions and the place of the current process in it. The latter needs a full description of the entire graph on every participating node, which may become prohibitively expensive at scale. This issue, that was apparent from the very beginning, got fixed in the MPI-2.2 standard (see Section 7.2.5.5).

7.1.5.5 Probing for Incoming Messages

There may be situations, albeit rather artificial, when the MPI application needs to find out if any incoming messages are there, and if so, whether they will fit into the respective application buffer. In this case, people use the blocking MPI_Probe() and the nonblocking MPI_Iprobe() calls. They work as if they were the MPI_Recv() and the MPI_Irecv() followed by the MPI_Test(), respectively, but they do not receive any data and do not complete the request involved. Instead, they look into the envelope of the incoming message. When MPI_Probe() returns and MPI_Test() returns with the flag argument set to 1, the message source rank, tag, and count can be retrieved from the MPI_Status argument using the usual means. So far this does not look fancy, until you notice that according to the MPI standard, the following MPI_Recv() or MPI_Irecv() call must match the same message. This may become a problem if you have more than one fabric to serve or more than one thread to control, which explains the later addition of the thread-safe MPI_Mprobe() call and friends (see Section 7.3.4.2).

7.1.5.6 Message Cancellation

If a contest of the most cumbersome and absolutely useless MPI features were held, the MPI_Cancel() would score big if not win the crown outright. Why would an application want to issue a nonblocking send or receive request that needed cancellation afterwards? Get organized!

However, as sometimes happens, there was a great goodness in that the MPI_Cancel() was defined in the MPI-1 standard. In order to implement it, every MPI request I know of carried around an integer dedicated to the flag normally referred to as cancelled, and so did every MPI_Status object. A total waste of space, if you ask me, since just a bit would suffice, and even that was superfluous, just because the whole feature was intrinsically superfluous.

But when the time came to extend the MPI standard once again, this extra integer word came in handy. Long counts that we will consider in Section 7.3.2, if introduced mindlessly, would break binary compatibility between MPI-3 and all earlier standards: MPI_Status lives in user space, and so a larger count that has to be stored by the MPI_Status objects for the MPI_Get_count() function to work, would need an extra space that was not there. A forced rebuild of all MPI applications would ensue.

Would, but did not. How? Well, message counts are nonnegative by definition. And so, one lucky morning I came up with an idea to use one bit of the cancelled word for the actual cancellation flag and use the rest of this word for the high bits of the long count. The MPICH team kindly agreed to this hack, and thus MPICH ABI was born (Supalov & Yalozo, 2014).

Life is full of paradoxes, isn't it?

7.1.5.7 Signal Handling

This is possibly the most obscure part of the MPI-1 standard. A good quality MPI implementation is supposed to handle the most popular signals like SIGHUP, SIGTERM and SIGKILL in a way natural to the users. In practice, this means that pressing <ctrl>C on your keyboard should terminate the respective MPI job rather than only the mpiexec running on the control terminal. Unlike many other, rather esoteric features described in this section, signal propagation is really useful and worth careful attention, if a bit tricky.

7.1.6 Exercises

Exercise 7.1 What communication mode was implemented by the shared memory MPI subset in Chapter 2?

7.2 MPI-2 Features

The MPI-2 standard was a major leap forward compared to MPI-1. Apart from fixing and completing certain MPI-1 features in the MPI-1.2 standard, it added dynamic process management, one-sided communication, thread support, file I/O, and myriad other, smaller features. I was on the team that produced the first full MPI-2 implementation in the world (Bißeling, Hoppe, Latour, & Supalov, 2002). In my experience, conversion of an MPI-1 implementation to an MPI-2 version was always extremely cost-intensive, to the point of making this an almost futile exercise. Read on, and you will see why.

7.2.1 Dynamic Process Management

Dynamic connections come from the MPI_Spawn(), MPI_Spawn_multiple(), and MPI_Comm_join() calls (the latter is almost never used and hardly ever implemented and tested). If you compare this new situation to the lazy connection establishment, you will find substantial similarities. Indeed, that same machinery can be reused almost unchanged.

However, there is a catch or two here. First, you do not know in advance how many connections are going to be established in the end. This means that you need to look for them always and make place for the new connections either dynamically or up to some limit normally referred to by the attribute MPI_UNIVERSE_SIZE of the MPI_COMM_WORLD communicator.

Second, in case of a multifabric run, it is possible that a new connection will come via fabric that has not yet been used by this particular MPI process. This leads to the need to listen to quiescent fabrics as well and this is why, in some cases, there are recommendations to explicitly select fabrics to watch, or even tell the library that dynamic processes will not be used in this particular run at all. This helps, just as does a hybrid progress engine that takes care of the relatively slow and infrequent connection requests in background, without braking the main MPI thread.

7.2.2 One-Sided Communication

This is probably the best example of a very advanced and elegant feature that did not appeal to users at all. Why? Excessive generality. The feature was not focused enough from the start to get proper design. Indeed, if you try to define one-sided communication in a way that allows it to be mapped upon fast RDMA-capable hardware as well as point-to-point calls on top of TCP/IP, you need to go a long way instead of taking shortcuts that were possible when your semantics were a little more specific. This is somewhat akin to the situation with subset MPI versus full MPI that we have been dealing with in this book: sharp and focused assumptions lead to substantially more robust and simple design, and hence better performance. Indeed, the very presence of the fast collective operations in MPI-3 (see Section 7.3.1) is yet another indication that the original design did not reach its users.

For all their sophistication and power, I can recall only one (right, one) serious application that tried to use MPI-2 one-sided operations in experimental (yes, experimental) mode. I cannot recall whether this brought any performance gain, though. This is quite understandable: when you have to map one-sided operations upon point-to-point ones, or at least make sure that this is possible, plus support three more or less obscure synchronization modes that basically allow this (with the notable exception of the passive target communication that requires either some kind of extra agent to detect the incoming communication requests, or RDMA support), you cannot do very well at all.

On top of that, the semantics of the MPI-2 one-sided operations are so exotic that there are probably about three people in the world who do understand it to the point of being able to explain how it should work. However, doing this is not the goal here. The dry residue is this: if you have RDMA support on the target platform, you collect all RDMA requests until the next synchronization point and then complete the pending operations in order. If you have no RDMA support, you map all pending RDMA operations onto equivalent point-to-point operations on both sides—the origin and the target—and complete them again in order. In all cases you should pay close attention to the state in which the caches and the main memory remain once the operations are completed, especially if your target platform does not guarantee *cache coherence* in hardware.

7.2.3 Thread Support

Threads have always been second class citizens in the MPI world. This was especially true of the earlier standards and is getting better step by step in the progress toward MPI-3.1 and beyond. MPI-2 made the first tangible step by accepting thread existence explicitly and making provisions for using threads orthogonally to MPI.

There are four levels of thread support selected at the MPI initialization time: MPI_THREAD_SINGLE, MPI_THREAD_FUNNELED, MPI_THREAD_SERIALIZED, and MPI_THREAD_MULTIPLE. The former two are practically identical. However, in order to support the MPI_THEAD_FUNNELED mode, the MPI library must be built using thread safe underlying libraries and respective compiler options. This is necessary because threads other than the main one, in which the MPI library has to be called, may still be using those other supporting libraries, and there should be no collision between the threads in this case.

The `MPI_THREAD_SERIALIZED` mode is basically a half step beyond that, because the MPI library may be called by any thread, but only by one of them at a time, such that any MPI function starts and stops in the same thread and there is no other MPI function being called by any other thread during this time. However, remember that one thread may start a nonblocking operation, while another thread may try to complete it later on. This implies that MPI data structures need to be kept in the general, that is, non-thread specific memory.

The biggest challenge, however, is the `MPI_THREAD_MULTIPLE` mode. Here everyone can call anything at any time (provided this is still sane), and hence all internal resources of the MPI library, and all system resources it uses, need to be protected somehow from accidental reentry and the conflicts possible due to this. One could, of course, enforce behavior akin to the `MPI_THEAD_SERIALIZED` mode by protecting the entry to the MPI library by a single global lock. However, this would defeat the very purpose of the `MPI_THEAD MULTIPLE` mode, and hence can only be considered a partial and incorrect solution. A correct solution is to protect all internal data structures subject to potential multiple access, and make sure that the whole thing does not deadlock.

Modes of protection may be different. Most opaque objects, like communicators, will need an individual lock each for one writer and multiple readers, so that only one writer is allowed to modify the object, while many readers may not query the object features while the writer is busy, but may happily read the object contents in the absence of active writers. Whether a writer that needs access to the object involved should wait on all readers already in the critical section is a good question. Probably, yes, for MPI does not know partially completed calls—they either succeed or fail as a whole, with the exception of the `MPI_Waitany()`, `MPI_Waitsome()` and friends. However, those partial completions relate to the output arguments rather than the function itself, so they should follow the same one writer, many readers rules.

Some internal structures, like message queues, will also need to be protected. The same is true of the data structures that are used for data transfer, be that shared memory segments, sockets, or whatever else. Here, some protection will be provided by the operating systems, since system calls naturally serialize access to objects like sockets and such. However, many threads issuing conflicting commands to one socket will certainly confuse it completely on the sending side and lead to an unrecoverable mess on the other end. Hence, access should also be guarded by the MPI library itself.

As far as queues are concerned, it may be and usually is a good idea to exploit so called lock-free or lockless queues. Certain processor commands combined with the fitting data structures provide this functionality based upon elementary mutual exclusion built into said commands. There is typically no need and there is indeed no time available for heavy-weight POSIX mutexes, semaphores, and such. It makes no sense to press any extra ounce of juice out of the data transfer machinery, just to protect it by those slow and cumbersome constructs where a simple test-and-set, compare-and-swap, or even exchange operation will do. You will easily find several implementations ready for you to grab or at least provide a good source of inspiration, for example, in this article (Gropp & Lusk, 1997).

7.2.4 File I/O

Traditionally considered a major addition of the MPI-2 standard, this massive extra subsystem is actually almost never used. Why? Wrong design. And when I say "wrong," I do not mean that it is inelegant, powerless, or inexpressive. No, quite the contrary: it is very elegant, extremely powerful, and exquisitely expressive. But a description of the data to be written and read in terms of the etype and filetype turned out to be way too elementary for the result to be convenient to the MPI users, easily understandable to the MPI internals, and readily mappable upon the capabilities of the existing parallel file systems. It is as if instead of the MPI_Send() and MPI_Recv() you were given a set of obscure calls to define the communication pattern, start the exchange, and then complete it in so many ways. There are calls like that in the MPI, too, by the way, but there are also the MPI_Send() and MPI_Recv() everybody knows, loves, and actually uses.

In the MPI I/O, as it is casually referred to, there are no such high-level calls, in a sense. Yes, you can read and write data from and to files in many ways, but you have to describe the file structure in terms of several datatypes, and then let the MPI library figure out what exactly you wanted to do, and how to do this fast as well. In the past, parallel I/O was different: you had a set of *file I/O modes* that let you, for example, produce ordered output from all processes, or append a data item to the common file, or read data items in turn, and so on. This was very convenient, quite transparent, and rather easy to implement—but limited. So the MPI Forum decided to provide the MPI users with a whole toolbox of very low-level primitives in the hope that both they and the MPI libraries will be able to make sense of this heap.

Well, they did not. The clear sign of it is that there appeared additional packages like HDF (Hierarchical Data Format, 2017) that tried to plug up this usability and implementation hole. Maybe this was exactly what was intended, but then we would also expect to see extra packages built on top of all those very elementary point-to-point calls I have mentioned earlier. Well, there are none: the MPI_Send() and MPI_Recv() do their job well enough for that to be unnecessary. And so, the very elegant, powerful, and expressive design of the MPI I/O must be dubbed an utter failure from the usability and implementability points of view.

Due to this, do yourself a favor and never try to implement MPI I/O yourself, unless paid extravagantly for this. Take one of the standard packages, like the ROMIO (Thakur, et al., 2016), slam it on top of your basic and fast MPI, and be happy.

7.2.5 Other Features

Wait-wait-wait! Is this all? And what about extended collectives, generalized requests, and so many other bits and pieces that make a real MPI expert produce a volume dedicated to each of these features alone? Well, this is not the point of this book. The 10/90 principle is actually a double-edged sword: it makes you dedicate 90% of your time to those feature that are really used out in the field, but it also relegates the remaining 90% of features to the remaining 10% of the time you have. And so it goes that apparently major features, making up 90% of the MPI standard, are either mentioned in passing, like we do this, or take 900% (yes, nine hundred percent; or, more exactly, 910%) of the time to discuss and implement them.

7.2.5.1 Extended Collectives
These are collectives over intercommunicators. Nothing really special apart from working on two groups—local and remote—instead of just one for intracommunicators.

7.2.5.2 Generalized Requests
This is probably the fanciest feature of the MPI-2 standard. The idea was that by means of generalized requests, very advanced MPI users would be able to formulate their own operations and use them in their very advanced MPI applications. Well, I have never seen them used, even for the MPI-3 neighborhood collectives (see Section 7.3.4.3).

7.2.5.3 Attributes Revisited
Attributes typical of communicators in MPI-1 standard were added to other first-class opaque objects like files and windows. Other than a marginal function renaming and deprecating in the communicator case, this is again just a lot of nearly useless sweat, blood, and tears.

7.2.5.4 The Info Object
The introduction of the MPI_Info object marked capitulation of the MPI Forum in the face of the otherwise unmanageable complexity. As soon as a standard allows such a catch-all argument to be passed to selected functions, the role of the usual arguments is put in question. Of course, strict discipline in relation to the data that may be transferred this way and the level of its influence upon program behavior helps a little in keeping things organized, but let's be frank—this was a capitulation.

Extra spice was added by the fact that there had already been attributes attached to so many opaque objects (see Section 7.2.5.3). Why add yet another, substantially more flexible way of representing basically the same key/value data sets? Were attributes not general enough? Why did they survive then? Backward compatibility? Beg your pardon, here something apparently went very wrong indeed.

7.2.5.5 Distributed Graph Topology

Fixing an old issue with the virtual graph topology mentioned before (see Section 7.1.5.4), distributed graph topology introduced by the MPI-2.2 standard allowed graph description in local terms, thus making the respective call vastly more scalable. However, first implementations of the MPI_Dist_graph_create() used an MPI_Allgatherv() followed by the MPI_Graph_create() for simplicity, and I will not be overly surprised if most of them still do this.

7.2.6 Exercises

Exercise 7.2 Build subset MPI with thread safe libraries and see whether there is any performance difference.

7.3 MPI-3 Features

A lot of time passed between MPI-2 and MPI-3. If you remember Section 1.2.1, the intermediate steps (MPI-2.1 and MPI-2.2) were meant to fix the remaining issues and introduce backward-compatible features, respectively. Major additions to the MPI standard were all slated for MPI-3. In some sense, it fell short of these expectations, since the fault tolerance and treating threads as first-class citizens did not make the cutoff. Still, some features were big enough to qualify for a major standard version increment: fast one-sided operations, long counts, and nonblocking collectives were among them.

7.3.1 Fast One-Sided Communication

We dwelt upon the shortcomings of the MPI-2 one-sided operations in Section 7.2.2. In MPI-3, the design became substantially more focused and targeted those systems that could do RDMA, active messaging, or analogous operations efficiently in hardware. In addition, it extended the set of operations to include the most natural way of completing individual operations—via an individual MPI_Request entity rather than three more or less obscure synchronization methods defined in the MPI-2 standard. I hope that these new operations will find their excited users.

7.3.2 Long Counts

Addition of the long counts took quite a bit of effort. Nobody really wanted to double the number of calls that contain a count argument, traditionally sized as default 32-bit integer. However, the increasing dominance of the 64-bit platforms in the HPC and elsewhere, and the tendency of message size growth made inclusion of the long count support somewhat unavoidable. In the end, it was decided to go for a compromise and introduce a few calls manipulating long counts but leave the datatype system unchanged in its biggest part. Large messages were represented by multiple instances of lesser messages that in total would exceed the 2 GiB size limit effectively set by the 32-bit integer word.

Of course, this does not help inside the library where all counts need to be kept in 64-bit integers. Had the MPI Forum foreseen this eventuality, declaring all counts using a predefined datatype in both C and Fortran bindings, this would have saved a lot of time and effort spent working around this initial design decision.

7.3.3 Nonblocking Collectives

First time I heard of a nonblocking barrier I was a bit amused. Still, this makes sense, if you think of it, in the context of intensive numerical calculations that need to be synchronized across the processes involved. In this case, issuing an `MPI_Ibarrier()` in advance and waiting for it to complete once the calculations are over may indeed give you an extra ounce in performance, provided that your hardware and software does support so-called computation/communication overlap. However, there is a high price tag attached to this promising extension: a process may now be involved in more than one collective operation at a time. You can learn more about the consequences of this from the original description of this feature (Hoefler, Lumsdane, & Rehm, 2007).

7.3.4 Other Features

There were also myriad smaller features added, some representing very specialized nitpicks, some opening the door for further extensions in the MPI-4 timeframe.

7.3.4.1 Shared Memory Extensions

By the time of MPI-3, multicore processors became more than just commonplace. Many powerful SMP systems had also existed, in the time of MPI-1 and earlier, but they were somewhat exotic high-end creatures rather than something you might find casually purring in your lap. Now every plain vanilla laptop does just that.

Of course, MPI had to react to that. For starters, it made it easier to create communicators that belong to the same shared memory domain via the MPI_Comm_split_type() call with the predefined value MPI_COMM_TYPE_SHARED, and use shared memory efficiently in one-sided operations via the MPI_Win_allocate_shared() call. The implementation side of these extensions is rather trivial.

7.3.4.2 Matching Probe

This was one of those additions that are done with a single MPI application in mind. Imagine that you have a multithreaded application that probes and receives messages in many threads. In this case, as mentioned in Section 7.1.5.5, you have no guarantee that the thread that successfully probed for a message will also be able to receive it. Since the message has already made its way up the cache hierarchy on the probing thread, switching context to another thread will incur performance overhead. Hence, receiving on the same thread will work faster. Whether it really makes sense to resolve this issue by introducing a whole new group of calls around MPI_Mprobe() and add yet another opaque object MPI_Message is open to debate.

7.3.4.3 Neighborhood Collectives

Another sign of the scalability epoch presented itself in the form of neighborhood collective operations. Indeed, most numerical methods work very intensively on local exchanges between immediate neighbors in the process topology, and before the advent of the neighborhood collectives, you'd have to program these exchanges either by using plain point-to-point calls or by creating a lot of smaller neighborhood communicators and then synchronizing the MPI processes so that they do not block each other by entering different collective operations at the same time.

Instead of this, neighborhood collectives look up the virtual topology of the communicator involved and operate on the neighbors of the current process. Apart from this, there is no real difference between them and their big brothers. A good low-level implementation of the support calls will help you accommodate neighborhood collectives in a snap.

7.3.5 Exercises

Exercise 7.3 Propose an MPI device extension for one-sided operation on an RDMA-enabled platform.

7.4 Conclusions

Was it worth it to extend the MPI standard so much? I doubt it. I seriously doubt it, especially since this apparently adversely influenced the resulting MPI library performance. We will look closer into the reasons behind this in the next chapter.

Chapter 8
A Look Ahead

Any design, however successful it is—and MPI has been remarkably successful thus far—hits boundaries with time. Look at humans: we have been evolving for the past 7 million years, starting with ape-like creatures who left behind only their bones, tools, pictures, and some remote relatives running wild in the forests and savanna. Sometimes they are running wild in the cities, too. Oh well, no politics.

The point is: everything has to adapt or die, and the original design will fail if exposed to conditions too harsh to be accommodated. Imagine humans in space: a couple of years, and cosmic radiation will do its job on anyone. This applies to MPI, too. How far can it go under pressure exerted by extreme scalability, frequency wall, processor and memory latency gap, and especially competing interfaces?

In this chapter we will try to understand MPI limitations to better assess its past and future, including the (hopefully) coming MPI-4 standard.

8.1   Selected MPI-4 Features (En Route)

The original MPI design limitations as well as untamed creativity of the MPI Forum participants have manifested themselves fully during discussion of the features that failed to make the MPI-3 cutoff: fault tolerance and threads as MPI processes. There are also other, smaller topics still active in the ongoing MPI Forum discussions, but we will let those cook in their juices until they smell like something we mere mortals may taste, appreciate, and survive.

8.1.1 Fault Tolerance

There was no other topic in the MPI-3 timeframe, and even prior to that, that took so much time and effort as the fault tolerance. I can remember at least three credible attempts to get something into the standard, but my unassisted memories may be a little blurred at this point.

What's the problem? Well, the original MPI standard was all based on the "all or nothing" principle. You get a message or you get it not. You die due to an error or you have had no error at all. You abort the whole job or you keep running as a whole. With some minor semantic deviations that very few MPI implementors really cared about and even fewer MPI users ever exercised, these were the principles put into the very foundation of the MPI standard and hence most of the implementations. This preference is confirmed to a degree by the choice of the default error handler `MPI_ERRORS_ARE_FATAL`.

DOI 10.1515/9781501506871-008

Tangible faults, such as nodes going down and up again due to a rogue muon from Alpha Centauri causing unrecoverable multibit memory failure, or network segments becoming unreachable for a spell thanks to the sysadmin who tripped over a dangling wire, or a part of the machine failing *suddenly* due to insufficient cooling on a hot summer day—all these events will immediately and irretrievably kill all directly affected MPI jobs, and probably all those unaffected, and those still waiting in the job queue, and those about to be submitted at the moment of failure and a day after, too.

In Chapter 2 we had a whiff of the feeling what it might look like when errors were ignored. There, we implied the error handler `MPI_ERRORS_RETURN` and bravely ignored the MPI return value. Of course, this was done a little naïvely. What should a serious application do if it does expect some failures?

It is hardly possible to make any given application completely fault tolerant and give a guarantee that it will work in all possible situations. Remember that the algorithms rule? Well, a fault tolerant application should start with fault tolerant algorithms. If sudden loss of a certain data part (and this is what failure is basically about) may lead to the result getting far away from where it would be after a completely successful run, such an application is probably not going to fly, be it fault tolerant or not. Checkpoint/restart facilities are only good up to a certain scale, and interaction with them has not been incorporated into the MPI standard yet.

A good algorithm, however, will be able to withstand a minor data loss, and a good program and MPI library will be able to steady themselves after a single-node failure. What about the failed process? What happens if it is addressed by some other process via point-to-point, one-sided, or even collective operation? Should this failed process be restarted somehow or just ignored? How will the normally dense MPI rank sequence behave in either of these cases? Will there be a reordering, a hole left, or a new process taking the place of the failed one? In some cases, a solution is relatively simple. For example, when an application is structured using the master-slave pattern, the slaves communicate only with the master(s), and there are no collective operations used, the master notified of a slave failure via a respective MPI error code may easily avoid addressing this process in the future. In one of the ultimately unsuccessful attempts to get some fault tolerance into the standard, I proposed to add this scenario to a sequence of increasingly complicated use cases to be supported by MPI, and we even implemented and documented it. However, shortly after this the wind changed, and this promising discussion had to be restarted on completely different terms.

We have yet to hear of a real breakthrough in this area, hopefully, without the `MPI_Info` objects and communicator attributes being used in the process.

8.1.2 Threads as MPI Processes

We have already mentioned MPI thread support, shared memory extensions, and thread safe probing before (see Sections 7.2.3, 7.3.4.1, and 7.3.4.2, respectively). Those MPI features were intended to make MPI coexist peacefully with threads. Treating threads as MPI processes goes a good step farther than that.

There are at least two issues to address here. The first one is how to address the threads that have historically never had any representation in the MPI standard. In order to send a message to a thread while keeping the rest of the MPI standard unchanged, one has to assign an MPI rank to a thread. This potentially leads to multiple complications. Upon several proposals, one of them emerged as a clear winner (Dinan, et al., 2013). It separated the notions of control and communication by introducing MPI endpoints to which both OS processes and threads could attach themselves at will. Then, threads become individually addressable in the respective *endpoint communicator*. Voilà!

The second issue is the memory locality. Threads share memory and must do extra work to enable their own, private memory. Processes, on the contrary, have their own memory at their complete disposal. Thus, a traditional MPI program needs to be changed when mapped upon threads treated as MPI processes. Fortunately, this can be done automatically with a little help of a clever compiler (US Patent No. 8,539,456,2013).

Of course, this is not the first time ever threads are going to play a more important role in the MPI. However, this is the first time they will do this with the blessing of the MPI Forum.

8.1.3 Exercises

Exercise 8.1 Propose an MPI checkpoint/restart interface.

8.2 MPI Limitations

Let's look at certain MPI features that are to some extent set in stone and can hardly be changed without breaking backward compatibility in a very noticeable way. We've scratched some of these issue in passing earlier in this book. Now it's time to examine them more thoroughly and see what lies beneath. This way we should be able to assess when quantitative corrections will have to give way to a qualitative change and thus possibly the end of the MPI that we know and love.

8.2.1 Dense Process Rank Sequence

One of the most fundamental MPI features is the dense process rank sequence, from zero to the number of processes in the respective communicator minus one. We have already seen that this causes substantial trouble as soon as failures are permitted to be non-fatal, when rank sequence perforce becomes noncontiguous, at least for a spell. And failures are unavoidable with scale, to the point of a checkpoint creation taking more time to complete than the mean time between failures.

Note that the dense rank sequence is not the only way of representing and address-ing job processes. Unix uses process identifiers (pid), PVM used task identifiers (tid), some other interfaces represent processes with opaque objects that can be thought of as handles. Storing these objects seems to be wasteful, but that is true only at first glance. Think of the reference counting used elsewhere in the MPI anyway, and of the memory used to represent processes (i.e., connections) inside the MPI library. And a typical opaque object—a handle—takes just 8 bytes on a modern 64-bit platform, i.e., the size of the virtual address. A rank takes 4 bytes and is easily deducible—if their sequence is dense.

So, when you calculate everything, you might end up with about as much memory spent and fewer troubles facing you than with the current seemingly very compact and implicit dense rank sequence.

8.2.2 Reliable and Pairwise Ordered Data Transfer

Another feature that can hardly be changed easily is the reliable and ordered message delivery. MPI is a sort of "shoot and forget" interface: once sent, a message is supposed to reach its destination, no matter what, unless it has been successfully cancelled at the source (see Section 7.1.5.6). The only other exception is the destination process dying prematurely in wait, which again means a failure that should in principle lead to the job committing *seppuku*, unless it is being finalized anyway. At the same time, every serious communication interface provides a means of setting timeouts on the respective calls, which comes in very handy in case of transient connection failures. And making an intrinsically unreliable communication medium, such as datagrams, carry messages reliably, essentially forces the MPI implementor to a (normally lame) recreation of the TCP protocol.

The same is true of the ordering of the messages being exchanged between any pair of MPI processes. Conditional on the tagging, message A1 must always come after message A2 that was sent later than A1, period. And what if the network routing has just switched to a faster path, and A2 as well as many later messages alongside it become available at the destination earlier than A1? Well, they will have to be buffered away by the MPI library until better times. Frankly, this is counterproductive, espe-cially if the underlying medium is intrinsically unordered, as many modern networks are, and the application algorithm can withstand unordered delivery. If the MPI user really cared of the message ordering, they would use message tags to enforce this, right? Many of them do this anyway, just to be on the safe side.

Can this be fixed using MPI_Info objects and communicator attributes? Sure. Will this be elegant and natural? No. And what kind of programs and performance gains are going to be possible then? Especially compared to a more compact interface that does not guarantee reliability and ordering by default? Think of our earlier MPI subsets. You are fully entitled to a comparable improvement expectation here, as well.

8.2.3 Message Progress Rules

Message progress is one of the worst understood areas of the MPI standard. This can be partly explained by the distributed nature of the respective definitions, which is normal to the MPI standard. Its sections 3.5 ("Semantics of Point-to-Point Communication"), 3.7.4 ("Semantics of Nonblocking Communications"), 3.8.1 ("Probe"), 3.8.2 ("Matching Probe"), 5.12 ("Nonblocking Collective Operations"), 11.4 ("Memory Model") and 11.7.3 ("Progress") in Chapter 11 ("One-Sided Communications"), 12.2 ("Generalized Requests"), 12.4 ("MPI and Threads"), 13.6.3 ("Progress") and 13.6.5 ("Nonblocking Collective File Operations") in Chapter 13 ("I/O") all contain definitions, explanations, illustrations, and references to the progress rules necessary for understanding this notion.

In a nutshell, progress means that all MPI communication calls must proceed without the application doing anything to this end, and that a temporarily blocked operation may not block other pending operations that are independent of it. This does not exclude the possibility of a deadlock, but a program that deadlocks over a fully standard-conforming MPI implementation is deemed incorrect anyway.

The reality of the MPI implementation is somewhat different. Excluding the rare cases of availability and full functionality of the asynchronous and hybrid progress engines that we have considered before (see Sections 5.2.4.3 and 5.2.4.4, respectively), all known MPI implementations do require the MPI application to kick the progress engine indirectly once in a while by calling the respective MPI functions. Most of the time this works. However, consider the following fully legal example (see Listing 8.1):

Listing 8.1: Nonblocking synchronous send paradox

```
//Sending process            // Receiving process
MPI_Issend(…,&req);          MPI_Irecv(…,&req);
…                            …
MPI_Wait(&req);              MPI_Wait(&req);
```

If the MPI_Wait() on the sending side is separated from the MPI_Issend() by a long computation without any intervening MPI calls, this program may block for an arbitrarily long time in most MPI implementations. For example, if the receiving side does not post the respective receive request before the sending process posts its synchronous send request, the sending process will not be able to fully process this message until the MPI_Wait() has been reached. Why? If you recall one popular way of implementing synchronous sends we have described in Section 7.1.1—the rendezvous protocol—even though the late coming MPI_Irecv() will be able to see the "ready-to-send" control message coming from the sender and reply to it by sending the "ready-to-receive" control message back, there will be no agent on the sending side to start the actual data transfer until the progress engine is kicked by the MPI_Wait() call there. There is no such agent on the receiving side either until the MPI_Wait() is called there.

In some sense, the (almost never fully implemented) progress rule stands in stark contrast to the absence of fairness guarantee in the MPI standard. What will happen if the progress rule is relaxed? Well, we have investigated this throughout this book. What we did in the first, very limited MPI subset (see Section 2.1) was to replace all standard MPI_Send() calls with synchronous MPI_Ssend() calls semantically and to bang completely on the progress rule because synchronous blocking calls did not need it at all. What we did in the second subset (see Section 3.1) was to allow non-blocking communication to a degree but to keep the blocking nature of the underlying progress engine. Again, the progress rule was not satisfied in this case. In some sense, the unlimited progress rule may be the root of many performance and implementation issues of the MPI standard.

8.2.4 Exercises

Exercise 8.2 Find yet another MPI design limitation and consider the repercussions of removing or at least relaxing it.

8.3 Beyond MPI

Is there life beyond MPI? The answer depends on whom you ask. MPI gurus will vehemently defend their baby and proclaim that MPI is here to stay and satisfy all the most exquisite needs and whims of the technical HPC community forever. In some sense, MPI is slowly achieving status comparable to that of the ever-evolving Fortran has among the programming languages. Authors of alternative interface A or B will counter that their baby is better suited for a particular task at hand and is deemed to take over once the world becomes a better place than it has been so far. Who is right?

Well, there is the way things ought to be, and then there is reality. Because of this, MPI will of course keep evolving, possibly taking into account some ideas from this book, since MPI subsetting can be deployed in several ways. Here, we have chosen the simple path of restricting the implementation from the outset and hoping (not in vain) that there will be applications fully satisfied by it and the resulting performance. One can imagine more complicated solutions, like dynamic detection of the working MPI subset either by means of static analysis or by dynamic tracking, and selecting either a specific library or a set of specific code paths inside a bigger library in pursuit of the best possible performance. If the subsetting idea takes root, and why shouldn't it, this topic will be investigated and exhausted fully in an astronomically short time. Doing this is not the purpose of this book. It provides a demonstration of what is possible with very simple and inexpensive means as soon as you go and shoot several sacred cows.

However, some things will not be able to change even then. One of them is the list of function arguments. In the simplest subset we have considered, only two arguments would be really necessary in a program running on two processes: the user buffer address and its length. The rest could be deduced from the context: the destination or source is the other process, the datatype is `MPI_BYTE`, the tag is not necessary, and the communicator is superfluous, not to mention the `MPI_Status` and the return code that are going to be ignored anyway. So, we should get two arguments instead of six in `MPI_Send()`, and of seven in `MPI_Recv()`. And if we only send zero size messages, we can have zero arguments, with the respective simplification of the code paths below the interface boundary. If you recall how function arguments are treated down below, you will see that we will need at most two registers instead of six or seven, and no register for the return value. For argument lists longer than four or five arguments, this will also mean that we do not really need any stack space, if you talk about modern RISC processors rather than the outdated CISC monsters that rule the world now by lucky chance and mighty installed base.

This was only a technical example indicating that there are some things no redesign will wipe away easily, apart from introducing yet another set of special MPI calls dealing with the two-process situation specifically along the lines mentioned above or mapping unnecessary MPI complexity upon much simpler calls by macros that basically throw away most if not all of their arguments. However, beyond this, there are higher level considerations that speak in favor of a more compact, single-minded interface based on a certain paradigm, be that DMA or RDMA, or something completely exotic (take Portals, for one), or something else we do not yet know about.

Let's consider some of the current challenges facing the HPC world and see how MPI fits in and whether it makes the best possible fit there.

8.3.1 Scalability, Mobility, and IoT Challenges

If there is a name to the epoch we live in, it's Scalability. After the qualitative leap of electricity coupled with cybernetics, we are currently going through a dialectically unavoidable phase of quantitative expansion that will end up with yet another qualitative leap, be that optics or quantum computing. This tendency spans the high-, medium-, and low-end parts of the computing universe.

PetaFLOPS machines are becoming commonplace (top500.org, 2018). The next scalability step is ExaFLOPS, i.e., 10^{18} floating point operations per second. Traditionally, MPI has been considered the best possible fit for those monster computers. Is it? Hardly. Apart from bringing the existing application code base with it, it has very few features that make it amenable to this level of scalability. And even the application base may be more of a liability than of an asset, since algorithms necessary to address faults and transient errors, data loss, and cosmic radiation are going to be completely different from those popular in the pre-ExaFLOPS era. What you need for Exascale is a very limited set of calls that scale, scale, scale... well, that's it.

MPI is even less suited for the ever-growing world of mobile computing. I will not be surprised if the cumulative computing power of all smartphones and tablets in the world, if harvested properly, has already surpassed or is soon going to surpass the total computing power of the TOP500 list mentioned above. What is specific to this potential market that MPI Forum has proudly ignored so far? Well, relatively large latencies, relatively low (by the current standards) bandwidths, relatively high levels of faults. Again, no perfect fit for the MPI as currently defined.

And then come the smart fridges. Not as powerful as smartphones, but probably more numerous if combined with smart air conditioners, smart washing machines, smart frying pans, and so on. If they ever take part in SETI@home (University of California, 2018), they might yet discover extraterrestrial life on our behalf. Most likely, it will be them talking to alien smart fridges. And now imagine a Chief Smart Fridge in Charge coping with something like `MPI_Reduce_scatter()`...

8.3.2 Frequency Wall and Memory Lag

Another issue to address is the so-called frequency wall. Using current technologies and not going to live in liquid nitrogen or exchange quanta of light, you are not going to get much higher than, say, 5 GHz processor frequency. Shortly after that, the heat dissipation will approach that of the core of a nuclear reactor, and that will be that. At the same time, the working frequency of the main memory reaches its limit at about a third of that, while latencies differ by three orders of magnitude in favor of the processor. The main problem is then to feed data from the main memory into the main processor that runs away as a fighter jet ahead of a pedestrian.

What to do about this? Let memory and processors fuse, and keep most if not all of the data to be processed in the network where it properly belongs, just like money changing hands rather than tucked under a mattress. Is MPI suited for this? No. What is? Well, we do not know yet.

8.3.3 Programmability Barrier and Stack Crawling

And then comes the complexity, the obtuse complexity of the MPI interface multiplied by the complexity of the algorithms multiplied by the complexity of concurrent execution. One can do little about algorithms, still less about the intrinsic pitfalls of the parallelism, but one can and actually must reduce the complexity of the interfaces. Otherwise the network drivers start crawling up the program stack and assume the place of the MPI. Take the OFA libfabrics. It's MPI in all but name. MPI calls implemented on top of this presumable low-level interface are, well, trivial: some error checking, some argument preprocessing, and then a call down in the hope all is well there. Is it? In what part? The point-to-point, the collectives, the one-sided, or something else? What will you use today, sir?

The point is that passing visible complexity from above down under is not going to decrease the complexity of the whole, rather it will increase it. And the slow-down and all that matters will reappear down there, too.

8.3.4 Exercises

Exercise 8.3 Propose an interface that can take on MPI in the target markets mentioned above or elsewhere.

8.4 Conclusions

So, what will happen to MPI? A new, very simple yet fast interface will emerge, to be ignored at first, then to be laughed at, then to be fought against, and finally to win (Christensen C. , 1997).

I hope to have contributed to the unavoidable with this book.

Bibliography

Argonne National Laboratory. (2017). *MPICH*. Retrieved from www.mpich.org: https://www.mpich.org/

Balaji, P., Buntinas, D., Goodell, D., Gropp, William, Krishna, J., Lusk, E., & Thakur, R. (2012). *PMI: A Scalable Parallel Process-Management Interface for Extreme-Scale Systems*. Retrieved from http://www.mcs.anl.gov/papers/P1760.pdf

Beowulf.org. (2002). *Beowulf.org: Overview - FAQ*. Retrieved from www.beowulf.org: http://www.beowulf.org/overview/faq.html

Bißeling, G., Hoppe, H.-C., Latour, J., & Supalov, A. (2002). Fujitsu MPI-2: Fast Locally, Reaching Globally. *Recent Advances in Parallel Virtual Machine and Message Passing Interface, 2474*, pp. 401-409. Linz, Austria. Retrieved from https://www.researchgate.net/publication/221597662_Fujitsu_MPI-2_Fast_Locally_Reaching_Globally

Christensen, A. L., Brito, J., & Silva, J. G. (2004). The Architecture and Performance of WMPI II. *Recent Advances in Parallel Virtual Machine and Message Passing Interface, 3241*, pp. 112-121. Retrieved from https://link.springer.com/chapter/10.1007/978-3-540-30218-6_21

Christensen, C. (1997). *The Innovator's Dilemma: When New Technologies Cause Great Firms to Fail*. Retrieved from http://www.claytonchristensen.com/books/the-innovators-dilemma/

Cray. (2014). *hjelmn/xpmem*. Retrieved from Github.com: https://github.com/hjelmn/xpmem

DelSignore, J. (2010, October 11). *The MPIR Process Acquisition Interface Version 1.0*. Retrieved from mpi-forum.org: http://mpi-forum.org/docs/mpir-specification-10-11-2010.pdf

Dinan, J., Balaji, P., Goodell, D., Miller, D., Snir, M., & Thakur, R. (2013). Enabling MPI Interoperability Through Flexible Communication Endpoints. *Recent Advances in the Message Passing Interface*, (pp. 13-19). Madrid, Spain. Retrieved from http://www.mcs.anl.gov/~thakur/papers/endpoints.pdf

Eisler, M. (2006, May). *RFC 4506 - XDR: External Data Representation Standard*. Retrieved from tools.ietf.org: https://tools.ietf.org/html/rfc4506.html

George, W. L., Hagedorn, J. G., & Devaney, J. E. (2000, May-June). *IMPI: Making MPI Interoperable*. Retrieved from www.nist.gov: https://www.nist.gov/sites/default/files/documents/itl/math/hpcvg/j53geo.pdf

Goglin, B., & Moreaud, S. (2013, February). KNEM: A generic and scalable kernel-assisted intra-node MPI communication framework. *Journal of Parallel and Distributed Computing, 73*(2), pp. 176-188. Retrieved from https://www.sciencedirect.com/science/article/pii/S0743731512002316

Goldberg, D. (1991, March). What Every Computer Scientist Should Know About Floating-Point Arithmetic. *ACM Computing Surveys, 23*(1), pp. 5-48. Retrieved from http://pages.cs.wisc.edu/~david/courses/cs552/S12/handouts/goldberg-floating-point.pdf

Gropp, W., & Lusk, E. (1997, January). A high-performance MPI implementation on a shared-memory vector supercomputer. *Parallel Computing, 22*(11), pp. 1513-1526. Retrieved from https://www.sciencedirect.com/science/article/pii/S0167819196000622

Hempel, R., Hoppe, H.-C., & Supalov, A. (1992, December 17). *PARMACS 6.0 Library Interface Specification*. Retrieved from PARMACS 6.0 Library Interface Specification: http://citeseerx.ist.psu.edu/viewdoc/summary?doi=10.1.1.47.2244

Hierarchical Data Format. (2017). Retrieved from Wikipedia.org: https://en.wikipedia.org/wiki/Hierarchical_Data_Format

Hoefler, T., Lumsdane, A., & Rehm, W. (2007). Implementation and Performance Analysis of Non-Blocking Collective Operations for MPI. In I. C. Society/ACM (Ed.), *Proceedings of the 2007 International Conference on High Performance Computing, Networking, Storage and Analysis*,

DOI 10.1515/9781501506871-009

SC07, presented in Reno, USA, Nov. 2007. Reno, USA. Retrieved from http://htor.inf.ethz.ch/publications/img/hoefler-sc07.pdf

IEEE 802.3 ETHERNET WORKING GROUP. (2017). *IEEE 802.3 ETHERNET WORKING GROUP*. Retrieved from www.ieee802.org: http://www.ieee802.org/3/

InfiniBand Trade Association. (2017, February 22). *InfiniBand Trade Association*. Retrieved from www.infinibandta.org: http://www.infinibandta.org/

Intel Corporation. (2011, April). *DAPL IB collective extensions*. Retrieved from Openfabrics.org: https://www.openfabrics.org/downloads/dapl/documentation/dat_collective_preview.pdf

Intel Corporation. (2018). *Introducing Intel® MPI Benchmarks*. Retrieved from https://software.intel.com/en-us/articles/intel-mpi-benchmarks

Kanevsky, A., Skjellum, A., & Rounbehler, A. (1997, January 2). *Real-Time Extensions to the Message-Passing Interface (MPI)*. Retrieved from http://citeseerx.ist.psu.edu/viewdoc/summary?doi=10.1.1.36.1387

Li, S., Hu, C., Hoefler, T., & Snir, M. (2014, December). Improved MPI collectives for MPI processes in shared address spaces. *Cluster Computing, 17*(4), pp. 1139-1155. Retrieved from https://link.springer.com/article/10.1007/s10586-014-0361-4

Magro, W. R., Poulsen, D. K., Supalov, A. V., & Derbunovich, A. B. (2010, Januar 5). *USA Patent No. 7,644,130*. Retrieved from http://patft.uspto.gov/netacgi/nph-Parser?-Sect1=PTO2&Sect2=HITOFF&p=1&u=%2Fnetahtml%2FPTO%2Fsearch-bool.html&r=17&f=G&l=50&co1=AND&d=PTXT&s1=supalov.INNM.&OS=IN/supalov&RS=IN/supalov

Mellanox Technologies. (2017a). *Messaging Accelerator (MXM)*. Retrieved from Mellanox Technologies: http://www.mellanox.com/page/products_dyn?product_family=135&menu_section=73

Mellanox Technologies. (2017b). *Unified Communication X (UCX) Framework*. Retrieved from Mellanox Technologies: http://www.mellanox.com/page/products_dyn?product_family=281&mtag=ucx

Mellanox Technologies. (2017c). *Fabric Collective Accelerator (FCA)*. Retrieved from Mellanox Techologies: http://www.mellanox.com/page/products_dyn?product_family=104&mtag=fca2

Message Passing Interface Forum. (1994, May 5). *MPI: A Message-Passing Interface Standard*. Retrieved from mpi-forum.org: http://mpi-forum.org/docs/mpi-1.0/mpi-10.ps

Message Passing Interface Forum. (1995, June 12). *MPI: A Message-Passing Interface Standard*. Retrieved from mpi-forum.org: http://mpi-forum.org/docs/mpi-1.1/mpi1-report.pdf

Message Passing Interface Forum. (1997a, July 18). *MPI-2: Extensions to the Message-Passing Interface*. Retrieved from mpi-forum.org: http://mpi-forum.org/docs/mpi-2.0/mpi2-report.pdf

Message Passing Interface Forum. (1997b, July 18). *MPI-2 Journal of Development*. Retrieved from mpi-forum.org: http://mpi-forum.org/docs/mpi-jd/mpi-20-jod.ps

Message Passing Interface Forum. (2008a, June 23). *MPI: A Message-Passing Interface Standard Version 2.1*. Retrieved from mpi-forum.org: http://mpi-forum.org/docs/mpi-2.1/mpi21-report.pdf

Message Passing Interface Forum. (2008b, May 30). *MPI: A Message-Passing Interface Standard Version 1.3*. Retrieved from mpi-forum.org: http://mpi-forum.org/docs/mpi-1.3/mpi-report-1.3-2008-05-30.pdf

Message Passing Interface Forum. (2009, September 4). *MPI: A Message-Passing Interface Standard Version 2.2*. Retrieved from mpi-forum.org: http://mpi-forum.org/docs/mpi-2.2/mpi22-report.pdf

Message Passing Interface Forum. (2012, September 21). *MPI: A Message-Passing Interface Standard Version 3.0*. Retrieved from mpi-forum.org: http://mpi-forum.org/docs/mpi-3.0/mpi30-report.pdf

Message Passing Interface Forum. (2015, June 4). *MPI: A Message-Passing Interface Standard Version 3.1*. Retrieved from mpi-forum.org: http://mpi-forum.org/docs/mpi-3.1/mpi31-report.pdf

Metcalfe, V., Gierth, A., & et al. (1996). *Please explain the TIME_WAIT state*. Retrieved from Programming UNIX Sockets in C - Frequently Asked Questions: http://www.softlab.ntua.gr/facilities/documentation/unix/unix-socket-faq/unix-socket-faq-2.html#time_wait

Message Passing Interface Forum. (2017, February 5). *MPI Documents*. Retrieved from http://mpi-forum.org/docs/

Nagle, J. (1984, January 6). *Congestion Control in IP/TCP Internetworks*. Retrieved from Network Working Group: https://tools.ietf.org/html/rfc896

Nazarewicz, M. (2012, March 14). *A deep dive into CMA*. Retrieved from LWN.net: https://lwn.net/Articles/486301/

Netlib.org. (2017). *BLAS (Basic Linear Algebra Subprograms)*. Retrieved from netlib.org: http://www.netlib.org/blas/

OFI Working Group. (2017). *Libfabric Programmer's Manual*. Retrieved from Libfabric OpenFabrics: https://ofiwg.github.io/libfabric/

Open MPI Consortium. (2017). *Open MPI: Open Source High Performance Computing*. Retrieved from www.open-mpi.org: https://www.open-mpi.org/

Openfabrics.org. (2017). *OFED Overview*. Retrieved from Openfabrics Software: https://www.openfabrics.org/index.php/openfabrics-software.html

OpenMP Consortium. (2017, February 5). *Specifications - OpenMP*. Retrieved from http://www.openmp.org/specifications/

PCI Special Interest Group. (2017, February 7). *Specifications | PCI-SIG*. Retrieved from Specifications | PCI-SIG: https://pcisig.com/specifications/pciexpress/

Pješivac-Grbović, J., Angskun, T., Bosilca, G., Fagg, G. E., Gabriel, E., & Dongarra, J. J. (2007, June). Performance analysis of MPI collective operations. *Cluster Computing, 10*(2), pp. 127-143. Retrieved from http://www.netlib.org/utk/people/JackDongarra/PAPERS/coll-perf-analysis-cluster-2005.pdf

PVM: Parallel Virtual Machine. (2017). Retrieved from PVM: Parallel Virtual Machine: http://www.csm.ornl.gov/pvm/

Rabenseifner, R. (2000). Automatic MPI counter profiling. *Proceedings of the 42nd CUG Conference*. Noorwijk, The Netherlands. Retrieved from http://citeseerx.ist.psu.edu/viewdoc/download?doi=10.1.1.68.6687&rep=rep1&type=pdf

Sandia National Laboratories. (2017). *Portals 4.0*. Retrieved from Portals Network Programming Interface: http://www.cs.sandia.gov/Portals/portals4.html

Squyres, J. (2014, October 29). *The "vader" shared memory transport in Open MPI: Now featuring 3 flavors of zero copy!* Retrieved from Cisco Blogs: https://blogs.cisco.com/performance/the-vader-shared-memory-transport-in-open-mpi-now-featuring-3-flavors-of-zero-copy

Stackoverflow.com. (2013). *How to ssh to localhost without password?* Retrieved from Stackoverflow.com: https://stackoverflow.com/questions/7439563/how-to-ssh-to-localhost-without-password

Stevens, W. R. (2012). *Unix Network Programming, Volume 2: Interprocess Communications (2nd Edition)*. Prentice Hall.

Stevens, W. R., Fenner, B., & Rudoff, A. M. (2003). *Unix Network Programming, Volume 1: The Sockets Networking API (3rd Edition)*. Addison-Wesley Professional.

Supalov, A. (2003). Lock-Free Collective Operations. *Recent Advances in Parallel Virtual Machine and Message Passing Interface, 2840*, pp. 276-285. Retrieved from https://link.springer.com/chapter/10.1007/978-3-540-39924-7_40

Supalov, A. (2009, May 26). *USA Patent No. 7,539,995*. Retrieved from http://patft.uspto.gov/netacgi/nph-Parser?Sect1=PTO2&Sect2=HITOFF&p=1&u=%2Fnetahtml%2FP-

TO%2Fsearch-bool.html&r=19&f=G&l=50&co1=AND&d=PTXT&s1=supalov.INNM.&OS=IN/
supalov&RS=IN/supalov

Supalov, A. (2018). *Inside the Message Passing Interface: Creating Fast Communication Libraries.* De
Gruyters. Retrieved from https://www.degruyter.com/view/product/488201

Supalov, A. V. (2009, July 28). *USA Patent No. 7,567,557.* Retrieved from http://patft.uspto.
gov/netacgi/nph-Parser?Sect1=PTO2&Sect2=HITOFF&p=1&u=%2Fnetahtml%2FP-
TO%2Fsearch-bool.html&r=18&f=G&l=50&co1=AND&d=PTXT&s1=supalov.INNM.&OS=IN/
supalov&RS=IN/supalov

Supalov, A. V. (2010, December 21). *USA Patent No. 7,856,348.* Retrieved from http://patft.
uspto.gov/netacgi/nph-Parser?Sect1=PTO2&Sect2=HITOFF&p=1&u=%2Fnetahtml%2FP-
TO%2Fsearch-bool.html&r=15&f=G&l=50&co1=AND&d=PTXT&s1=supalov.INNM.&OS=IN/
supalov&RS=IN/supalov

Supalov, A. V. (2011, June 21). *USA Patent No. 7,966,624.* Retrieved from http://patft.uspto.
gov/netacgi/nph-Parser?Sect1=PTO2&Sect2=HITOFF&p=1&u=%2Fnetahtml%2FP-
TO%2Fsearch-bool.html&r=12&f=G&l=50&co1=AND&d=PTXT&s1=supalov.INNM.&OS=IN/
supalov&RS=IN/supalov

Supalov, A. V. (2014, May 13). *US Patent No. 8,725,875.* Retrieved from http://patft.uspto.
gov/netacgi/nph-Parser?Sect1=PTO2&Sect2=HITOFF&p=1&u=%2Fnetahtml%2FP-
TO%2Fsearch-bool.html&r=5&f=G&l=50&co1=AND&d=PTXT&s1=supalov.INNM.&OS=IN/
supalov&RS=IN/supalov

Supalov, A. V., Chuvelev, M. V., Dontsov, D. V., & Truschin, V. D. (2016, September 20). *USA
Patent No. 9,448,863.* Retrieved from http://patft.uspto.gov/netacgi/nph-Parser?-
Sect1=PTO2&Sect2=HITOFF&p=1&u=%2Fnetahtml%2FPTO%2Fsearch-bool.
html&r=1&f=G&l=50&co1=AND&d=PTXT&s1=supalov.INNM.&OS=IN/supalov&RS=IN/supalov

Supalov, A. V., Sapronov, S. I., Syrov, A. A., Ezhov, D. V., & Truschin, V. D. (2010, August 3).
USA Patent No. 7,769,856. Retrieved from http://patft.uspto.gov/netacgi/nph-Parser?-
Sect1=PTO2&Sect2=HITOFF&p=1&u=%2Fnetahtml%2FPTO%2Fsearch-bool.
html&r=16&f=G&l=50&co1=AND&d=PTXT&s1=supalov.INNM.&OS=IN/supalov&RS=IN/
supalov

Supalov, A. V., Truschin, V. D., & Ryzhykh, A. V. (2012, October 2). *USA Patent
No. 8,281,060.* Retrieved from http://patft.uspto.gov/netacgi/nph-Parser?-
Sect1=PTO2&Sect2=HITOFF&p=1&u=%2Fnetahtml%2FPTO%2Fsearch-bool.
html&r=10&f=G&l=50&co1=AND&d=PTXT&s1=supalov.INNM.&OS=IN/supalov&RS=IN/
supalov

Supalov, A. V., Van der Wijngaart, R. F., & Whitlock, S. J. (2013, Septeber 17). *USA
Patent No. 8,539,456.* Retrieved from http://patft.uspto.gov/netacgi/nph-Parser?-
Sect1=PTO2&Sect2=HITOFF&p=1&u=%2Fnetahtml%2FPTO%2Fsearch-bool.
html&r=7&f=G&l=50&co1=AND&d=PTXT&s1=supalov.INNM.&OS=IN/supalov&RS=IN/supalov

Supalov, A., & Yalozo, A. (2014, June). 20 Years of the MPI Standard: Now with a Common Application
Binary Interface. *The Parallel Universe*(18), pp. 28-32. Retrieved from https://software.intel.
com/sites/default/files/managed/6a/78/parallel_mag_issue18.pdf

Supalov, A., Semin, A., Klemm, M., & Dahnken, C. (2014). *Optimizing HPC Applications with Intel
Cluster Tools.* Apress. doi:10.1007/978-1-4302-6497-2

Superuser.com. (2014). *Find which HT-cores are sharing a physical core from /proc/cpuinfo.*
Retrieved from Superuser.com: https://superuser.com/questions/331308/find-which-ht-cores-
are-sharing-a-physical-core-from-proc-cpuinfo

Thakur, R., Rabenseifner, R., & Gropp, W. (2004). *Optimization of Collective Communication
Operations in MPICH.* Retrieved from http://www.mcs.anl.gov/~thakur/papers/ijhpca-coll.pdf

Thakur, R., Ross, R., Latham, R., Lusk, R., Gropp, B., & Choudhary, A. (2016). *ROMIO: A High-Per-formance, Portable MPI-IO Implementation*. Retrieved from Argonne National Laboratories: http://www.mcs.anl.gov/projects/romio/

The Ohio State University. (2018). *MVAPICH: MPI over InfiniBand, Omni-Path, Ethernet/iWARP, and RoCE*. Retrieved from MVAPICH :: Home: http://mvapich.cse.ohio-state.edu/

top500.org. (2018). *TOP500 Supercoputer Sites*. Retrieved from https://www.top500.org/

Truschin, V. D., Supalov, A. V., & Ryzhykh, A. V. (2012, August 14). *USA Patent No. 8,245,240*. Retrieved from http://patft.uspto.gov/netacgi/nph-Parser?-Sect1=PTO2&Sect2=HITOFF&p=1&u=%2Fnetahtml%2FPTO%2Fsearch-bool.html&r=11&f=G&l=50&co1=AND&d=PTXT&s1=supalov.INNM.&OS=IN/supalov&RS=IN/supalov

University of California. (2018). *SETI@home*. Retrieved from https://setiathome.berkeley.edu/

Vo, A. (2013, December 12). *The MPI Message Queue Dumping Interface Version 1.0*. Retrieved from mpi-forum.org: http://mpi-forum.org/docs/msgq.5.pdf

Wikipedia.org. (2018a). *Berkley sockets*. Retrieved from Wikipedia: https://en.wikipedia.org/wiki/Berkeley_sockets

Wikipedia.org. (2018b). *POSIX Threads*. Retrieved from wikipedia.org: https://en.wikipedia.org/wiki/POSIX_Threads

Wikipedia.org. (2018c). *HAL 9000*. Retrieved from Wikipedia.org: https://en.wikipedia.org/wiki/HAL_9000

Wikipedia.org. (2018d). *OSI Model*. Retrieved from www.wikipedia.org: https://en.wikipedia.org/wiki/OSI_model

Wikipedia.org. (2018e). *Network File System*. Retrieved from https://en.wikipedia.org/wiki/Network_File_System

Appendix A
Solutions

In this appendix you will learn how to answer all those nagging questions the easy way.

A.1 Overview

Exercise 1.1 Examine your computer system and clarify how many nodes it has, what node types are available, and by what network(s) they are connected.

The Lenovo P50 I am writing this book on is a full-featured single-node engineering workstation. The cluster I am using for test runs involving networking is so special that I do not feel that I can describe it here.

Exercise 1.2 Examine your computing nodes and find out what kind of processor(s) they exhibit, how many cores and threads they run, what memory type and what caches they use, and what networking and graphics capabilities they have.

The Lenovo P50 laptop mentioned in Exercise 1.1 has the following characteristics:
- CPU: Intel Xeon E3-1535M v5 (8 MB Cache, up to 3.8 GHz)
- Graphics: NVIDIA Quadro M2000M 4 GB
- Memory: 64 GB (16 x 4) DDR4 2133 MHz ECC SoDIMM
- Storage: 500GB HD 7200RPM + 2 x 512 GB SSD PCIe-NVMe
- Display: 15.6" 4K (3840 x 2160), IPS
- Network: Wi-Fi 802.11n, Gigabit Ethernet

Exercise 1.3 Read the latest MPI standard. I mean it: you will need this to proceed sensibly beyond this point.

The latest MPI standard 3.1 can be found at (Message Passing Interface Forum, 2015) alongside all earlier standards. If you do not want to read them, you can buy an audiobook. This was a joke.

Exercise 1.4 Explain the difference between the blocking and nonblocking communication. Is blocking local or non-local?

Blocking communication calls return control to the application as soon as the user provided buffer can be reused in other operations. Nonblocking communication does not guarantee that. Blocking communication operation may be local or nonlocal depending on the communication mode requested. For example, buffered `MPI_Bsend()` is local, because the message contents are buffered away by the MPI implementation in order to get back to the application as soon as possible. Contrary

DOI 10.1515/9781501506871-010

to this, synchronous `MPI_Ssend()` is nonlocal, because it depends upon the receiving side sending an acknowledgement back to the sending side once a matching receive operation has been posted on that side. Standard `MPI_Send()` is usually a combination of those plus more.

Exercise 1.5 Find out what the standard point-to-point communication mode is comparable to in your favorite MPI implementation. In particular, find out whether it is buffered or not, and if so, to what degree.

You may have figured out by now that I prefer MPICH. In this package, the standard communication mode is a combination of buffered, eager, and synchronous message transfers.

In addition to the buffered and synchronous modes described briefly in Exercise 1.4, MPICH may use the so-called eager protocol for sending away messages up to a certain length. Such a message may cause an unexpected receive operation on the other side, during which the data will be buffered away by the MPI implementation until the point that a matching receive operation is posted.

Beyond this threshold, synchronous message transfer takes over. It can only be started when the receiving side has posted a matching receive operation. This may lead to extra delay on departure but will save the internal buffer space on the receiving side. In other words, this is a trade-off between the message latency and the memory consumed.

Finally, buffered message transfer may kick in if the eager or synchronous transfers take too long to arrange—for example, if the respective channel is busy pumping another, earlier piece of data. In this case MPI will store user data away temporarily on the sending side, just to let the application proceed.

Exercise 1.6 Clarify whether your favorite MPI implementation provides nonblocking collective operations and how they compare in performance to the blocking ones, and whether they support any computation/communication overlap.

MPICH supports nonblocking collective operations. Depending on the package configuration and the communication fabrics in use, it may or may not provide computation/communication overhead. A good way to do this is to use the Intel® MPI Benchmarks (Intel Corporation, 2018) that provide a set of tests specifically targeted at answering this question.

Exercise 1.7 Examine corresponding MPI documentation and source code to see whether one-sided communication is mapped upon the point-to-point communication or supported natively, and if so, over what communication fabrics.

It depends. First of all, some fabrics may *only* provide point-to-point communication, in this case point-to-point mapping may be required. Second, the way in which one-sided operations are implemented may be altered during package configuration time. Finally, as far as the so-called fast collective operations introduced by the MPI-3 standard are concerned, they will probably be implemented directly for those fabrics that support them, and not implemented otherwise.

Exercise 1.8 If you are unable to give articulate answers to any of the above questions, read the answers in this Appendix, then go back to Exercise 1.3 and repeat until done.

I had to do this a couple of times, just to make sure I had the correct information. Did you?

Exercise 1.9 Find out what operating system is installed on your computer, whether it is up-to-date, and what support for the threading, process binding, memory locality control, interprocess communication, and networking it offers.

I run two kinds of Linux on the laptop mentioned above. For program development and quick testing, I use Ubuntu 16.06 LTS. For heavy weight lifting, like the shared memory benchmarking, where the GUI and excessive daemons may get in the way big time, I use Debian 8.4 in the plainest configuration available without the trouble of building the kernel myself: no GUI, no servers, rudimentary networking, as few daemons as possible. Both systems are close in provided functionality that is used throughout this book.

Exercise 1.10 Define a sensible MPI subset a step beyond one that includes collective operations and nonblocking point-to-point communication.

The next sensible addition depends on your goals. Adding MPI_Sendrecv(), MPI_Waitall() for its implementation, and MPI_Testall() for symmetry looks good to me. Whether you need the MPI_Waitsome(), MPI_Waitany(), and their nonblocking pendants is open to discussion.

Exercise 1.11 Examine the code management utilities offered and choose one you like most. You will probably want to use it throughout this book, so be careful when choosing between the cvs, svn, git, and something else.

I decided not to use any code management tool for reasons stipulated in Section 1.4.2. You are of course free to choose you own way.

Exercise 1.12 Experiment with the program building tools, including the C and the Fortran compilers, and find out what level of the respective language standard they support. If you have both GNU and other compilers, compare them using standard benchmarks like SPECint, SPECfloat, and Stream.

I used OS-provided GNU compilers and the associated tool chain throughout this book on the aforementioned laptop. For Ubuntu, this was gcc (Ubuntu 5.4.0-6ubuntu1~16.04.4) 5.4.0 20160609, for Debian this was gcc (Debian 4.9.2-10) 4.9.2. In both cases I show the essential part of the output produced by the gcc --version command.

Exercise 1.13 Try out the MPI libraries that come with your OS distribution (if any) and make yourself comfortable with the respective MPI commands like `mpicc`, `mpif77`, and `mpiexec` (or `mpirun`).

I've tried both MPICH and Open MPI provided by the respective OS in additional binary packages. However, I found that using a freshly built MPI based on the latest stable source available was a better bet, for many reasons including but not limited to freshness, full control over the configuration, and the good ol' feeling of seeing lots of cryptic output.

This way I used MPICH 3.2 and Open MPI 2.0.2 and 2.1.0 for most measurements in Chapter 2 (Shared Memory). Both libraries were evolving as the book was being written, MPICH reaching version 3.2.1, and Open MPI progressing to major version 3 at least. Some of the issues observed by me and highlighted throughout the book were reported or at least presumed to have been fixed in those later versions. Unfortunately, I did not have time to verify that information.

A.2 Shared Memory

Exercise 2.1 Define a couple of MPI subsets a bit wider than the one described above but stopping short of adding nonblocking communication.

The easiest and most natural thing to do will be adding support for the `MPI_STATUS_IGNORE` constant. Another interesting option is adding support for the `MPI_ANY_TAG` and `MPI_ANY_SOURCE` wildcards. Finally, one can imagine a situation in which a limited send-to-itself operation might be useful.

Exercise 2.2 Define a subset or two still narrower than the one described above. Think about message length, tag values, and other entities.

A sensible thing to do, from the performance point of view, would be a limitation put on message size. Another possible simplification would be completely ignoring the tag value.

Exercise 2.3 Check out bigger alignment offsets (up to and including 64) in the refactored naïve shared memory benchmark. Also recheck offsets from 8 down on your system, process the data and see whether there is any difference in performance. Use the best offset detected in the rest of your exercises.

Based on the results obtained offline, I settled on the 8-byte offset.

Exercise 2.4 Redo Exercise 2.3 using a different (hopefully, even better optimizing) compiler. If its `memcpy(3)` function performs noticeably better, use this compiler from now on. If data transfer performance demonstrates any dependency upon the alignment, adjust the offset accordingly going forward.

I decided not to pump down several gigabytes of data to get Intel Compiler available under free trial license. I remember vaguely that it was not allowed to use it in performance measurements other than your own, and I have no legal department supporting me to recheck the license once again. The full compiler was way too expensive.

Exercise 2.5 Study the test programs provided in the source code archive and determine whether there are any corner cases and error conditions that have not been verified. Add a respective program or programs to the set in this case and extend the Makefile accordingly.

Well, we have not done any checking of the MPI return error codes so far. This will be corrected in Chapter 4.

Exercise 2.6 Look more closely into the library and decide whether the current implementation is really doing only what is necessary and nothing more—even after code streamlining. If you find superfluous code segments that are executed more than once without necessity, reformulate the library code accordingly and redo the testing and benchmarking.

I did this and a bit more in version 4 of the source code (see shm/mpi/v4/mpi.*).

Exercise 2.7 Find out the optimum size of the BUFLEN constant on your system. Take care of trying suitable non-power-of-2 values as well, minding the paging and alignment requirements. Use the identified best value from now on.

Based on the data collected thus far, I settled upon the 32 KiB buffer size.

Exercise 2.8 Add support for the MPI_STATUS_IGNORE status placeholder to the original MPI_Recv() implementation presented in this chapter. Use NULL as the respective constant value in the mpi.h header. Change the ping-pong benchmark to use this feature and find out whether this brings any noticeable performance improvement. Use conditional compilation throughout to make this feature selectable at build time.

I did not observe any performance difference (see shm/mpi/v4/mpi.*).

Exercise 2.9 Add support for the MPI_ANY_TAG tag wildcard selector to the MPI_Recv() implementation. Use -1 as the respective constant value in the mpi.h header. Change the ping-pong benchmark to use this feature and find out whether this addition brings any noticeable performance overhead compared to the original implementation. Pay attention to running both the MPI_ANY_TAG and specific tag selection benchmarks using your new code.

See Section 7.1.2.1. I did not observe any performance difference.

Exercise 2.10 Add support for the MPI_ANY_SOURCE source process wildcard selector to the MPI_Recv() implementation. Use -1 as the respective constant value in the mpi.h header. Change the ping-pong benchmark to use this feature and find out whether this addition brings any noticeable performance overhead compared to the original implementation. Pay attention to running both the MPI_ANY_SOURCE and specific process selection benchmarks using your new code. Do not forget to try running more than two processes.

See Section 7.1.2.2 . Performance difference stayed within about 2% in this simple case.

Exercise 2.11 Get rid of the double buffering altogether by restricting the maximum MPI message size to BUFLEN. Benchmark the resulting implementation to find out how much additional performance this brings in the small message area compared to the implementations created so far.

There is no need to change MPI library code to get the answer. Just look at the timings obtained using the naïve shared memory benchmarks sm1, sm2, and sm3.

Exercise 2.12 Restrict the MPI implementation even further to pass only zero-sized messages, possibly without paying attention to the tag value. Find out how much performance this adds on top. Imagine your users will only be sending zero-sized messages (for example, for synchronization purposes). Does it make sense for them to use a full-blown MPI implementation?

Again, to get the numbers, comment out the memcpy(3) in the naïve shared memory benchmarks sm1, sm2, and sm3. You will save one function call. That's all.

Exercise 2.13 Try using clock_gettime(3) instead of gettimeofday(2) in the MPI_Wtime() function. Does this influence the timings in any way?

It does: they become more fine-grained and hence more precise. Note that you may have to link your program against the librt library to use this function.

Exercise 2.14 Compare shared memory performance of the latest versions of MPICH and Open MPI to versions 3.2 and 2.0.2, respectively, used in this chapter. Is there any difference? Why?

There should be no difference here if there was no material change to the way the code works.

A.3 Sockets

Exercise 3.1 Define a couple of MPI subsets a bit wider than the one described above.
See Exercise 1.10.

Exercise 3.2 Define a subset or two that are still narrower than the one described above. Think about message length, tag values, and other entities.

See Exercise 2.2.

Exercise 3.3 Redo the intranode benchmarking done so far over sockets on your system. How does it fare versus my test laptop?

I cannot help you here.

Exercise 3.4 If you have access to an Ethernet-based cluster, do the measurements over this network as well as intranode, and compare your results to mine. Do they essentially match?

Again, this is your particular situation, but I doubt you will find much difference beyond the expected difference between my 100-Mbit and your GigE configurations.

Exercise 3.5 If you have access to cluster with InfiniBand or another network, find out how sockets are implemented over that network, and redo the measurements done above and compare the performance results with the intra- and internode data obtained so far.

Alas, I did not manage to get access to such a cluster. What about you?

Exercise 3.6 Would it not be better to put the MPI_Send() and MPI_Recv() operations into the transfer completion calls?

No. The point at which the message request is injected into the system does not determine the order in which it will be received in the presence of the tag matching. It does, however, determine the order in which consequent messages are registered with the system, so that messages with the same tag will conceptually arrive to a specific receiving process. If you put the MPI_Send() and MPI_Recv() into the transfer completion calls, you may easily spoil this ordering, since completion calls are not required to come in the same order as the respective MPI_Isend() and MPI_Irecv() calls.

Exercise 3.7 How can one map the MPI_Test() upon the MPI_Wait()?

Just rewrite the MPI_Test() to use MPI_Wait(). Nobody will notice the difference because the underlying progress engine is actually blocking.

Exercise 3.8 Implement the buffered short message receive scheme outlined in Section 3.4.6. Do you notice any difference in latency now?

Indeed, short message latency was down by 15% due to this change (see source code in sock/mpi/v4/mpi.*). At least, this is what my quick benchmarking told me.

Exercise 3.9 Compare socket performance of the latest versions of MPICH and Open MPI to versions 3.2 and 2.0.2, respectively, used in this chapter. Is there any difference? Why?

There should be no difference unless someone really screwed up.

A.4 Extensions

Exercise 4.1 Define a couple of MPI subsets a bit wider than the one described above.

You can add more collective operations, for one. An extension to more communication modes is also possible. There are quite a few communicator management calls that can be built largely on top of the MPI_Comm_split(). Finally, a full Fortran binding, or even Fortran 90 and Fortran 2008 bindings are all within reach.

Exercise 4.2 Define a subset or two that are still narrower than the one described above. Think about message length, tag values, and other entities.

See Exercise 2.2.

Exercise 4.3 Estimate the effort needed to implement the extensions outlined in this chapter. Once done with the implementation, compare reality to your forecast. Does it look like the "double and round up" rule works for you?

It did this for me.

Exercise 4.4 Do you feel lost in all these short program names? If so, introduce more descriptive names and see that happens.

I did not need this. Ever. The less letters to type in, the better.

Exercise 4.5 Correct the MPI_Bcast() by using a zero-size message to close the ring (see Section 4.5.3.2).

Just replace the respective message exchange by zero-sized MPI_Send() and MPI_Recv() (see ext/v5/src/coll.c).

Exercise 4.6 Implement the MPI_Allreduce() directly using the ring pattern.

The requires change is rather trivial: you just pass the data around, performing the required operation on your part and the buffer you get (see ext/v5/src/coll.c).

Exercise 4.7 Make the MPI_Barrier() strongly synchronizing this way or another (see Listing 4.43). Consider controlling the manner in which this is achieved, if at all, at run time.

A macro STRONG_BARRIER will do for starters (see ext/v5/src/coll.c). This leaves some room for further extensions, basically selecting the algorithm to apply.

Exercise 4.8 Improve the MPI_Gather() by sending and receiving only the required portions of the buffer along the ring (see Section 4.5.3.5).

This is a bit of a hassle, because you need to figure out the number of the send and receive operations on each process individually (see ext/v5/src/coll.c). After all, noncontiguous data buffers are not such a bad idea, huh?

Exercise 4.9 Implement the MPI_Allgather() directly using the ring pattern.

This is just an extension to the MPI_Gather() done in Exercise 4.8, blended with the MPI_Allreduce() done in Exercise 4.6 (see ext/v5/src/coll.c).

Exercise 4.10 Translate the unified point-to-point benchmark b.c into Fortran and compare performance of both incarnations.

Stopping short of extending the Fortran bindings and using C preprocessor for selecting different code paths, I implemented the ping-pong, ping and pong patterns (see ext/v5/test/b.f) and observed no difference in performance compared to the C version.

A.5 Optimization

Exercise 5.1 Implement the MPI_Sendrecv() call. What other call(s) would you add as well?

To feel comfortable, you will need an MPI_Waitall() capable of processing at least two requests. Mapping this function upon two MPI_Wait() calls would suboptimal, however, because you would impose a certain ordering upon the operation completion then (see opt/v4/src/xfer.c).

Exercise 5.2 Optimize the MPI_Bcast() using the horseshoe pattern and the pipelining idea (see Section 5.1.6.2.1).

You just need to split the buffer into chunks and follow the pattern for all of them in turn (see opt/v4/src/coll.c).

Exercise 5.3 Implement the MPI_Reduce() using the tree pattern.

Well, this is relatively easy, but may require prior separation of the tree management functions defined elsewhere (see opt/v4/src/coll.c).

Exercise 5.4 Implement the MPI_Allreduce() using the hypercube pattern and the MPI_Sendrecv() call.

This will look almost like the ring algorithm done in Exercise 4.6, with the changed number and order of operations (see opt/v4/src/coll.c).

Exercise 5.5 Describe a lock-free shared memory MPI_Barrier().

Allocate an extra array of the MPI_COMM_WORLD size for the synchronization variables at device initialization time. On entry into the optimized MPI_Barrier(), set your synchronization variables to one, then count the variables set to one for all the ranks of the current communicator. As soon as this number gets to match the communicator size, set your synchronization variable to zero and count the appropriate variables set to zero until their number reaches the size of the communicator involved. Thus, you will do the fan-in first and fan-out next, making sure nobody runs away and starts a new barrier operation prematurely. If you have a nonblocking progress engine handy, kick it once in a while as well while counting the variables, just to make sure that your pending point-to-point operations do not get stuck while you sit in the barrier.

Exercise 5.6 Add sched_yield(2) into the shared memory path and measure the slow-down if this system call is issued at every iteration of the progress engine.

You will see the time needed to do the respective system call. Whatever the actual values obtained, this will kill your latency, right?

Exercise 5.7 Implement the -ppn option instead of the MPI_RANKSTEP environment variable in the mpiexec command. What takes precedence if both the variable and the option are used at the same time?

If both ways are implemented, the option should take precedence. In addition to this, it may be advisable to output a warning to the user in this case (see opt/v4/bin/mpiexec.ppn).

Exercise 5.8 Kick the progress upon enqueueing a new send request rather than before that. What happens to performance? Why?

In principle, messages should start getting out earlier, if only just. This means that the respective completion calls will have nothing to do (see opt/v4/src/xfer.c). However, the blocking nature of our current progress engine will make all sending calls effectively blocking, and this may lead to new errors.

Exercise 5.9 Like Exercise 5.8, but for the receive request. Same questions.

In principle, there should be fewer unexpected messages. This means that the respective completion calls will have nothing to do (see opt/v4/src/xfer.c). However, the blocking nature of our current progress engine will make all calls receiving effectively blocking, and this may lead to new errors.

A.6 Tuning

Exercise 6.1 Recall the double buffer size tuning from Section 2.3.6.4. Was this static or dynamic, platform- or application-specific tuning?

Static and platform-specific by the purpose, and static and application-specific by the execution, since only the unified point-to-point benchmark was used.

Exercise 6.2 Inspect the implementation of the MPI_Cart_create() and MPI_Graph_create() calls in your favorite MPI implementation. Do they reorder process ranks for better performance?

Judging by the man pages, MPICH does not do this.

Exercise 6.3 Implement statistics gathering mode that will collect data on the operations and buffer sizes used per process involved and output this information on request.

In order to keep the memory space occupied under control, I split the buffer size accounting into buckets: small, medium, and large, with appropriate limits preset and thus potentially controllable. Data is collected and output only if the MPI_STATS environment variable is set (see opt/v5/src/).

A.7 And the Rest

Exercise 7.1 What communication mode was implemented by the shared memory MPI subset in Chapter 2?

Synchronous.

Exercise 7.2 Build subset MPI with thread safe libraries and see whether there is any performance difference.

Compile your code with the -fpic option set and use the -shared option to produce your shared library out of the object files so obtained. The times have changed: I did not notice any difference in performance now.

Exercise 7.3 Propose an MPI device extension for one-sided operation on an RDMA-enabled platform.

If you stay within the paradigm proposed in this book, you will have to introduce two more queues—one for the messages to be put to a target process, and one for the messages to be gotten from an origin process. The rest of the job will be covered by the progress engine going through these queues as it does through the pending send and pending receive queues.

You might also want to experiment with using the same send queue for the put requests, and the same receive queue for the get requests. Does this work out?

A.8 A Look Ahead

Exercise 8.1 Propose MPI checkpoint/restart interface.

Let us introduce calls like MPI_Prepare_for_checkpoint() and MPI_Recover_after_restart(), for getting ready for a coming checkpoint done by means outside the MPI library, and for making sense of the mess upon restart, respectively (see Listing A.1):

Listing A.1: Checkpoint/restart interface

```
int MPI_Prepare_for_checkpoint(MPI_Comm comm);
int MPI_Recover_after_restart(MPI_Comm comm);
```

Both calls are collective over the communicator provided. Individual backup/restore can be expressed by using the MPI_COMM_SELF.

Exercise 8.2 Find yet another MPI design limitation and consider the repercussions of removing or at least relaxing it.

I cannot come up with any other fundamental design issue, or I would have described it in the text. Do you have a better idea? Please let me know over LinkedIn or directly at alexander@supalov.com.

Exercise 8.3 Propose an interface that can replace MPI in one of the target markets mentioned above or elsewhere.

Let us consider the following interface (see Listing A.2):

Listing A.2: Basic subset

```
int LOH_Init(int subsets,int threads);
int LOH_Finalize(void);
LOH_Pid LOH_Get_pid(LOH_pid pid);
LOH_Size LOH_Send(LOH_Pid pid,void *pbuf,LOH_Size size);
LOH_Size LOH_Recv(LOH_pid pid,void *pbuf,LOH_Size size);
```

This is a basic, highly optimized subset zero for static environments and blocking point-to-point communication targeting HPC, mobile, and IoT markets.

Active subsets are determined by the subsets mask argument of the LOH_Init() call. Other subsets can be selected and combined at will (see below). The basic subset cannot be deselected.

Thread safety level of the library as a whole is determined by the number of threads indicated at initialization time: zero for none, one for at most one thread using the library at any one time, more than one otherwise.

All integer valued calls return a negative value in case of error, or a nonnegative value otherwise. Data transfer and data query calls return the real size of the buffer sent out, received, or filled by the function. This size cannot exceed the size of the buffer defined by the respective size argument. The LOH_Init() returns the selected subset mask that may differ from the requested one.

Note that the current process or thread always has index zero in the system process id list returned by the LOH_Get_pid() call. The lists returned to different threads and processes are thus not required to match, nor there is any need to return a full pid list of the whole job unless explicitly requested. In particular, in server-client configurations, a client will normally get just two process ids: its own as number zero, and server's as number one.

The initial process and thread startup, usage, and synchronization are strictly application and platform matters. A process id can equally refer to an OS process or a thread inside any process. None of the failures are fatal. Fault recovery and other hassles are also all up to the application.

For dynamic environments of subset LOH_SUBSET_BLOCKING_DYNAMIC, the following call is added (see Listing A.3):

Listing A.3: Blocking dynamic subset

```
int LOH_Join(LOH_Jid jid);
```

This call is blocking. A thread can participate in at most one job at a time, including the initial job with job id zero. It leaves the original job by joining another job. If a thread wants to return back to the original job, it has to explicitly rejoin that job. Upon joining, a new call to the LOH_Get_pid() will return the process ids of the current job formatted as usual.

For nonblocking point-to-point communication of subset LOH_SUBSET_NONBLOCK-ING_PT2PT the following calls are added (see Listing A.4):

Listing A.4: Nonblocking point-to-point subset

```
LOH_Req LOH_Start_send(LOH_Pid pid,void *pbuf,LOH_Size size);
LOH_Req LOH_Start_recv(LOH_Pid pid,void *pbuf,LOH_Size size);
LOH_Size LOH_End_send(LOH_Req req,LOH_Timeout timeout);
LOH_Size LOH_End_recv(LOH_Req req,LOH_Timeout timeout);
```

All calls in this subset are nonblocking as long as the timeout of the respective completion call is set to zero. Timeout set to -1 blocks the call until completion. Pointer valued calls return (void *)0 in case of error, or a valid pointer otherwise.

Nonblocking one-sided subset LOH_SUBSET_NONBLOCKING_1SIDED is defined analogously to the nonblocking point-to-point subset (see Listing A.5):

Listing A.5: Nonblocking one-sided subset

```
LOH_Req LOH_Start_get(LOH_Pid pid,void *ptarget,void *pbuf,LOH_Size size);
LOH_Req LOH_Start_put(LOH_Pid pid,void *ptarget,void *pbuf,LOH_Size size);
LOH_Size LOH_End_get(LOH_Req req,LOH_Timeout timeout);
LOH_Size LOH_End_put(LOH_Req req,LOH_Timeout timeout);
```

The attending type and constant definitions are as follows (see Listing A.6; conditional compilation constructs related to the subset selection in effect are skipped for clarity):

Listing A.6: Declarations for an HPC platform

```
#define LOH_SUBSET_BLOCKING_DYNAMIC 1
#define LOH_SUBSET_NONBLOCKING_PT2PT 2
#define LOH_SUBSET_NONBLOCKING_1SIDED 4

typedef void *LOH_Pid;
typedef long LOH_Size;

typedef void *LOH_Jid;

typedef void *LOH_Req;
typedef double LOH_Timeout;
```

If a target platform cannot handle long counts and double precision arithmetic efficiently, the attending declarations can be changed as follows (see Listing A.7):

Listing A.7: Declarations for an IoT platform

```
typedef void *LOH_Pid;
typedef int LOH_Size;

typedef void *LOH_Jid;

typedef void *LOH_Req;
typedef int LOH_Timeout;
```

Timeouts expressed for an HPC platform in terms of fractions of a second will have to
be converted to the tick numbers of the platform, if such a need ever arises, since zero
will stay a valid value.

That's it. Further subsets can be added later if the need arises. By explicitly
selecting them, you will give the library a chance to tune itself up without any need
for static or dynamic analysis: but you know best what you are going to use in the
program, right?

You may ask about message tags. Well, if you need them, add them.

Index

DOI 10.1515/9781501506871-011